CONNECTING PEOPLE, IDEAS, AND RESOURCES ACROSS COMMUNITIES

International Series on Technology Policy and Innovation

General Editors:

Manuel V. Heitor, Secretary of State for Science, Technology, and Higher Education, Portugal

David V. Gibson, IC^2 Institute, The University of Texas at Austin, Texas

Pedro Conceição, Deputy Director, Office of Development Studies, United Nations Development Program

CONNECTING PEOPLE, IDEAS, AND RESOURCES ACROSS COMMUNITIES

Edited by
David V. Gibson, Manuel V. Heitor, and Alejandro Ibarra-Yunez

Purdue Unversity Press
West Lafayette, Indiana

Printed in the United States of America.

ISBN 978-1-55753-448-4

Library of Congress Cataloging-in-Publication Data

Connecting people, ideas, and resources across communities / edited by David V. Gibson, Manuel V. Heitor, and Alejandro Ibarra-Yunez.
p. cm. -- (International series on technology policy and innovation)
Includes bibliographical references and index.
ISBN-13: 978-1-55753-448-4 (alk. paper) 1. Telecommunication. 2. Telecommunication--Social aspects. I. Gibson, David V. II. Heitor, M. V. (Manuel V.), 1957- III. Ibarra Yunez, Alejandro.
TK5101.C676 2007
621.382--dc22

2006032650

Contents

Foreword

IC^2 Institute and the LBJ School of Public Affairs at The University of Texas at Austin were pleased to co-sponsor along with the Conference Chair EGADE-ITESM, Monterrey, Mexico, and the Instituto Superior Tecnico, Lisbon, Portugal, the 7th International Conference on Technology Policy and Innovation (ICTPI) that was held in Monterrey, Mexico, June 10-13, 2003. This event provided an important international forum to discuss how academic, business, and governmental policies and actions can best facilitate regional innovation and economic development with the theme "Connecting People, Ideas, and Resources Across Communities."

The ICTPI conference occurred at an important time frame in Mexico's political history as Vicente Fox, an opposition candidate, defeated the almighty ruling PRI party and won the presidency in July, 2000. A peaceful transfer of power occurred when Fox was sworn in as President of Mexico on December 1, 2000. Most recent times have faced challenges both for Mexico and its quest to more deeply integrate into a North American competitive region. Biggest challenges are that a new 2006 election in Mexico has resulted in a contested result and a partisan divided congress. On the bilateral front with the US, after-9/11 security policy has taken precedence over other pressing agendas for North America, such as convergence in critical infrastructures, trade facilitation with a secure border, and deeper integration of energy, technology, and finance.

One can assume that the Mexican governing elite has become genuinely plural. This change has occurred as access to higher education had risen to a 15 percent matriculation rate compared to only 2 percent in 1960; including an increase of women accounting for 48 percent of the students enrolled in and graduating from college. These increases will help establish both new talent and the opportunity for new policy implementations—which would seem to set the stage for Mexico to attain a higher degree of technology innovation, at the same time that challenges of our times need to be embraced.

The June 2003 ICTPI conference focused on community connectivity facilitated by technology to explore new economic development opportunities: Both for and in Mexico and for and to the global community. Mexico is not alone in its attempt to put aside barriers and challenges (both old and new) to develop and accelerate its economic success. This is a timeless and common struggle of all the nations of the world. Successful national role models do exist—and were profiled in this conference—such as Korea's one generation climb from a rural nation to its present status as one of the most technologically developed nations that successfully exports to global markets its value-added technology innovations.

The 7th ICTPI also had a strong Latin American focus and featured a high percentage of women participants including the topic of "Women in Science" in a round table discussion. To attain a well-balanced view of any subject, it is important for women to have an equal voice—and perhaps most especially on the topic of connecting people, ideas, and resources across communities—the conference theme. Empirical literature has demonstrated the centrality of women as auditor and watchdog of the state. As such, women present a vital voice to matters of policy which must be observed in times of great change such as Mexico's ongoing governmental transition. The political transition in Mexico is far from complete and what the current regime transitions *to* will be determined to a large extent by what the regime is transitioning *from.* It is certain that the world, and especially Latin America, has been watching Mexico's recent political and economic transitions with the hope that a new success story will arise. Time is needed to shape new policies, review the results and provide critical improvements. Yet, if a nation can make dramatic transformation within the space of one generation, such as Korea's recent history, this is as a brief moment on the international scale.

The following body of scholarly work presents discussions on topics that are vital to government at all levels and for all nations. I hope you enjoy reading *Connecting People, Ideas, and Resources Across Communities.*

Dr. Victoria E. Rodriguez
Vice Provost and Dean of Graduate Studies
College of Graduate Studies
The University of Texas at Austin

Acknowledgments

The chapters included in this volume are based on select presentations of the seventh International Conference on Technology Policy and Innovation (http://in3.dem.ist.utl.pt/confpolicy), which was held in Monterrey, Mexico, June 10-13, 2003. The three sponsoring academic institutions of the conference were The Instituto Superior Técnico (IST), Lisbon, Portugal, through the Center for Innovation, Technology and Policy Research, IN+; The University of Texas at Austin, including the LBJ School of Public Affairs and IC^2 Institute; and the Monterrey Institute of Technology (ITESM), Mexico.

We thank the Organizing Committee of the seventh International Conference on Technology Policy and Innovation. We are also grateful to those who so effectively managed the conference, in particular the staff at the Graduate School of Business Administration and Leadership (EGADE) of the Monterrey Institute of Technology.

We thank the chapter authors for sharing their insights and perspectives on a range of important topics regarding science and technology policy and innovation.

Finally, the editors are especially grateful for the dedicated and excellent publication effort including Miguel Silveiro, Publications Coordinator at the Center for Innovation, Technology, and Policy Research, Technical University of Lisbon, Portugal and Katie Chase, copyeditor, for successfully bringing this volume to publication.

David V. Gibson
Manuel V. Heitor
Alejandro Ibarra-Yunez

1

Introduction: Connecting People, Ideas, and Resources Across Communities

Alejandro Ibarra-Yunez, David V. Gibson, and Manuel V. Heitor

The objective of connecting people, ideas, and resources across communities in the knowledge-based economy is complicated by increasing global competition as well as the reality that required know-how and infrastructure are asymmetrically distributed across regions worldwide. The evidence seems clear that through enhanced networks within regions, production chains are extended toward increased competitive positions of economic and social actors. However, the economics of networks calls for compatibility and minimum standards to optimize positive externalities. Some key conditions include the areas of financial markets and physical infrastructure such as ports, roads, and international borders and telecommunications, as well as coordination and compatibility in business practices, labor markets, and legal and regulatory settings.

This volume is a collection of select papers presented at the Seventh International Conference on Technology Policy and Innovation, where 129 researchers from academia, business, and policymaking sectors from 31 countries came together at EGADE-ITESM, Monterrey, Mexico, June 10-13, 2003. The venue for the conference, the northern capital and industrial powerhouse of Mexico, has deep roots in the Mexican economy as well as strong economic, education, and government linkages with the U.S. economy in general and the Texas economy in particular under the North American Free Trade Agreement (NAFTA). This volume is organized under the following headings: (1) Promoting Institutional Change, (2) Building Capacity, (3) Fostering Connectivity, (4) Measuring and Modeling for Improved Understanding, and (5) Learning from Case Studies.

This volume centers on the need to connect ideas with resources and people and it discusses the requirements for capacity building and connectivity, putting in perspective new challenges and opportunities that have arisen in the context of the emerging global learning society. This introductory chapter and the concluding section of the book, on case studies, make special reference to the challenges of Latin America in its quest for enhanced competitiveness and technology upgrading. On the one hand, institutional and infrastructure change within Latin America has led to technological upgrading and regional leapfrogging. For example, Chile's committed policies of the 1980s and 1990s for improved institutions and infrastructure is an excellent example of improving conditions and developing new capacities for the creation of capital. On the other hand, Latin America, in general, faces considerable challenges as a result of lagging in its development of infrastructure for wealth creation when compared with other emerging regions in Asia and Mediterranean Europe as well as new participating economies in Central and Eastern Europe.

In economics there is an underlying question concerning the challenge of (1) how to facilitate networks that connect people, ideas, and resources, and (2) how active government technology policy should be to facilitate such connectivity. According to economic theory, positive network externalities arise when products from connected firms create value for consumers, or if more consumers use the same goods or complementary ones. In order to extend the benefits of externalities, strategic moves of firms in technology competition can coexist with efforts toward creating complementary goods and services across them, in forms that can extend from tacit complementarities, to subcontracting, to consortia. Externalities can be firm-specific or extend to an entire industry. For example, firm-specific benefits are accrued by a company if its standards are widely used and industrywide benefits arise if a technological standard is shared by suppliers via the same or complementary goods or services. The technology base can also be strategic, such that firms seek to maintain incompatible technologies as a means of market differentiation. While some standards are mandated (Tirole, 1993), at times standards are difficult to implement as was the lack of agreement on the Code Division Multiple Access project in Korea, and the quest of rulings under the International Mobile Telecommunications 2000/Global System Mobile (IMT-2000/GSM) technologies in mobile communications between Europe, Asia, and the United States.

The fact that emerging economies often lag concerning technology-based wealth creation calls for increased research and understanding on critical issues. On the one hand, international practices can become the norm for emerging economies to adhere to and such international integration is argued to promote technology transfer and adaptation by firms in developing areas necessary to enhance their competitive position. For such industrial leveraging, learning and adaptation are key along with linking with firms that are technology leaders (Richard, 2004). On the other hand, government action can attempt to guarantee competitive physical capital, financial resources, labor training programs, and institutional infrastructures. While it is argued that economic development implies moving low-productivity activities toward high-productivity, many emerging countries suffer from governmental lack of understanding or immobility, not only because of insufficient resources but also from the lack of coordinated technology policies that involve many secretariats or government ministries.

Many emerging economies have been unable to foster sufficiently competitive environments. In the case of Latin America, as a result of inferior macroeconomic policy, fiscal resources for public investment have diminished from around 35% of GDP in the in the early 1980s to around 25% in 2000. There is an absence of policy paradigms for acceptable boundaries to government action as seen in the contrast of government programs in Southeast Asia, China, Central Europe, and Africa. Regarding manufacturing value added per capita, and manufacturing exports per capita, Asian economies including China lead the developing world, followed by Latin America and the Caribbean. Some countries such as Mexico, Costa Rica, Malaysia, the Philippines, and Thailand have exports of medium- and high-technology products. According to Richard (2004), cases where technology upgrading has been apparent in exports include Argentina, Brazil, Costa Rica, and Mexico in Latin America; China, Hong Kong, India, Malaysia, The Philippines, Singapore, South Korea, Taiwan, and Thailand in Asia; and Saudi Arabia, Turkey, and South Africa.

Chile and Hungary are good examples of technology adaptation arising from connecting domestic and international economic activities through foreign direct investment (FDI), industrial integration, and licenses. However, FDI is a necessary but not a sufficient condition for deepening productive networks, generating domestic research and development (R&D), and moving to higher levels of sophistication in productive processes. Countries linking to international value chains often participate marginally as assembly operations of multinational corporations due to a focus on cheap labor or abundant natural resources. With this orientation, connecting resources without ideas—lack of technology upgrading or market expansion through better infrastructure—will result in a constrained economic base. The goal should be to expand opportunities for new entrepreneurs and providers of goods and services and for this to occur, government effort is often needed to provide competitive infrastructure, such as quality education systems and smart infrastructure as well as highways/railroads, telecommunications, energy, and finance at competitive levels.

Global value chains require industrial systems where firm boundaries are redefined and extended. Such value chains are generally not linear in the connections of economic agents, but work in more complex forms. Knowledge networks, business practices, licensing, market and product intelligence, and promotion need to be internalized. A technology-based environment of coordination should emerge, where investment in both the benefits and costs of the network are calculated. Compatible and standardized integrated systems require business leaders to extend their decisions from within the firm, to members of the coalition, to other coalitions, and to sources of knowledge and human capital such as technology centers, universities, and laboratories.

It is clear that in emerging economies such as Latin America, limited resources and the narrow scope of technology policy ministries and government agencies necessitate education for increased coordination given a wider scope of involvement. This includes education and retraining programs, information-gathering and dissemination, the application of subsidies with objectives of economic efficiency and attraction of new firms, fostering entrepreneurship, and international cooperative agreements. Regulatory agencies need to regulate for the market, but be active rather than reactionary to conflicts arisen from predatory conduct of companies.

Latin American economic development policymakers suffer in comparison with their European counterparts as well as policymaking efforts in the United States and Asia. It is clear that enhanced coordination and cooperation are needed among Latin American policymaking authorities at federal, regional, state, and local government levels while focusing on the promotion of industry, university, and government research centers as well as the needs of large firms and small and medium-sized enterprises (SMEs) and to enhance infrastructure development that connects people with ideas and resources. The focus cannot be a short term; what is needed is the development of systems of networks that evolve over time for the long term, leading to the creation of wealth and high-value jobs.

CONNECTING IDEAS AND GOVERNMENT INSTITUTIONS

Global commodity chains require well-structured industrial systems. This makes connecting of businesses, large and mainly small and medium enterprises, of strategic importance. Some lay strategists assume that linkages of the various production levels are linear. However, empirical evidence shows that rather than linear value chains, competitive firms relate to others in complex ways, either circular, starlike forms, or lattice or matrix forms. By the complex system of interactions, firms enhance demand-driven economies, supply-driven economies, contracts, and licenses, all of which move decision makers to new forms of investment: high capital and technological investment, high investment in norms and standards to improve the network, and high investment in the links of the network. Sometimes this is accomplished through intermediaries such as technology laboratories, universities and research centers, industry chambers, or government agencies. Through connections, new business architectures and high-innovation networking place challenges not only to firms (endogenous innovation is never purely endogenous to a firm), but to support infrastructures: government agencies, universities and research centers, industrial parks, physical infrastructure layouts, the financial system, and the legal environment.

In analyzing the state of technology approaches by firms and government agencies in Latin America, it seems apparent that firms and agencies coincide in a rather outdated vision of technology upgrading and policy, which is demand driven, in use by the United States and Europe in the middle of the 20th century. During the 1970s, an interactive model was tried as a more integral model that assumed that linkages to improve competitiveness are not linear. The model has become multifaceted and systemic in Europe and leading Asian economies, but seems difficult to attain for Latin America (Chaminade and Roberts, 2003).

Company decision makers often pursue technology upgrading via benchmarking. Upon investment decisions, then they frequently assume price competition or development of niches. This is a typical two-step economic model of oligopoly competition. Even if useful, a question is whether a firm can do better than the two-step model. The answer is yes, if self-centered decisions and anchoring give way to decisions around networking with suppliers, product developers, information and communication technology options, and other subcontracting. Obviously, vertical and horizontal relations are costly. This is why information infrastructures are critical. International institutions such as the Organization for Economic Cooperation and Development (OECD) or United Nations Industrial Development Organization (UNIDO) have been at the forefront of promoting gathering of technology and scientific information, creating of databases for emerging networked businesses, implementing training, and norms and procedures. This has been pursued since the early 1990s. Some countries' agencies such as CNR/ISTAT in Italy, CNRS/LATAPSES in France, or the IFO Institute in Germany have systemized studies on innovation. Whereas CNR/ISTAT encompasses all aspects of social, economic, and technology statistics and research, CNRS/LATAPSES concentrates on networking and information on new technologies for experts and firms. IFO/CESIFO depends on university networks related to business strategies and economic analysis related to technology and innovation.

National, transnational, and subregional centers profusely exist in the European panorama that are not clearly replicated in emerging economies such as those in Latin America. For example, the National Science and Technology Council (CONACYT) in Mexico provides relevant information on technologies, scientific reports, national guidelines for firms, and promotes and funds scientific research related to business applications. However, retraining programs, venture-capital promotion, and networking are not part of their emphasized activities. In the case of Brazil, the Ministry of Science and Technology (MCT) and the National Institute of Metrology, Standardization, and Industrial Quality (INMETRO) both generate relevant information and promote standards in industry as well as accrediting scientific laboratories. This does not emphasize critical aspects of competitiveness of Brazilian firms mainly related to networking. On its part, Costa Rica has the Ministry of Science and Technology (MICIT) that promotes shared decisions between the central government, firms, and the scientific community, mainly in providing infrastructure.

Connecting firms arises from facilitation, but also from competitive forces and survivability of companies. The link between SMEs and large firms is frequent but it is not a necessary or sufficient condition for deepening competitiveness. SMEs account for a high share of R&D. For example, the figure is 65% for Italy, 50% for Greece and Ireland. In contrast, SMEs' share in R&D expenditures in the United States is less than 15%, while in other EU economies the figure is around 25% (OECD, 2004).

On the part of large firms, they are important in establishing large-scale innovations, structuring markets, leading in business practices, and coordinating SMEs. Often large and smaller firms are complementary, rather than having large dominant firms coexisting with marginal or residual SMEs up to situations of market foreclosure. Complementarities do not necessarily mean, however, that deepening production chains is attainable *ipso facto*. Action can arise from large firms extending supplier-client contracts, or from SMEs trying to "bandwagon" on an expanding sector. Sometimes government support programs are needed, when access to funding, skills, information on markets, or market intelligence is not easily available. Most often, SMEs are encapsulated away from competitive strategies of leading firms. Some even are impeded from efforts in innovation and technology upgrading, or even replicating business practices via technology investment (process, product, or infrastructure).

COMPATIBLE INDUSTRIES

Turning to the question of firm coordination, cooperation, and complementarity, Economides in the economics field sheds light on the theory and empirical evidence on networking (Economides, 1995, 2003; Economides and Flyer, 1997). A pair of vertically related industries form a one-way network where their products are complementary, where consumers demand the composite good or service. In order for the interlinkage to be stable, compatibility of these industries is needed, via technology convergence, standardized goods, goods with some type of physical or chemical component characteristics shared between the partners, financial soundness of the partners, and finally commercialization, logistics, and business practices. For many products, however, compatibility needs to be achieved through a third-party oversight of standards. This is why a holding company in a technology consortium or the government technology agency is of paramount importance. Additionally, it is possible that firms in a vertically related industry might seek incompatibility, such as proprietary designs or refusal to connect with some partners, as a means to gain and keep market power.

For economic research, studies in the 1970s and 1980s concentrated in models of two industrial partners, being then either vertically related or network industries. Also, research was dedicated to situations where each network was owned by one firm. In the 1980s, partly as a result of the divestiture in AT&T, research focused on issues of interconnection and compatibilities. Economic theory and empirical analyses then aided in the study of proprietary networks, coordination-cooperation among independent firms, and clusters. Part of the empirics of networks proved that external economies, compatibility, risk sharing, and even switching costs for consumers, generate cost reductions and market expansion needed in a globalized world. Moreover, cost reductions have transformed some industries because new services and new technologies arise to expand the product space of entire industries.

In Figure 1.1, A_1 and A_n would coordinate either directly or via an intermediary or switch S; Bs would do the same. This amounts to independent and noncompatible industries. If interconnection is sought between A_1 and B_1 it would be accomplished in the form ASB, or ASTB, where T is a transferer (of technology, standards, etc.). An integrated compatible network would be ASB or ASTB; a partial integration would occur as A_1SA_n and STB_1 or STB_n. C_1, C_2,... would be entering firms to the network that can be accommodated or else foreclosed.

Figure 1.1
An Example of Networking with and without Foreclosure

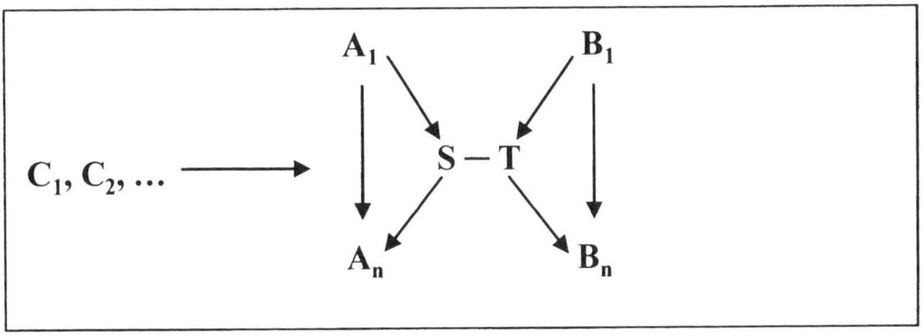

How many firms can coexist in a network is an unsettled question. If coordination, compatibility, and interconnection costs rise more than the benefit of market expansion for partners, or if the above costs are higher and more complex than staying out of an expected partnership, then an entering firm, say an SME, will be excluded from the network. This is quite similar as a condition for a coalition. Its stability will depend on two conditions typical in the new industrial organization discipline: a "participation constraint" and an "incentive compatibility constraint." In countries where promotion policies, information and infrastructure, or legal environments are primitive, costs of coordination, interconnection, and compatibility would increase. At the same time, large firms in trying to stay competitive infrequently seek compatibility with network partners but only possibly temporal subcontracting. New entrants and SMEs, even if seeking participation, would not find it advantageous to participate in a network (the incentive compatibility constraint would not be granted). In case the network is a vehicle for technology upgrading, then excluded firms would become less technology competitive, underinvest, or disappear from the market.

Some industries are characterized with high network externalities. Examples are telecommunications, transport systems, financial intermediation, computer-operating systems; industries where a packaged good is sold, such as the fashion textile industry, furniture, and housing features design; and industries heavily dependent on technical standards, such as automobiles, pharmaceuticals, or the aircraft industry. Other industries are low in network externalities, such as cases where horizontal differentiation of goods is more prevalent, or when technology is nonproprietary. In such cases, coalitions do not prevail, and viability of competition depends on the cost structures and size of the various players in competition. According to Economides and Flyer (1997), when externalities are strong, entry of new firms into a coalition has small impacts on sales, prices, and profits, but participant (incumbent) leading firms are less prone to make technical standards available to newcomers. Within the coalition, a feature is that leading firms would want to establish their standards, hence value chains will be characterized by firms of different sizes, market power, and prices. However, industry output is greater under compatibility than in other equilibrium with some incompatible firms (Katz and Shapiro, 1985).

NETWORKS IN TECHNOLOGY TRANSFER

Research and development and the carryout of technology transfer and applications can be rather cumbersome for firms seeking to upgrade their competitive position. In order to reduce the costs of technology upgrading, various approaches can be followed, which include member firms of a coalition but also out-of-coalition facilitators. One such case is that of technology experts, mostly from government agencies but also from risk assessment organizations, that identify firms, mostly SMEs, most capable to benefit from a transfer of technology. An involved process would entail contacting the heads of the company, providing information on the benefits of the transfer, offering advice and contacts, helping in approaching the origins of the transfer, such as universities, consultants, laboratories, then aiding the firm in finding sources of funds from financial agents, angels, venture capitalists, or government financing. Finally the advisor would monitor the project implemented by the firm. Many cases exist in Europe of this type of technology transfer network, but this is not the case in Latin America. No parallel program in Latin America is similar to the European Business Network (EBN), with the above objectives.

Technology centers are a second means of technology transfer, such as the CRITTs in France, whose aims are the development of enhanced technologies at the subregional level in that country at the same time that it looks for emerging technology areas such as robotics, microelectronic mechanical systems (mems), and nanotechnology. One of the main objectives of this type of facilitator is to bring together scientists, engineering schools, and public research laboratories with industrial application centers and chambers of commerce. Additionally centers such as the CRITTs act as intermediaries and intermediate also partially fund technology developments. While these types of centers exist increasingly across the world, in the case of Latin America such centers exist only to network university-industry efforts.

Technology incubators are also very active in many parts of the world. Some operate to aide newly established high-technology companies in the beginning facets of operation, after which they move to industrial parks. Start-up advice and capital are key for incubators, as well as an important role of coordinating various firms to join in clustering efforts. A problem area with incubators is that often they assume that coordination-cooperation-standardization is a natural and optimal business structure. Given the brief analysis in the preceding section on the economics of networks, it is clear that the asymmetric equilibrium and proprietary *versus* nonproprietary technical standards and the width of externalities calls for a broader approach by many technology incubators.

Much more difficult to create is another form of facilitation structure, which is that of technology parks. Examples exist in Europe such as the Südniedersachsen Park of the University of Göttingen in Germany, the Bangalore/Tarnataka Technology Park on software development, HKSTP in Hong Kong, the Barcelona Science Park on biomedicine, the Bioindustry Park Canavese in Italy for biotechnology, the Los Alamos Research Park, The Technology Park of Western Australia, among others. In Latin America, some salient parks are the Itaipu Technology Park in Brazil, Technology Park of São Paulo, The Parque Tecnologico de Antioquia of Colombia in life sciences, and the Technoparque Internacional of Panama specializing in transportation systems and information technologies. There are less than 150 technology parks as such in the world, mainly because of high investment needs.

In general, networks exist because the complex coalitions formed are strategic rather than purely operational. This is why involved technology parks tend to specialize in some technology or scientific development. Now, as argued here, many SMEs in the world and knowledge centers navigate the competitive pressure with little or no support or network. This sets the challenge to widen policies, but also reaching better contacts, technology and strategic information, and infrastructure conditions and resources.

CONNECTING RESOURCES

Networks exist at various levels. The overall economic approach is that of production networking. However, there are networks related to market servicing, and information transfer networks. All in all, value chains are complex and as argued, infrequently linear. Given the complex nature of technology-related networks, SMEs are less prone to enter into complex coordination games as larger firms. So there is a challenge for SMEs and also the awareness that networking to improve technology positioning is a long-term, step-by-step action.

What are the main variables that are critical resources to competitiveness, and then competitiveness of a network?

On physical infrastructure for competitiveness, firms in the United States, Europe and Asia that are located far from technology centers and markets still have an easy access not only to the multifaceted and varied forms of facilitators of technology upgrading, but have ready access to information and communications technologies, ports-highways-railroads, warehousing, and market facilities. In stark contrast SMEs in emerging economies such as in Latin America, are culturally far from markets and technological developments due to insufficient and lagging infrastructure resources: ICT, ports-highways, other physical capital, financial and venture capital, and education. Some countries have been decisively upgrading these infrastructures, but many are lagging behind.

On ICTs, it is clear that teledensity and access to the Internet is concentrated in main countries. Medium- to high-income emerging economies are catching up and face growth rates in their ICTs in annual averages of more than 30%. In deregulation and privatization, telecommunication infrastructures accounted for the highest percentage of all world privatizations during the 1990s. Along with this process, specific or sectoral regulators were created from scratch in many countries. Other sectoral regulators, such as energy, electricity, banking, and insurance, and even in ports, emerged also in developing economies. Taking infrastructure participants not only as agents with the same challenges for technology development, but also as critical elements of competitiveness for other sectors, after a period of infrastructure privatization and regulatory change that boosted investment and competition, emerging economies of the world, and patently in Latin America, are now facing an industry that has moved from an increasing number of competitive players in the 1990s to consolidation and concentration of services and infrastructure in few firms in 2003. As a result, prices have reduced their pace of decline, investment and servicing in noneconomic areas or small towns has tapered off, and introduction of new technologies and product lines has declined (Ibarra, 2004).

Regulations to improve competition and governance structures of infrastructure agents need to be strong, transparent, and aimed at moving the economy to higher levels of competitive positions. Urgency is compounded with new market structures to make regulators and promoters of improved physical infrastructures slow and myopic. Some countries have succeeded more than others, mainly because regulators have been able to redefine the boundaries and complexities of the telecommunication industry (Melody, chapter 2 in this volume). The same can be observed in the dynamics and restructuring of ports and transport hubs (Brooks, 2004). Participants in these networked physical infrastructures include leading firms, multinational corporations (MNCs), SMEs, technology centers, and, obviously, regulators. At least a clear vision toward competition should be evident for advancement of the technology agenda. It also seems apparent that technology experts, technology centers, incubators, or parks alone cannot lead in the networking. Reform of infrastructure sectors places an additional cost in that most reforms are irreversible or difficult to reverse as a recent World Bank report (2004) emphasizes. Subsidies and support programs do not generally reach poorer and underinvested areas, which calls for renewed efforts for increased coverage.

On venture capital, it is a small share of all investment resources, but has proven to be the main source of funding for new technology-based firms. Whereas venture capital was concentrated in the United States, it was less in Finland, Ireland, or Switzerland. In Canada, for example, 80% of new technology-based companies, in their early stages, were funded with venture capital, but in Spain less than 20% if start-ups were financed with venture capital (OECD, 2004).

Now, internationally originated venture capital flows are also noted. For example, U.S. venture capital is largely invested in Europe or Asia, while little is absorbed by economies in Latin America. On their part, R&D resources and activities are internationalized at a slower pace than production or subcontracting. It is through product and processing specifications that innovation and R&D are invested in contracting parties. Economies like South Korea, Singapore, and Taiwan in Asia; Mexico, Brazil, and to lesser extent Costa Rica in Latin America, are open economies with large intraindustry and intrafirm operations that facilitate technology transfer. Some subsidiaries of MNCs, in establishing operations in developing economies, carry out design and development of product lines. Statistical information is mixed on whether affiliates or domestic suppliers of MNCs carry out more R&D than purely indigenous firms. Basic infrastructure is also uneven in two possibly comparable regions, as shown in Table 1.1.

Table 1.1
Telephone and Water Access in Urban and Rural Areas: Latin America and Central Europe/Middle East in the 1990s

Telecommunications	Percent	Water	Percent
LAC urban	40	LAC urban	77
LAC rural	6	LAC rural	39
ECA urban	69	ECA urban	80
ECA rural	45	ECA rural	29

Note: LAC = Latin America and the Caribbean; ECA = European Central Asia
Source: World Bank (2004), from Table 6.1 and database.

There is another critical resource that conditions coordination-cooperation among aspiring firms to technology upgrading. It is the legal setting, observance, and protection of intellectual property, but extends to legal conditions to start a business or leave a market. This is especially critical in many countries of Latin America. For example, NAFTA pushed Mexican authorities to upgrade and operationalize the country's intellectual property protection measures around 1994; Colombia and Chile had modern property protection systems since the mid-1970s, but faced administrative problems; Brazil upgraded its intellectual property rights protection only in the beginning of 2001. On ease of opening a business or leaving a market, problems in emerging economies and especially in Latin America are well documented by Djankov et al. (2000), where calculation of number of procedures, official time and cost show high costs in most countries.

OVERVIEW OF THIS VOLUME

Part I—Promoting Institutional Change—leads off with "From Telecom Reform to the E-economy" by William H. Melody. This chapter emphasizes that sustained growth in information economies requires investments in human capital as a high-priority policy tool of governments to enhance the micro-economic performance of specific economic sectors, for building competitive advantage in regional and global markets, and for enhancing individual income and well being. Chapter 3, "The Co-evolution of Technology and Institution in the Korean Information and Communication Industry," by Jae-Yong Choung and Hye-Ran Hwang, emphasizes the importance of organizational and institutional change in accord with technological capability accumulation. The authors characterize Korean innovation in information technology while tracing how technology and institutions co-evolved throughout the development process. This section of the book concludes with "Solving Defense Problems Across the Borders: Military and Industrial Reforms in Europe's New Security Environment," by Ioanna Boulouta. This chapter reviews complex issues of the transformation of European defense structures after the end of the Cold War given rapid advances in technology. To do this Boulouta brings together technical, political, economic, financial, and legal sectors to examine their interactions and concludes with the belief that it is not technological issues that offer the greatest cross-border challenges but finding sufficient political will and leadership.

Part II—Building Capacity—begins with a chapter by Philip E. Auerswald on "Emerging Technologies for Change: Mobilizing Invention and Entrepreneurship for People in Poor Places." Auerswald emphasizes that while the positive long-term impact of technology-based innovation on human welfare is evident, these benefits are unequally shared. Access to such basic human needs as clean water, medicine, and education are still out of the reach of much of humanity. The defining feature of the 21st century, an efficient and interconnected global economy and rapid technological advances, are remote at best and threatening at worst. In chapter 6, "Industrial Modernization in Transformation Economies: Analysis of Factors and Strategies," Tobias Schauf emphasizes the need for improvement in organizational and management tools and education for systemic economic planning so that middle and top management get a clear understanding that modernization and innovation need to be considered routine and that information systems and planning procedures need to be developed to accomplish these ends. Cristina Casanueva in "Human Capital, Technology Adoption, and Export Performance in Mexico's Manufacturing Industry (1989-1999)" identifies trends in the practices introduced by manufacturing companies, excluding the special case of "maquiladoras," within the context of the totally new conditions arising from the opening of the Mexican economy. In support of the human capital hypothesis, Casanueva's statistical analysis shows a positive and significant correlation between education and productivity throughout the full set of Mexican industries, including less technologically complex firms, at both at the beginning and end of the decade. Chapter 8 concludes Part II of the book with a discussion on "The Use of Nontraditional Policy Tools to Support Technological Innovation and Economic Growth." The authors, João Pedro Taborda, Pedro Conceição, and José Rui Felizardo, look at the practice of "offset processes" as a policy tool to foster innovation in developed and developing countries. These authors use case data on the purchase of weapons systems developed by developed and developing countries in terms of offset compensation programs.

Part III—Fostering Connectivity—begins with "Emerging Technology Trade Triangle: Japan Joins Mexico, the United States, and Canada," by R. Ray Gehani and Rashmi A. Gehani. The authors propose a dynamic open system model to examine Mexico's evolving relationship with Japan in the context of Mexico's trade with the European Union and North American Free Trade Agreement. Chapter 10 on "Participation of 'The Periphery' in International Research Collaboration Based on Norwegian Experiences" by Kaja Wendt examines strategies to reduce Norway's scientific marginality as a basis for economic growth. Internationalization of research is considered at system, institutional, and individual levels of analysis. In Chapter 11, "Innovation Clusters and Cooperation Networks to Foster Technology-Based Firms," Carlos Quandt and Luiz Márcio Spinosa present an approach for the development of small and medium-sized technology-based firms in Paraná, Brazil. The cases illustrate how the competitiveness of local software producers is enhanced by support of entrepreneurial activity and technology development as well as through information exchange and capacity building among local agents. Network links with international institutions need to be integrated into the processes of knowledge creation, sharing, and building institutional and business relationships. Kari Laine and Matti Lähdeniemi in "Effective Model for Higher Education and Industry Interaction" argue that in knowledge-driven economies there is a growing need for deeper and more productive interaction between higher education and industry. They look inside institutions of higher education to get a better understanding of the requirements for successful collaboration such as incentives and strong institutional interaction. The goals are long-term partnerships where trust, commitment, and mutual benefit can be achieved. The final chapter of Part III is on "The Interface Role of Virtual Team Project Success: A Study in the IT Sector" by Lauro Noboru Hassegawa and Roberto Sbragia. These authors investigate the associations between local and remote interface aspects in Brazilian international virtual teams in the success of information technology projects. The chapter emphasizes the importance of classical elements of project management such as human, technical, and organizational interfaces as well as intervening variables such as team size and project complexity.

Part IV—Measuring and Modeling for Improved Understanding—begins with "Does Income Distribution Affect Innovation?" Apiwat Ratanawaraha empirically examines the hypothesis that income distribution affects innovation. He suggests that, in general, countries with more equal income distribution spend more on innovative activity, produce more innovative outputs, and are more productive in creating innovations than are those with less income distribution. Chapter 15 by David V. Gibson and Pablo Rhi-Perez on "At the Crossroads for Binational Development: Cameron County, Texas, and Matamoros, Mexico" provides a methodology for assessing the assets and challenges of cross-border regions to accelerate technology-based growth. Demographic, survey, and cluster data were analyzed for constructing a binational roadmap. The final chapter of this section is on "Innovation and Knowledge Sharing Across Public and Private Sectors: The U.S.-Brazil Sustainability Consortium," by John Motloch, Pliny Fisk, Rodolpho Ramina, and Pedro Pacheco.

Part V—Learning from Case Studies—begins with Chapter 17 by Carlos A. Góngora-Caamal and Enrique Díaz de León López on "Venture Capital Investments on New Technology-Based Ventures in Mexico." These authors conducted interviews and a survey in Mexico to find that in both the public and private sectors there is increasing awareness of the importance of venture financing and that, while still relatively low, venture capital investment is increasing. Gabriela Dutrénit and Alexandre O. Vera-Cruz in "Technological Capability Accumulation in the 'Maquila Industry' in Mexico" present an analytical framework to analyze the levels of technological capability accumulation of three maquiladoras. They conclude with a list of generalizable factors including the observation that local accumulation is a necessary but not sufficient condition for global firms to transfer activities to Mexico. In Chapter 19, "A Successful Experience of Innovation and Technological Learning in the Automobile Industry: The Tremec-Chrysler Case," Salvador Padilla Hernández and María de la Luz Martín are concerned with the factors that explain the success of technological and productive learning processes. To do this the authors studied two Mexican automobile firms while focusing on the suppliers of spare parts, noting how this affects the technological performance of users and producers. Beatriz C. Fialho, Lia Hasenclever, and José M. C. Mello in "The Geography of Innovation in the Pharmaceutical Industry: Assessing Implications for Developing Countries" use the case of Brazil to research challenges of production, trade, and R&D in the pharmaceutical industry. They conclude that developing countries need to overcome challenges related to internal macroeconomic conditions, unequal wealth distribution, and poor sanitary conditions as well as problems associated with technological dependence. The final chapter of Part V moves to Korea to look at success factors of the Code Division Multiple Access (CDMA) R&D Project that had the objective of securing the competence of the Korean mobile telecommunications industry. Authors Joong Ick Ryu and Heung Deug Hong ascertain an expanded list of success factors of a large-scale national R&D project that include strategies for technology selection and acquisition, motivation of policymakers, clarity of goal setting, and coordination and cooperation among participants.

REFERENCES

Brooks, M.R. (2004). "The Governance Structure of Ports," *Review of Network Economics*, 3 (2), June.

Chaminade, C. and Roberts, H. (2003). "Fostering Innovation in SMEs: When Internal and External Networks Matter, A European Experience," paper presented at the Seventh International Conference on Technology Policy and Innovation, Monterrey, Mexico [www.egade.sistema.itesm.mx/monterrey2003].

Djankov, S., La Porta, R., Lopez-de-Silanes, F., and Schleifer, A. (2000). "The Regulation of Entry," NBER working paper 7892. Cambridge, MA: National Bureau of Economic Research.

Economides, N. (1995). "The Economics of Networks," *International Journal of Industrial Organization*, 14 (2), March.

Economides, N. (2003). "Competition Policy in Network Industries: An Introduction," working paper [www.stern.nyu.edu/networks].

Economides, N. and Flyer, F. (1997). "Compatibility and Market Structure for Network Goods," working paper [http://raven.stern.nyu.edu/networks].

Ibarra, A. (2004). "Estructura del Mercado de Servicios de Telecomunicaciones y su Impacto en la Industria de Equipos y Componentes en Mexico (Structure of the Telecom Service Market and its Impact on the Equipment and Components Industry in Mexico), EGADE, Monterrey, February.

Katz, M. and Shapiro, C. (1985). "Network Externalities, Competition, and Compatibility," *American Economic Review*, 75 (3): 424-440.

OECD (2004). *Science, Technology, and Industry Scoreboard 2003: Towards a Knowledge-Based Economy*. Paris: OECD.

Richard, F. (2004). "Industrial Development Report 2002/2003, Competing through Innovation and Learning," United Nations Industrial Development Organization, Vienna.

Tirole, J. (1993). *The Theory of Industrial Organization*. Cambridge, MA: MIT Press.

World Bank (2004). *Reforming Infrastructure: Privatization, Regulation, and Competition*. Washington, DC: The World Bank.

PART I:
PROMOTING INSTITUTIONAL CHANGE

2

From Telecom Reform to the E-economy

William H. Melody

INTRODUCTION

An ever more popular theme in the scientific literature, government policy documents, and the popular press is that technologically advanced economies are in the process of moving beyond industrial capitalism to information-based economies. This transformation is expected to bring profound changes in the form and structure of economic, social, cultural, and political systems. European Union (EU) integration and the North American Free Trade Agreement (NAFTA) are just two illustrations of this ongoing restructuring of economies and societies.

This transformation is being driven by the development and pervasive application of information and communication technologies and services (ICTS). The electronics, computer, telecommunication, media, and information-content industries constitute a trillion dollar-plus global industry sector. It is the fastest growing sector of the global economy and is expected to remain so for the foreseeable future. Most national governments are counting on these industries to provide the primary stimulus to their future economic growth. This chapter examines the transformation process and the major forces driving it, paying particular attention to the interplay between changing technologies, markets, and government policies and regulation.

MAJOR FORCES DRIVING ECONOMIC TRANSFORMATION

The primary forces driving the transformation of national, regional, and global economies are dramatic changes in technologies, policies, and markets—the combination of the development and increasingly pervasive applications of ICTS on the one hand, and the worldwide movement to market liberalization and deregulation on the other. The conversion of telecommunication (telecom) networks and all forms of communication and information content to digital standards is creating an electronic network foundation that facilitates exchanges and transactions of all kinds. Electronic commerce and the next-generation Internet represent the next step in this process. Together with liberalized markets and reduced barriers to trade, this will ensure that the 21st-century information economy is primarily an international, or even global economy.

In an agricultural economy, land is the most valuable resource attracting investment capital. In an industrial economy, manufacturing plants, machinery, and other forms of physical capital are the focal point of investment activity. In the information economy, the expectation is that people will be the central resource attracting investment because information is essentially produced, stored, and applied by humans. Whereas the industrial economy was an era of physical capital with labor employed to facilitate its needs, the information economy is expected be an era of human capital with investment in the skills, competences, and capabilities of people being the central activity.

This suggests that the information economy can provide for a considerably higher level of human development than the industrial economy, for the conversion of what we know as the "labor force" into information or knowledge workers, and for a significant expansion in investment in education, training, research, and development—the major formal knowledge-generating and -distribution activities. It also suggests the possibility for a more widespread distribution of the wealth generated in the knowledge economy because the human resources attracting this increased investment are also workers and consumers.

Therefore, the most important elements of the new information economy will be:

1. the development and use of advanced high-speed (i.e., broadband) telecom networks—the information infrastructure—for electronic commerce and related next-generation Internet activities;
2. continued major market deregulation and reregulation at both national and international levels;
3. the increased generation and use of information as both an economic resource and a product exchanged in markets; and
4. a much greater emphasis on the role of human capital as the principal producer, repository, and disseminator of information.

This chapter focuses primarily upon points 1 and 2, which are necessary to prepare the ground for the productive development of points 3 and 4.

THE INTERPLAY AMONG TECHNOLOGIES, MARKETS, AND GOVERNMENT POLICIES

The path of both past and future development of the ICTS sector and the economy is shaped by the interrelations among changing technologies, markets, and government policies and regulations. Changes in each of these elements have had a major impact upon the industries and upon the other elements in a complex, but identifiable manner. This is illustrated in Figure 2.1. Note that the arrows connect each element—technologies, markets, and policies—to the ICTS sector industries, not to each other. This is to illustrate that the primary influences of technologies, markets, and government policies have not been directly upon one another (although on some occasions they have), but rather they have been mediated through the mix of influences that have shaped the development of the industries, individually and collectively. This includes, of course, other influencing factors than those being examined here, including changing social, cultural, and political factors. But here we focus on the primary determinants: technologies, markets, and government policies.

Figure 2.1
Key Factors Shaping Media Development

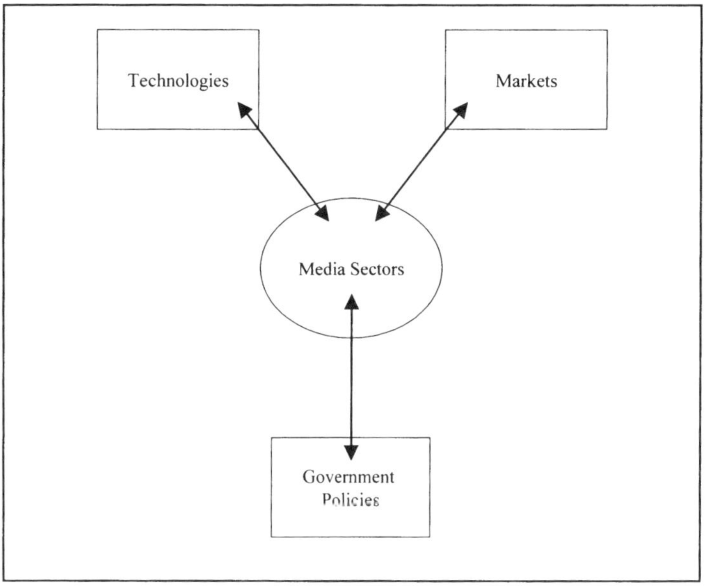

Note also that the arrows run both ways. Each element influences developments in the ICTS industries and in turn is influenced by them. In a dynamic environment, like the one we are examining here, there are no simple, direct, cause-and-effect relationships. There are schools of determinist thought that claim industry development is driven primarily by technological change, economic structures, or government policies. For example, technological determinists often suggest a linear and sequential model of change. Technology changes drive industry and market changes, which in turn require changes in government policies that have become obsolete.

But they neglect to ask what stimulated, or made possible, the technology changes. A little research generally shows it was sometimes changes in government policies and regulations, sometimes changes in industry markets, and sometimes independent inventions and innovations. The reason future changes cannot be predicted with any accuracy is that the circumstances spawning change could come from anywhere. Some illustrations of how the interplay between technologies, markets, and policies has influenced changes in the ICTS sector industries will demonstrate the process.

New Uses of the Spectrum

Modern communication makes heavy use of sending signals through the air, as we observe with cell phones, satellites, and more recently WiFi hot spots for Internet access. These are all developments that came to fruition in the marketplace during the latter part of the 20th century. The spectrum is the natural resource of the airwaves over which communication signals can be sent, if one has the appropriate transmission and receiving equipment tuned to the correct frequencies and there is no signal interference. Technological advances during the latter part of the 20th century have provided a revolution in our capabilities to communicate over the spectrum. What made these technological improvements possible, and encouraged their widespread development and use in the media industries today?

Many inventors and innovators have been experimenting with using the spectrum ever since the first radio signals were sent by Marconi in 1894, and radio stations were broadcasting for most of the 20th century. But a major technological advance occurred during World War II when the British invented radar to detect enemy planes as a result of wartime research and development. In fact, a great many advances in communication technology have come from military R&D for war or defense purposes. In these instances, technological advance is driven primarily by government defense policy objectives.

After World War II, telecommunication equipment manufacturers, working with the telephone monopolies of the day—the British Post Office and AT&T in the United States, among others—made further improvements that enabled the telephone operators to use landline microwave towers, strung across the country about 30 km apart, to transmit higher volumes of telephone traffic at lower costs than could be obtained from buried cable. This provided a stimulus for long-distance telephone calls and the expansion of the radio microwave networks.

At the same time, many firms were attempting to build on the same technical knowledge base to transmit television signals, which required far greater communication capacity (i.e., bandwidth) than voice phone calls. In addition to the local transmission of signals to TV sets, it came to be recognized that the much higher capacity transmission channels on radio microwave systems could carry television signals across the country. At this stage, it was industry responding to major new market opportunities that was driving this next stage of improved spectrum-based technologies.

But there would be no market opportunities unless there was agreement on the precise frequencies that would be used for television signal transmission within localities and for intercity signal transmission. Governments had to adopt policies and regulations with respect to the use of the spectrum for these distinct forms of communication that would make these developments not only possible, but also economically viable and capable of scalable expansion as the industry grew. This was government responding to perceived technological and market opportunities by setting policies and regulations that fostered further innovation in the technology and increased market opportunities. Clearly it was the constructive interplay of technology, markets, and policies that shaped these developments.

A similar story can be told about communication satellites. The early R&D and the early satellites were developed by the U.S. military for defense purposes. Commercial satellites required both continuing R&D from industry in response to a perceived profitable market, and the government policy allocation of appropriate spectrum frequencies that would facilitate development. International satellites required international government agreements through the International Telecommunication Union (ITU) and the World Administrative Radio Conference (WARC) for satellite orbit locations and frequencies.

The development of the mobile phone industry followed a slightly different path. The technological innovation was driven primarily from industry in pursuit of a perceived new market. But this was facilitated greatly in Europe by government policy facilitating the development of common technical standards for global system mobile (GSM) phones. Although this was led by familiar names like Nokia, Ericsson, Motorola, and other manufacturers, it was all built upon the continuing innovation in the microelectronics industry that permitted mobile phones (and a great many other devices) to shrink from suitcase to hand size. But it also depended on government policy not only to allocate appropriate spectrum frequencies, but also government regulation requiring the monopoly telecom operators to interconnect their networks with the new mobile operators at reasonable prices. However, in this case, both the industry and the governments dramatically underestimated the explosive market development, and as a result growth has been restricted somewhat by the failure of national governments and international agencies to allocate enough spectrum to accommodate greater competition and market growth.

This was not a major problem with respect to second-generation (2G) mobile market development, but it has contributed significantly to the failures of licensing policies for third-generation (3G) mobile services that have delayed development and implementation for years. After allocating insufficient spectrum for competitive 3G development, many governments, led by the United Kingdom and Germany, chose to assign 3G licenses using monopoly auctions designed to maximize revenue to the government at the expense of new network and service development. In many countries, government policy and regulation, which had done so much to foster and facilitate mobile network and service development in the past, has created an enormous, unnecessary, and inefficient barrier to 3G development in the future.

The Internet

It is frequently claimed that the Internet has been developed entirely outside the bounds of government policy. Rather it has been driven by vigorous competition in the microelectronics and computing industries that has been producing wave after wave of technological improvements. This is the classic example of the competitive market model of economic theory working magnificently. Although one must admire the rapid and dramatic improvements in these technologies, and the technical and services innovations that have been applied in a variety of different ways, it is useful to recall some of the key developments in its evolution.

What is now known as the Internet has developed from a U.S. Department of Defense concern about enhancing security by being able to distribute information rapidly to different points in a decentralized defense communication network. The initial development in the early 1970s (the Advanced Research Projects Agency or ARPA project) connected research labs at leading U.S. universities to permit experimentation. It became possible for industry to participate in this project when the U.S. communication regulator, the Federal Communications Commission (FCC), liberalized its regulations to allow firms other than the telecom monopoly (AT&T) to provide services over the telecom network. This was not done for purposes of the Defense project, but rather in response to a changing general market environment in the sector, which suggested to the FCC that widespread benefits could be realized if entrepreneurial firms could provide a variety of communication services over the telecom network. And this was all made possible because the FCC had been fostering new digital communication developments since the late 1960s.

When the Defense project grew well beyond Defense requirements, this new "data communication network" was handed over to the U.S. National Science Foundation, the government agency that funds the major share of R&D at universities and independent labs. Finally in 1995, Internet management was handed over to industry and it has been driven primarily by commercial interests in its further development. Nevertheless, R&D and innovation on various aspects of Internet development are taking place in universities, public and private research labs, government agencies, and corporations around the world. And discussions are under way in many national and international fora about the policies and regulations that will be necessary to overcome current problems and constraints, (e.g., spam, security, privacy) and facilitate the growth of e-services on the next-generation Internet. Clearly future Internet development will be shaped continuously by the dynamic interplay among technologies, markets, and policies.

THE MULTIPLE DIMENSIONS OF ICTS CONVERGENCE

For the foreseeable future the pace of change and transformation in the ICTS sector and the development of information economies will be influenced heavily by the next stages of interplay among technologies, markets, and policies with respect to ICTS convergence and its applications throughout the economy.

The convergence of telecom, information technology, and information content has been a recurrent theme of discussion in industry, government, and academia for a long time. Initially the focus was almost entirely on technological convergence. Then the scope of the analysis was widened to include the convergence of industries and markets, and particularly the restructuring of the telecom services sector as networks became increasingly digitalized. In more recent times, the scope has widened still further to focus on the development of new electronic services. Here the growth of value-added and Internet services were the first to capture attention; then the commercialization potential of the Internet became apparent. Now electronic commerce and converging media in a multimedia environment are recognized as forces leading to the upgrading and expansion of the fundamental capabilities of the telecom network.

To understand how convergence processes are actually developing, and their implications, one must look more closely at precisely what is converging, how, where, and with what effects. One must identify and examine the multiple dimensions of convergence. Some of the main characteristics of convergence are the following:

1. Convergence should not be seen only as a single-dimension technological issue. It applies to changes in the industries, markets, policies, and regulations as well. There are significant differences among them, and other dimensions may be even more important than technology in determining the directions of development.
2. Within technology, the primary technologies converging are generally classified as telecom, computing, and content technologies. But rarely do we see evidence of the full convergence of all three types of technologies. Combinations of partial convergence of two types is more typical (e.g., digital networks or CD ROMs). Each has very different implications for network and services development. Competition among alternative possible directions of convergence is more likely to be the trend in future years, with a particular focus on the extent of control of so-called intelligent networks that will reside with the network facility operators, independent protocol platform managers, Internet service providers (ISPs) and customers through the personal computer (PC) itself.
3. Industry and market convergence may play a more influential role than technology in determining the dominant convergence trends. But it is not simply a matter of mergers, strategic alliances, and joint ventures. The problem of identifying good partners for specific types of convergence is difficult and requires an understanding of latent market and demand characteristics, as well as industry and organizational cultures. The key is demand-driven, not supply-driven convergence.
4. Policy and regulatory convergence will set the framework for both industry and technological convergence. Despite the plethora of global information infrastructure and information society policy statements, there has been little follow through yet by most governments. The real policy and regulation being developed is industry

specific (e.g., telecom, broadcasting, ICTS). It is unclear at this stage what shape a set of convergence policies might take, despite the fact that a number of countries have recently established converged regulatory authorities. However it is clear they must build upon and be consistent with telecom policy and regulation. In turn, telecom policy and regulation is in the process of preparing the ground for convergence and the implementation of information society policies.

The convergence process is illustrated in Figure 2.2. Not too long ago, the telecom, computing, and content industries were quite distinct and separate in terms of the technologies they used, their capabilities, their industry economics, and their government policies and regulations. If another industry, say banking or transport, wanted to apply these services to improve their own businesses, these would have to be separate and distinct applications that they did themselves or with consultants.

Figure 2.2
The Major Dimensions of Media Convergence

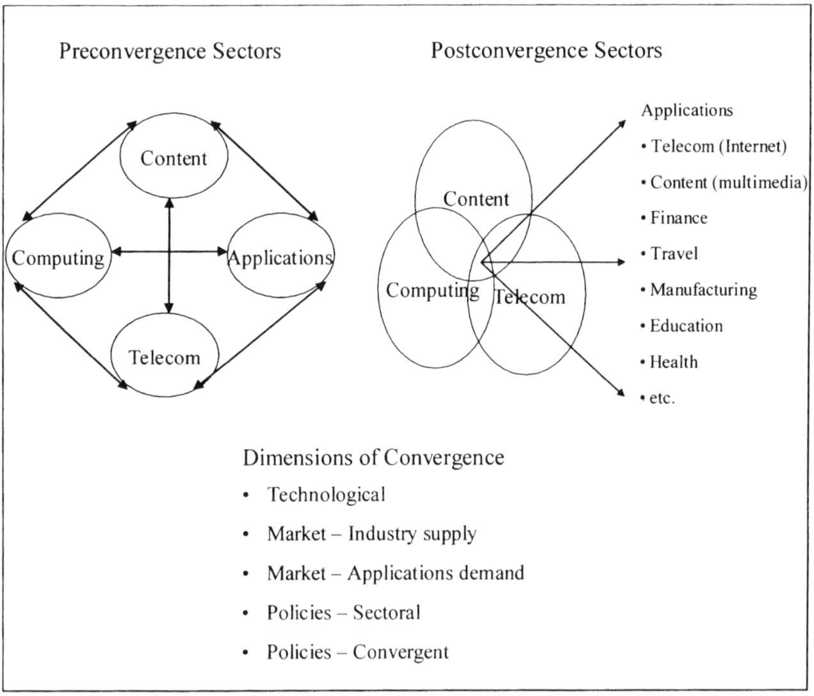

Convergence, as illustrated on the right side of Figure 2.2, involves a series of changes in technologies, the industries, the nature of their services, and in general government policies and regulations, which have permitted innovative applications of new converged services in many other industries, government agencies, and other organizations, as well as by individuals. The increasing application of digital communication standards from the computing industry in recording and playing media content (e.g., CDs for music, books, and documents of all kinds) has brought convergence between computing and content. The gradual adoption of digital standards in the telecom network has brought convergence between computing and telecom. The development of digital communication terminals with appropriate software permits digital content to be sent over the digital telecom network. The personal computer connected to the telecom network to send e-mail, reports, pictures, music, and videos is a good example of low-level convergence among computing, telecom, and content.

Achieving full technical convergence is much more complex than suggested by these illustrations. It involves continuing technological improvement within each of the industries and across them. In the early years of the 21st century, priority areas of attention for further technical improvement include increasing the capacity of the local connections to the network from narrowband (telephone) to broadband (video); improving the capability of mobile networks to handle data, provide broadband connections, and easy access to the Internet; and providing voice telephone calls over the Internet. The ultimate goal, of course, is to be able to provide any form of communication, delivering any form of content, from any location, seamlessly over the telecom network.

But as noted above, convergence involves much more than technical issues. Technological improvements get implemented in the real world only if they also pass an economic test. They must be capable of providing services or products that people want that are affordable. New services based on converged technologies must be developed and marketed. Since these bring together firms from each of the converging industries, and some outsiders as well, they must adapt to the changing environment and reposition themselves in a new converged ICTS industry sector. This requires experimentation and is risky. This is a period of dynamic change classified as "creative destruction" by the economist Joseph Schumpeter. Some firms go out of business, and mergers and acquisitions are common. A period of industry adjustment is at present under way and there is no indication that the markets will be stabilizing anytime soon. This industry and market convergence process is indicated at the bottom of Figure 2.2.

Convergence also raises issues for policy and regulation. Clearly the appropriate policies and regulations applied independently to computing (basically competitive production of hardware and software), telecom (reasonable access to telephones), and content (programming and access to information) must be reassessed in the new environment. Policies and regulations becoming obsolete must be removed as they may provide barriers to progress. Revised policies and regulations in light of the movement toward convergence can facilitate and prepare the ground for a converged future communication environment.

However, many experts believe there is much more to convergence than a major restructuring of the telecom, media content, and computing industries as a new ICT industry sector. These converged services are expected to provide enormous benefits when they are applied effectively in other sectors of the economy. In fact, it is expected that electronic communication and information services will be a major new resource that will lead to a restructuring of all other sectors. E-commerce will change the way that markets function, including trading and banking relations. E-government will change the way governments relate to citizens. The economy will become an E-economy. The limits of markets, identified by Adam Smith at the beginning of the age of the industrial economy as fundamental to specialization and economic growth, will be pushed back to global dimensions in many industries. This is illustrated by the list of applications industry sectors on the right side of Figure 2.2.

This convergence trend is seen by many people as a basis for the development of what are now increasingly called knowledge economies and information societies. Although it is undoubtedly premature to draw conclusions about what future information societies will really be like, it is clear that the communication industries will play an even more important role in future economies than in the past. It will be even more important that people have access to electronic communication and information services than they have had in the industrial economy. The matter of universal access to communication opportunities and to information is likely to become more important if people are to be able to participate effectively in the economy and society. Concerns have already been highlighted about "digital divides" both within and between countries. Fashioning policies and regulations that will respond to universal needs and public interest goals in the new economy will be a challenging task.

ENHANCING THE INFORMATION INFRASTRUCTURE

The transformation of the voice telecom network into an electronic information infrastructure is requiring a series of distinct technical enhancements. The first is a conversion from analog to digital standards to permit the supply of all forms of communication content over the same facilities network. This is largely complete in most countries. The second is an expansion of communication carrying capacity in the network to accommodate the demands of more sophisticated services, such as multimedia, and the dramatically increased volumes of communication and information transfer required in an information economy. This broadband capacity is being provided in intercity and international networks, and within the business districts of many cities by fiber optic cable, supplemented by cable TV, multidensity fiber (MDF), mobile, and satellite systems. The uncertain element at the moment is the local loop connection to residences and small organizations now supplied with copper wire connections. Although the copper wire capacity is continuously being enhanced by improved asymmetric digital subscriber line (ADSL) applications, many experts believe fiber to the home or new wireless connections will be necessary to meet future demands, which would require an enormous amount of new investment.

The discussion around the enhancement of capacity to date has revolved primarily around the need for bandwidth in communication networks and the capability of sending more communication/information over a given bandwidth, with South Korea setting the leading standards. However, the successful supply of electronic commerce and multimedia services requires much more than that. It requires a system of network management of communication signals that will provide an enhanced information infrastructure. This is the system of standards, protocols, and rules relating to functional elements such as numbering, domain naming, and routing that determine conditions of access, quality of service, and network capability. It will determine the extent to which the enhanced information infrastructure will be publicly accessible (i.e., the extent to which it will be treated as a universal service). This is illustrated as Layer 2, services infrastructure, in Figure 2.3.

Figure 2.3
Generic Information Infrastructure for the Provision of Network Capacity

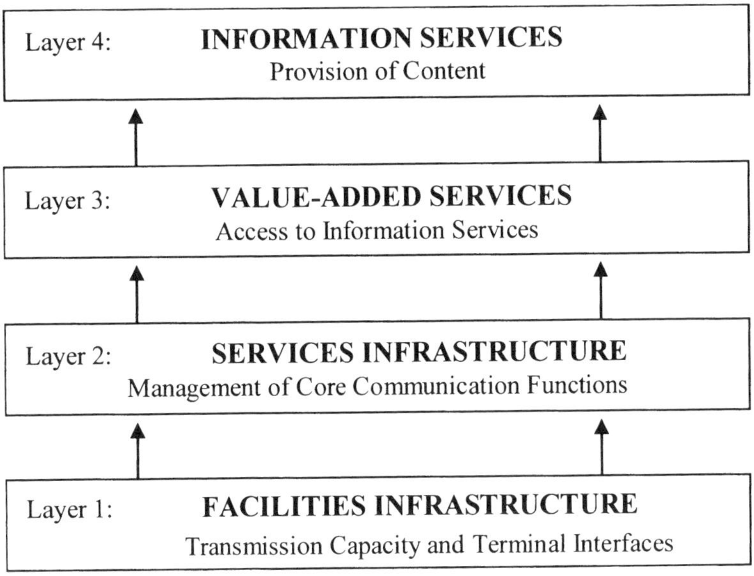

The specific generic characteristics of network services in Layer 2 are in the process of being defined. Layer 2 is an important area where the telecom and information technology sectors converge. Debates relating to numbering (telecom) and domain naming (information technology) over classification, assignment, prioritization, portability, flexibility, and the like will shape the enhanced network in important ways. The specific characteristics of security, privacy, and other quality dimensions of the new services platform will be determined by ongoing debates around technical, service, and public policy issues. The result will be the information infrastructure services platform on which electronic commerce and multimedia public services will develop.

Policy debates on these issues are currently focusing on Voice over Internet Protocol (VOIP), which is seen by some experts as finally bringing the possibility of complete technical convergence where voice phone calls can be just another Internet service. But the U.S. Federal Communications Commission, the European Union, and other policy authorities have noted there is much more to be considered than technical convergence. Phone calls over the Internet are not substitutes in every respect for calls over the telecom network, and raise a number of issues relating to service characteristics, quality, and policy. These include such things as emergency services and access in emergency conditions, network integrity and availability, the impact on national numbering, controls of extraterritorial providers, discriminatory treatment vis-à-vis obligations imposed on traditional providers, location of service provisioning, lawful intercept/access, number portability, termination payments, standards, universal service/access obligations, emergency calls under electricity failure, and impact on existing providers. Until technical convergence is achieved, the economic and policy issues are a second-order business. But once technical convergence is demonstrably achievable, the economic and policy implications become top-priority matters that are then subjected to intensive examination. With VOIP we have now reached that stage.

E-APPLICATIONS

The rate at which electronic commerce and multimedia services develop will depend upon successful applications in a variety of different sectors of the economy, ranging from banking and finance to local government, as well as a range of economic activities including teleworking and teleshopping. As these and other applications develop, increased attention will be directed to the capabilities of individuals, particularly in residences and small organizations, which are fundamental to the development of business to consumer E-commerce and other E-activities. This is illustrated in Figure 2.4, which shows the expanding involvement of individuals in their residences, participating in new network services in an increasing number of ways.

Figure 2.4
Next-Stage Limiting Factors in Information Economy Development: Sector-Specific Applications and Individual Capability

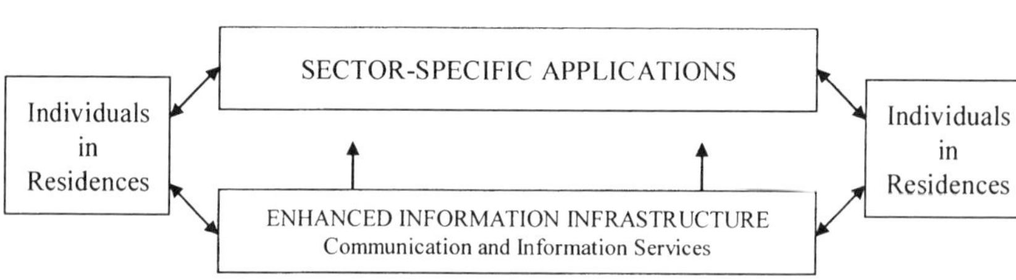

In an information economy where service production, employment, and consumption become increasingly dependent on enhanced network services, the capability and skill of individuals will become the most significant factor constraining market development. These will be needed to create customers capable of active participation in electronic commerce markets, workers capable of efficient participation in the production of enhanced services, citizens capable of participating in government services delivered electronically, and small-business suppliers capable of keeping pace with their large-firm customers.

Experimental trials around the world since the mid-1990s have demonstrated that investments in state-of-the-art technologies and services are not enough. There must be far more investment both in understanding consumer needs and in enhancing the consumer skill base before there will be widespread acceptance of the new E-services. There is increasing evidence that the pace at which the new technologies and services are driving the process of transformation to an information economy depends primarily on the pace of productive investment in human capital (i.e., the skill base of labor, management, consumers, and policymakers).

The evolving networks for electronic commerce and multimedia will become increasingly interdependent networks. The pace of development will be determined by the "weakest link" or limiting factor in the network. This will soon be recognized to be the individual capability of people participating primarily in their residences. The boundaries of electronic commerce and multimedia will be determined by the evolving capability of customers.

CONCLUSION

The processes of telecom reform and convergence in the development of the information infrastructure for new information or knowledge economies tend to focus attention on the factors limiting the next steps in its development. Initially attention was directed primarily to technological questions relating to facility network capability (digitalization) and capacity (bandwidth). The development of electronic commerce and multimedia has shifted the focus to the shaping and definition of the enhanced services network (i.e., the standards, protocols, and rules for the network management of platform services on the enhanced information infrastructure).

This, in turn, is bringing into the picture more clearly the importance of (1) developing successful sector-specific service applications, and (2) increasing the capabilities of individuals to participate efficiently and effectively as customers assessing and buying services, as workers supplying services, and as citizens exchanging information with and receiving services from their governments. As the topical issues of digitalization and broadband are resolved in the next few years, the pace of development of the E-economy will be determined less and less by technological considerations, and more and more by the breadth of individual capabilities that we are coming to recognize as human capital. Sustained growth in future information economies will require investments in human capital as a high-priority policy tool of governments—for enhancing the microeconomic performance of specific economic sectors, for building competitive advantage in regional and global markets, and for enhancing individual income and well-being.

BIBLIOGRAPHY

Economides, N. and Encaoua, D. (1996). "Special Issue on Network Economics: Business Conduct and Market Structure," *International Journal of Industrial Organization*, 14 (6).

Freeman, C. and Louçã, F. (2001). *As Time Goes By: From the Industrial Revolutions to the Information Revolution*. Oxford: Oxford University Press.

Henten, A., Samarajiva, R., and Melody, W. (2003). "Designing Next Generation Telecom Regulation: ICT Convergence or Multisector Utility?" WDR Final Report 0206. Washington, DC: *info*Dev, World Bank; and Lyngby, Denmark: LIRNE.NET. http://regulateonline.org/2002/dp/dp0206.htm.

Henten, A. and Skouby, K.E. (eds.) (1998). *Commercialisation of the Internet*. Lyngby: Center for Tele-Information.

International Telecommunication Union (ITU) (2003). *Trends in Telecommunication Reform 2003*. Geneva: ITU.

Mansell, R. and Steinmueller, W.E. (2000). *Mobilizing the Information Society: Strategies for Growth and Opportunity*. Oxford: Oxford University Press.

Mansell, R. and Wehn, U. (eds.) (1998). *Knowledge Societies: Information Technology for Sustainable Development*. Oxford: Oxford University Press, published for the United Nations Commission on Science and Technology for Development.

Melody, W.H. (1999). "Human Capital in Information Economies," *New Media and Society*, I (1).

Melody, W.H. (1999). "Telecom Reform: Progress and Prospects," *Telecommunications Policy*, 23, January.

Melody, W.H. (2002). "Designing Regulation for 21st Century Markets." In E. Miller and W. Samuels (eds.), *The Institutionalist Approach to Public Utility Regulation*. East Lansing: Michigan State University Press.

Melody, W.H. (2002). "Policy Implications of the New Information Economy." In M. Tool and P. Bush (eds.), *Institutional Analysis and Economic Policy*. Dordrecht, Netherlands: Kluwer.

Melody, W.H. (2003). "Preparing the Information Infrastructure for the Network Economy." In G. Madden (ed.), *World Telecommunications Markets: International Handbook of Telecommunications Economics*, Vol. III. London: Edward Elgar.

OECD (2003). *OECD Communications Outlook*. Paris: OECD.

Schumpeter, J.A. (1950). *Capitalism, Socialism and Democracy*, 3rd ed. New York: Harper.

Shapiro, C. and Varian, H.R. (1999). *Information Rules: A Strategic Guide to the Network Economy*. Boston: Harvard Business School Press.

Sheehan, P. and Tegart G. (eds.) (1998). "Working for the Future: Technology and Employment." In *The Global Knowledge Economy*. Melbourne: Victoria University Press.

Trebing, H.M. (1997). "Emerging Market Structures and Options for Regulatory Reform in Public Utility Industries." In W.H. Melody (ed.), *Telecom Reform: Principles, Policies and Regulatory Practices*. Lyngby: Technical University of Denmark.

World Bank (1998). *Knowledge for Development. 1998/99 World Development Report*. Washington, DC: World Bank.

3

The Co-evolution of Technology and Institution in the Korean Information and Communication Industry

Jae-Yong Choung and Hye-Ran Hwang

INTRODUCTION

Korea has been recognized as one of the successful followers, especially in the information industries. Korean industry also showed outstanding performance in the technology-intensive sector such as DRAMs (Dynamic Random Access Memories), TFT-LCD (Thin-film transistor/Liquid crystal displays), CDMA (Code Division Multi Access) handset. There have been several studies focusing on the technological development process of private and public sectors in Korea. Studies were performed in a broad range from the optimistic view of technology accumulation strategy (Hobday, 1995) and organizational learning at the firm level (Mathews and Cho, 1999) to pessimistic diagnosis, pointing to the narrow knowledge base and sticky specialization in technology development (Ernst, 2000). Despite these studies based on "capability" and "learning" approaches, few studies paid attention to the organizational and institutional change in accordance with the technological capability accumulation.

As Nelson (1994) suggests, firm strategy, supporting systems for innovation, industrial structure, and innovations are strongly interdependent at both sectoral and national levels. Also, in the context of developing countries, the organizational routines and institutional framework were established and evolved as the technological capability became enhanced. At the firm level, as latecomer firms enhance their capabilities, the organizational structures also evolve to meet the advanced technological needs. The role of public institutions also changed from direct supporters to generating industrial technology to developing generic and enabling technology.

During the technological catch-up process, Korean innovators adopted dual approaches to develop technological capabilities, both relying on foreign sources of technology and simultaneously achieving minor innovation in the design and manufacturing processes. Throughout the technological capability-building process, private and public sectors also developed working routines and organizational structures for learning and generating technological knowledge.

The main purpose of this chapter is to characterize the Korean innovation system in information technology (IT) areas. Success factors to catch-up in selected technology areas will be defined through the case study of TDX (Time Division Exchange), DRAMs, and CDMA. The main questions are: What were the strategies and learning mechanisms in this technology-intensive sector? and How did the institutional framework evolve throughout the development process?

CONCEPTUAL FRAMEWORK: CO-EVOLUTION OF TECHNOLOGY AND INSTITUTION IN THE DEVELOPMENT PROCESS

The generation of new technologies by developing countries relates to the wider issue involved in the transition from labor-intensive development to technology-intensive competitive advantage. At the center of this transition are the technological capabilities of innovators in developing countries to generate new technological knowledge. The term "technological capabilities" can be defined as "the knowledge and skill which firms need to acquire, assimilate, adapt, change and generate new technology" (Ernst, Mytelka & Ganiatsos, 1994). "Technological capabilities" are classified into the following two categories according to their depth, breadth, and direction-technology-using capability and technology-generating capability.

Technology-using capabilities relate to the knowledge and skills required to use existing technology in order to produce certain products. They encompass such activities as production management, production engineering, and repair and maintenance of physical capital. *Technology-generating capabilities*, on the other hand, refer to the knowledge and skills needed to manage and generate technology. They include adaptive engineering, improvement of product design/process, new product design, new product development, and new production process. The distinction is useful in order to understand how some developing countries are more successful than others in the transition to technology-based competition.

In order to upgrade from technology-using to technology-generating capabilities, the positive co-evolution process between technological development path and institutional/organizational arrangement has to be required. As indicated, in the developing countries' case, technological aspects divide into both technology-using and technology-generating aspects. In the institutional aspect, it is possible to indicate learning mechanisms, organizational arrangement, and the strategies of main innovators. As the technological needs from industry upgrade, the learning mechanisms, organizational arrangement, and strategy should be adapted to meet the changing needs.

Usually, in the initial stage of industrial development, the focus of technological development resides in the acquisition of existing technology and assimilation into the local environment. In this stage, the object of organization and institution resides in the fast learning and assimilation of existing knowledge through various mechanisms like licensing and implementation in the local environment.

As the technological capability accumulated, the technological needs from the industrial sector upgraded toward the generation of new technological knowledge to produce new products and processes. The generation of new knowledge requires a complex process of interrelationships among innovators than in the case of the learning and implementation of existing knowledge. The focus of technological development in this stage resides in the search for high-technological opportunities and integration of new technological resources with existing in-house capabilities. As a result, the learning mechanisms and organizational arrangement focus on the search for new technological seeds and coordination among various knowledge sources. The conceptual framework is summarized in Figure 3.1.

Figure 3.1
Co-evolutionary Process of Technology and Institutions

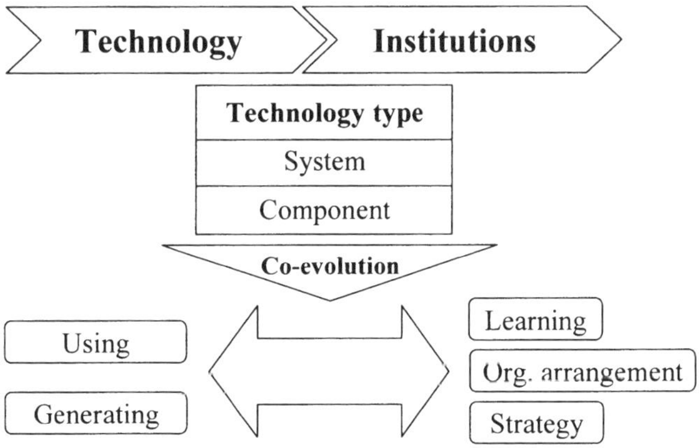

CASE EXAMPLES OF TECHNOLOGICAL CAPABILITY ACCUMULATION IN THE KOREAN INFORMATION AND COMMUNICATION SECTOR

The following three cases—TDX, DRAMs and CDMA—represent the technological strategy, managerial routines and institutional arrangements for innovation in the Korean IT sector since the late 1970s. In this section the stylized facts of these representative technology development cases will be displayed to understand the mechanisms of technological catch-up and characteristics of innovation systems. In particular, the pair case of TDX/CDMA and 4M DRAM/16/64M DRAM cooperative development programs showed the changing mechanisms of learning and organizational/institutional arrangement in two different technology development stages, using and generating.

TDX System Development (1976-1991)

Background

The development of the TDX project started in 1976 to resolve the telephone backlog. It usually took more than a year for telephone line installation and to accumulate technological capability in the digital switching system to foster the domestic telecommunication equipment industry. In addition, the government expected that enhancing domestic switching technology could generate an import-substitution effect, which heavily relied on import of foreign telecommunication equipment. The Korean government formed the Telecommunication Development Task Force (TDTF) in 1976. The TDX project consists of three phases of research and development: preliminary digital exchange development, TDX-1 series, and TDX-10.

The development switching system requires a huge resource mobilization in terms of R&D manpower and investment. The TDX-1 series called for US$57.5 million and even the development of the TDX-10 required US$142.7 million. It was a government-driven large-scale project, in which the government Electronic Technology Research Institute (ETRI), Korea Telecom, and four major manufacturing companies participated.

Technology Development Process

Since the base of technological capability was weak, the stage development model was adopted for technology development process. The preliminary development phase focused on technology import and assimilation. Korea Telecommunication Corporation (KTC) and Lucky-Goldstar (LG) imported production technology for an analog switching system from BTM.[1] In the first place, Korean firms imported the technology in a turnkey base. However, the condition of imported technology focused on the training of Korean engineers for assimilation of imported technology and technology acquisition for important parts and components such as chips.

The secondary stage of development, started in 1982, focused on technology assimilation and in-house development. According to the parallel approach, both technology import and in-house development was adopted for system development. In this period the organizational arrangement and working routines are set up for system development. The R&D organization for TDX was established in ETRI and, at the same time, the cooperative development framework between public research institutes and public/private companies was set up in this period. Three manufacturing firms, including Lucky-Goldstar, Dongyang and KTC participated in the development process. Particularly, the involvement of Korea Telecom (KT), the main user of the exchange system, speeded up the system development as it defined the working routines and coordinated R&D management processes. In 1984 ETRI succeeded in the development of the pilot system of TDX-1. After the pilot development, ETRI transferred the technology to the selected four private companies for commercialization. This included design technology, production technology, and operational technology through the transferring of technology documentation, technology training for engineers, and technology supports. However, the performance of TDX-1 was unsatisfactory due to the system instability. Complementing in-house development, ETRI imported the digital exchange technology on design technology and operational technology from Ericsson/Erifon. Based on the imported technology and previous in-house development experiences, ETRI developed the digital switching system in 1986.

The third phase of development started in 1987 and focused on the development of the large-scale exchange system, TDX-10. A cooperative development framework among public institutes, manufacturing firms, and main users in the previous development stage continued and was reinforced. In 1991 the TDX-10 system was developed and a commercial field test was begun.

Technology Learning Strategy and Organizational Structure

During the development process, the following three factors were important to achieve technological accumulation in digital switching system. First, the technology acquisition strategy in the beginning of technology development focused on fast learning. In the technology import from BTM, a large number of engineers were dispatched to acquire technological knowledge. Also in the case of Ericsson, 30 engineers were dispatched for more than two and a half years and 57 specialists from Ericsson were invited to train Korean engineers.

Second, the government-sponsored cooperative learning organization for system development had been very effective in responding to main users' needs and to diffusing technological knowledge to manufacturing firms. In the second phase of system development, the organizational arrangements for system development and working routines were settled. In the cooperative development framework, ETRI organized a whole process of system development and system design; KT proposed product specification and technological requirements; manufacturing firms developed subsystems through competition. The engineers from KT and other manufacturing firms played critical roles in the technology-diffusion process as technology liaison and system experts. The cooperative development organization was shaped to shorten the development period and maximize the diffusion effect to manufacturing firms in the commercialization. The cooperative development framework in the TDX project provides the organizational foundation to national projects that followed in the IT industry.

Third, it is important to indicate the user involvement in the development process. As in the advanced countries, main users in the development of complex product systems play a critical role in the system development as they provide technological knowledge, accumulated in the operating process.[2] Engineers from KT participated in the system development and later played a critical role in the commercialization process as a professional user. On the other hand, KT dispatched its engineers to foreign suppliers of digital switching systems[3] to learn the testing and evaluating exchange system. Based on the testing and evaluating knowledge from foreign vendors, KT defines exact specifications in the domestic digital switching system. It helped to define the technological specification and development goals in the development process.

CDMA (1990-1997)

Background

The case of CDMA system development represents another success story of the Korean system of innovation. As the need for mobile communication increased, the Ministry of Information and Communication (MIC) initiated the development of a digital mobile telecommunication system. In 1990, ETRI launched its digital mobile telecommunication system development project, which encompassed a mobile telecommunication switching system, a base station, and mobile phones. The large-scale national project continued to 1996 at a cost of US$77 million and the participation of 1,042 human resources.

The project consisted of three phases of development in cooperation with Qualcomm, a U.S. venture firm that had basic technology in CDMA. It includes system specifications, field testing, network design, basic structure, and design of the mobile handset. Based on the basic technology, which was acquired from Qualcomm and assimilated through the national-level cooperative project, four private firms developed their own commercialized versions. In 1994, the first pilot product was developed and the project was ended as the world's first CDMA-based commercial personal communication service (PCS) was launched in 1996.

The commercial success of the CDMA system blossomed from 1999. The export of the CDMA system reached US$10 billion in 2001. Samsung, the Korean leader company in the CDMA system, took first place (30% of market share) in the world CDMA handset market in 2001. The product range for export includes not only the CDMA handset but also CDMA network operating technology and consulting on the establishing CDMA network. It implies that the Korean CDMA industry deepened its knowledge base in a broad range of CDMA technology.

Technology Strategy

The technology acquisition strategy in the initial stage of system development displays an intriguing case in the technology management research. When Korean industry considered the development of a mobile system, the world standard was still dominated by the analog system and Time Division Multi Access (TDMA). MIC and chief engineers in ETRI paid attention to the CDMA technology, because it has a relatively low entry barrier, hence the possibility of catching up in a short period seemed high. Despite its high technological and market uncertainty, the Korean government selected CDMA technology to enjoy first-mover advantage in commercialization of the CDMA system.

The core technology of the CDMA system was imported from Qualcomm. When ETRI formed the cooperative research consortium, Qualcomm participated in the CDMA system development to provide technological knowledge. Qualcomm also enjoyed the benefits of testing and implementing its system design in the field. Despite the reliance on Qualcomm, the core technology still continues,[4] and the Korean manufacturer can enhance its technological capabilities to develop its core part, an MSM-electronic chip, in-house. In addition, the product range for export diversified to knowledge-intensive consulting on network operating and management.

Technology Learning Management and Organizational Structure

In 1990, ETRI launched a large-scale national project on its digital mobile telecommunication system development project. Various actors were involved: ETRI, designated domestic manufacturing firms, Qualcomm, and mobile service providers as shown Table 3.1.

Table 3.1
Technology and Institution: TDX and CDMA

	Strategy	Learning	Arrangement	Product
Using	• Technology import from foreign vendors (Ericsson)	• Turnkey-based →assimilation • User (KT) implementation	• Large scale with cooperating R&D program • Technology diffusion to private (components) sector • Key player of Public Research Inst.(ETRI)	TDX 1 Switching system: → TDX10 and CDMA
Generating	• Self-development of switching systems (TDX-10) • Cooperating development with foreign vendors (CDMA system)	• In-house development with cooperating R&D • User interaction for system development • Early entry to emerging standard	• Development via competition, division of labor, and coordination • Based on TDX management structure for CDMA development • Key player of Public Research Inst.(ETRI)	

During the development process, four manufacturing firms participated, including Samsung, Hyundai, Lucky-Goldstar, and Maxon Electronics. Among the participating firms, there was division of labor in terms of each subsystem. For example, Samsung and Hyundai developed the control station while LG and Hyundai developed the base station. The role of ETRI concentrated on the integration of the whole system and coordination of the development process.

DRAMs (1986-1993)

Background

The development of DRAMs in Korea has been recognized as a representative successful case in the developing countries' technological capability accumulation. The Korean semiconductor business is characterized by a highly concentrated market structure, focused on a highly specific product, especially DRAMs. This tendency toward concentration is related to the strategies of the *chaebols* that historically have focused on the mass production of standardized products. Since entering into the semiconductor business, the Korean *chaebols* have caught up with the world's most advanced companies in a short period and have become world leaders, at least in the DRAM area. Exports of Korean semiconductor firms increased from US$1 billion in 1985 to US$3 billion in 1988, and then to US$5 billion in 1990. Moreover, the share of Korean DRAM products in the world market grew from 15.7% in 1989 to 20% in 1993 and 29% in 1994.[5]

It is possible to distinguish two phases of technological development of DRAMs, technology-using to technology-generating. From the start to 4M DRAM development, the Korean semiconductor industry achieved a rapid learning process of its design and fabrication technology. The Korean leader firm developed 4M DRAM products based on their own design, although the level of technology still lags behind their major competitor in United States and Japan. From 16M DRAM onwards, the Korean semiconductor industry equipped technological capabilities compatible to world leader firms both in design and fabrication processes. It can be suggested that the Korean semiconductor industry entered into the technology-generating phase.

Technology Acquisition Strategy

In the early industry development process, ETRI has supported private firms in each phase of development. In the very first stage Korean Institute for Electronics Technology (KIET), the forerunner of ETRI, built up a substantial facility for, and considerable competence in, semiconductor technology. KIET conducted research on semiconductor technology, including design and process technology, for the purpose of supporting direct industrial production.

After the launch of the semiconductor industry, the public sector drive to the large-scale cooperative R&D program to learn basic technology and diffuse technological knowledge to private firms, as will be described in the next section. In parallel with public research and development, each *chaebol* firm licensed advanced technology from foreign firms and made consistent efforts for assimilation and adaptation. Three major semiconductor suppliers from the late 1980s developed their products based on in-house design. In product development, Korean *chaebol* firms, particularly the leader firm Samsung, continuously caught up to the 4M DRAM with advanced competitors, and achieved the leadership from the 64M DRAM onward. Korean follower firms, Hyundai and LG, also closed the time gap of product development and mass production. These data clearly show the technological generating capabilities of *chaebols* as they develop new products ahead of competitors.

As described, the main sources of learning shifted from 4M/16M DRAMs onward. In the initial stage of the semiconductor industry, three major semiconductor firms mainly relied on technology licensing from advanced foreign firms. From 4M DRAM, the Korean leader firm achieved its own design mainly based on in-house capability. As a means of acquiring technological knowledge, the large Korean firms developed various organizational arrangements, which included licensing agreements, close interaction with customers and suppliers, and in-house R&D. These organizational configurations were complex and improved as the firm's technological capabilities were enhanced. In contrast to the early stages when development relied on licensing agreements and major original equipment manufacturer (OEM)[6] partners to access technological knowledge, the major sources of technological knowledge became in-house R&D.

Organizational and Institutional Arrangements

Most of all, the government-initiated cooperative R&D program played a significant role in the initial stage of basic learning. During the 1980s, public institutes and the private sector continued to work toward achieving a consensus on how to develop technologies for semiconductors. Under the auspices of the Ministries of Science and Technology, Communication, and Trade and Industry, ETRI formed an R&D consortium in October 1986. This joint project was aimed at developing the 0.8μm process of 4M DRAM. Funding was divided at 57% (equivalent to 50 billion won) from the government and 43% (equivalent to 37 billion won) from three large companies. The consortium consisted of private firms—Samsung, LG and Hyundai—and seven universities with 11 projects, along with ETRI. The process consortium developed a 4M DRAM engineering sample in 1988 and 0.8μm process 4M DRAM technology by February 1989. In addition, 157 patents were applied for and 2,542 technical reports were prepared (ETRI, 1989). The 4M DRAM consortium had heavily concentrated in fabrication technologies, involving 62% of the total R&D manpower.

After the successful development of 4M DRAM, the government and the private sector continued to collaborate on further development of DRAM technologies. A 16M/64M DRAM consortium was formed immediately after the end of 4M DRAM development, and over a four-year period from April 1989 to

March 1993 it was involved in developing 0.5-0.6 μm 16M DRAM technology and 0.3-0.4μm 64M DRAM technology. In the 16M/64M DRAM consortium, the focus was on the development of material and equipment technology. Participation in the consortium increased to a total manpower of 1,400—with 19 universities involving 24 projects and with the two major public institutes, KRICT and KIST. In addition, major Korean semiconductor equipment/material suppliers also participated in this consortium. It implies that the object of the 16M/64M DRAM development consortium resides in the creation of new technological knowledge rather than cooperative learning of design and process technology as shown in the 4M DRAM consortium (see Table 3.2).

Table 3.2
Technology and Institution: DRAMs

	Strategy	Learning	Arrangement	Product
Using	• Development of 4M DRAM process technologies	• Previous technology capability on fabrication process • Licensing from foreign vendors • Knowledge diffusion via technical committee	• Coordination and project management by Public Research Inst.(ETRI) • Competition and cooperation	DRAM: 4M,16/64M
Generating	• Development of leading-edge Process →Design→Material (16M/64M DRAM)	• Basic and generic research • In house design • Knowledge diffusion via joint evaluation committee • New process development	• Coordination & project management, and material & equipment by Public Research Inst.(ETRI, KIST, KRICT) • Competition and cooperation • Material and equipment suppliers' (FST, Daebo, LG) participation • Strengthening basic research	

The role of ETRI was coordinating and management of participants to learn cooperatively generic and enabling technologies. Financial resources continued from the previous arrangement meant that 40% (equivalent to 75 billion won) came from the government and 60% (equivalent to 115 billion won) came from the three companies (ETRI, 1989). This case reveals that, as the private sector enhanced its in-house technological capabilities, the public sector transformed its role in technological development. The public-sector role was redefined from its role from direct supporter of industrial technology to coordinator of private technological activities, with the aim of conducting more pioneering research in newly emerging technology areas such as GaAs semiconductor and enabling technologies. Public institutions have therefore been a vehicle for semiconductor companies to collaborate to learn in key areas, and to access government funds for R&D and R&D management expertise.

KEY CHARACTERISTICS OF INNOVATION SYSTEMS IN THE IT SECTOR

The above three cases are summarized in Table 3.3. From the case studies, we can display common features of technological capability accumulation in the Korean information and communication sector. First, in the entry stage, public research institutes played active roles in acquiring technological knowledge and diffusion to the private sector. The cooperative R&D consortium was the main mechanism to assimilate foreign technology and to learn basic technology cooperatively. This helps to shorten the learning period in the first stage of development.

Table 3.3
Co-evolution Between Technology and Institution

		Strategy	Learning	Arrangement
System	Using	Existing system diffusion	Existing knowledge, training	System technology transfer to local
	Generating	Accumulation based on searching	Learning by searching	Joint system development with local and technology transfer
Components	Using	Existing Process diffusion	Learning by design and production	Joint learning of basic knowledge
	Generating	Integrating value chain	Learning by user-supplier interaction	• Basic research linkage • Back-end process firm participation

Second, the national R&D program was designed to encourage technology learning and commercializing of selected large firms in line with Korea's main economic growth strategy. The role of public institutes focused on the coordination of whole development process and system integration. It displays that the Korean innovation system in the IT sector is a large firm-centered structure under the coordination and integration of public research institutes.

Third, learning mechanisms and strategies have changed according to the upgrade of technological capabilities. In the initial stage, strategy and learning mechanisms focused on fast learning through technological acquisition from advanced foreign firms. As technological capabilities were enhanced, the main sources of learning relied more on in-house capabilities and strategies focused on the new technology creation through complex networking among various innovators, such as suppliers, basic research base, and technology-based small firms.

Fourth, it is found that cumulative learning occurs in both technology and organization. The working routine, institutional framework, and organizational structure was handed over to the other system development. The TDX system development project provides the institutional origin to following large-scale national R&D program. It is found that there were sharing of working routines, operating methods, and organizational configurations between generations and each case of large R&D programs.

Fifth is the interrelationship between technology sectors. Based on the TDX development, the technological capability in digital switching systems was acquired. It provided the technological base when ETRI developed the CDMA mobile telecommunication system. Hence the technology acquisition strategy in the CDMA system was different from the TDX development program. In the case of CDMA system development, the early stage of technology standard, developed by technology-based U.S. small firms, was adopted. Also, the relationship with the technology provider changed. Compared to the TDX system development case (importing technology in a turnkey base), in the CDMA system development program, Qualcomm worked cooperatively as a partner rather than a provider-recipient relationship. It implies that Korean innovators enhanced their technological capabilities up to entry in the early stage of technology.

IMPLICATIONS

Three representative cases in the Korean information and communication sector reveal that the institutional arrangement for learning at the national level is critical to reduce the technological gap with forerunners. A cooperative R&D consortium seemed to be effective to shorten the learning period and to diffuse technological knowledge to the private sector in the initial stage of development. In addition, through the cooperative R&D consortium, a domestic knowledge pool was shaped as diverse participants had a chance to learn cooperatively basic and generic technology.

It is also found that development experience in previous large-scale R&D programs provides the technological and managerial base to next-generation R&D programs. As shown in the above three cases, accumulated knowledge on the digital switching system in the TDX program transferred to the development of the mobile switching system in the CDMA program. As well as the technological base, the TDX program formed managerial routines and organizational framework for cooperative development, providing an institutional archetype to the national R&D program that followed. This implies that accumulation of technological and managerial experience is extremely important to upgrade innovative capabilities at the national level.

Finally, as technological capabilities were enhanced, learning mechanisms and institutional arrangements have changed to meet the advanced technological needs. It shows the co-evolutionary relationship between technology and institution.

NOTES

1. A subsidiary of ITT. ITT was acquired by Alcatel, a French telecommunication company.

2. M. Hobday (1998).

3. KT has purchased a foreign digital switching system as a monopoly user in Korea. Hence, KT took a advantage of bargaining power. In the purchase decision process, KT proposed the condition of training of domestic engineers to learn evaluation and testing exchange systems.

4. The CDMA handset manufacturer pays royalty to Qualcomm, 5.75% of sales per handset unit.

5. Federation of Korean Information Industries, Korean Information Industry White Paper 95.

6. OEM is a form of subcontracting, by which local firms, frequently in developing countries, produce under foreign brand names. OEM contracts allow high volumes and the reduction of production costs, based on low-cost labor.

REFERENCES

Ernst, D. (2000). "Catching-up and Post-crisis Industrial Upgrading: Searching for New Sources of Growth in Korea's Electronic Industry," East-West Center Working Paper.

Ernst, D., Mytelka, L., and Ganiatsos, T. (1994). "Export Performance and Technological Capabilities—A Conceptual Framework." In D. Ernst, T. Ganiatsos, and L. Mytelka (eds.), *Export Performance and Technological Capabilities: Lessons from East Asia*. Geneva: UNCTAD.

ETRI (1989). White Paper on TDX Development Project.

Hobday, M. (1995). "East Asian Latecomer Firms: Learning the Technology of Electronics," *World Development*, 23 (7): 1171-1193.

Hobday, M. (1998). "Product Complexity, Innovation and Industrial Organization," *Research Policy*, 26 (6): 689-710.

Mathews, J. and Cho, D.S. (1999). "Combinative Capabilities and Organizational Learning in Latecomer Firms: The Case of the Korean Semiconductor Industry," *Journal of World Business*, 34 (2): 139-156.

Nelson, R. (1994). "The Co-evolution of Technology, Industrial Structure, and Supporting Institution," *Industrial and Corporate Change*, 3: 47-64.

4

Solving Defense Problems Across the Borders: Military and Industrial Reforms in Europe's New Security Environment

Ioanna Boulouta

INTRODUCTION: Defense-Sector Transformations, a "Paradigm Shift"?

In 1962 Thomas Kuhn wrote the *Structure of Scientific Revolution* and fathered, defined, and popularized the concept of "paradigm shift,"[1] a revolutionary change where "one conceptual world view is replaced by another." This does not just happen; it is rather driven by agents of change. According to Donnelly we are now in the midst of such a "revolutionary change" driven by the new power balance that has emerged after the end of the Cold War and the collapse of the bipolar security system and by the rapid advances of technology.[2]

A study of history shows us that approximately every 50 years the world experiences such a revolutionary change in the nature of armed conflict, provoked by sociological, technological, or other factors. For example, information technology can act as a catalyst to such a revolutionary change. We are shifting from a mechanistic, manufacturing, industrial society to an organic, service-based, information-centered society. As a result, reliance on information technology has rendered our society very vulnerable to certain forms of terrorist attack and terms such as "cyberterrorism" have come to fruition today. More and more technology insertion in the armed forces seems inevitable while increases in technology impact globally.

The current paradigm change seems inevitable and, as a result, all nations around the world are faced with the great and urgent task of reassessing what constitutes security, what are the threats to security, and what should be the responses. The events of September 2001 have brought this into sharp focus. Therefore, military and industrial reforms will dominate the strategic landscape of the 21st century.

So what exactly is this transformation? What does the change involve? It is certainly about a new security scene, which generates new security threats and requires new responses. It is certainly about technology, but it is also about changing institutions, structures, doctrines, and ways of doing business.[3] Ultimately, it's about being able to deal effectively with the myriad threats of the 21st century.

But whatever the transformation may be, whatever these reforms may involve, three things are certain:

1. The delay between recognizing changing demands and the creation of appropriate structures to satisfy them often creates a security gap. The greater the gap the greater the danger the threats pose to security. Hence transformation need is urgent.
2. This transformation cannot happen automatically. It faces enormous resistance that is not due to human beings' natural resistance to change, but it is rather the result of large political and economic obstacles.
3. It takes the efforts of all nations to find appropriate ways to transform and to find solutions to their problems.[4] There are no ready answers. The development and recent progress of the European Security and Defence Policy (ESDP) promises to help nations in finding solutions to the challenges of transformation.[5]

But will European nations overcome the political and economic obstacles to allow ESDP to develop and meet its goals? ESDP may be a political reality today but there is considerable amount of work yet to be done before it becomes a practical reality.[6] As the European Union (EU) high representative J. Solana emphasized in his speech to the Institute of European Affairs in Dublin in May 2003: "We can each contribute in different ways to the development of a European security identity. We can opt not to participate in specific operations or specific measures. But we can none of us opt out of the search for security. The question for each of us is not *whether* but *how*."

THE NEW SECURITY ENVIRONMENT: RETHINKING SECURITY; NEW THREATS AND NEW RESPONSES

The end of the Cold War has had a profound impact on the old security thinking. "National security" was synonymous with "defense."[7] The threats of invasion and territory attack were common and have been the most important national security fears. War was seen in the context of East-West conflict. It would be massive, even total, and certainly would take place between militaries. Deterrence was the job of the armed forces, backed up by the threat of nuclear weapons. "Security" was measured largely in military strength. Today "security" has become a much broader issue.

For most European countries, security today is primarily measured in nonmilitary terms and threats to security are nonmilitary in nature. What is today under attack is no longer the territory of the state but the nature of the society. Security experts, at large, used to focus only on military issues and ignored such questions as domestic stability, the legitimacy of political institutions, the competence of political elites and their ability to guarantee public order, law enforcement, and economic welfare. Many of these threats have not traditionally been viewed as security matters. According to NATO's new security concept as agreed between the heads of member states in 1999 in Washington,[8] some of the 21st-century threats to security are incompetent government, corruption, organized crime, terrorism, smuggling, illegal migration, ethnic and religious conflict, proliferation of weapons, shortage of natural resources, and, of course, terrorism.9 These threats are more difficult to define than purely military ones and more difficult to counter.

In the old security thinking, business was not directly involved in national security.[10] Security, for the business world, was mostly protection against competition or fraud. The major focus has been on economic rather than on security issues. But today, big business may be the actual prime target of terrorism pursued with a political, rather than an economic, motive. Security has also become the major determining factor of foreign direct investment (FDI). For example FDI doubled in Poland after the country joined NATO.[11]

We have witnessed great changes in the security environment. These have triggered a new security thinking and made the need for new responses obvious.

TRANSFORMATIONS

The Need to Reform

Since every change begins in the minds of people, the need to understand the new concepts of security precedes any reforms. According to Donnelly, it is less than 10 years since the concept of national security has been properly understood in the new democracies of central and Eastern Europe, where "security" to most people meant the work of the secret police.[12] Today, new concepts of security are widely understood in Europe, paving the way for reforms, but, as noted earlier, the delay between recognizing changing demands and creating appropriate structures to meet them creates a dangerous security gap. The less a government makes adequate provision to reform in order to meet these new threats, the more serious will be the danger that the threats pose. Hence the need for reform is urgent.

Today many international organizations, including NATO and the European Union, are expected to evolve to meet nonmilitary threats to security. Meeting these new security requirements demands fundamental reform and large financial investments on national structures, international institutions, and systems of government.[13] As NATO Secretary General Lord G. Robertson emphasized in his Mountbatten Lecture (February 2001), "Security in Europe is still a work in progress."

Defense structures today generally reflect older approaches and concepts of security and are set up to deal with "defense" rather than "national security" issues. The situation became critical for Europe during the ethnic conflicts in the Balkans. The severe capability gaps and inefficiencies of European armed forces as well as the United States' outstanding supremacy in its military, technological, and industrial base became apparent.

The urgent need to reform has been obvious to all Europeans. There is a tendency today to believe that the most difficult issue is to understand the new security scene and to identify new threats and define new responses. Indeed, this is the first and most difficult step, but today it seems as we have gone through this step. The challenge for the present and the future is not whether we need to reform but how. The implementation of these reforms is what troubles most European countries today.

Reshaping the European Armed Forces

During the Cold War period most countries had relatively large armed forces based on conscription, large-scale mobilization, and designed to fight in defense of national territory. European members of NATO did not have to spend much to have credible defense since they relied on mutual support, particularly on U.S. support, and deterrence by nuclear backup. The threat was common for all states as was the response. Dominated by the Napoleonic concept of mass armies, procurement policies used to emphasize force size and structure rather than capability. For example, it was more important to buy more aircraft than the electronic warfare (EW) capability. After the end of the Cold War, most European countries reduced their defense budgets and force structures considerably. Conscription periods were shortened. Equipment was not upgraded. Training was cut back. European NATO armed forces under decreasing defense budgets allowed themselves to become dependent on U.S. "force multiplier" technologies, which are available because of coalition warfare.[14]

As a result, today Europe lacks certain up-to-date capabilities and cannot effectively deploy its forces out of area without U.S. support. One of the most unpleasant conclusions from the Kosovo war was that the United States provided 70% of the aircraft and 80% of total weapons delivered. Europeans could not perform in the framework of the chosen strategy without U.S. support, and the lack of command and control systems (C2), suppression of enemy air defenses (SEAD), and offensive electronic warfare (OEW) has not been less embarassing.[15]

Since then, the European Union states have set themselves the ambitious objective of constituting self-sustaining forces able to perform the full range of "Petersberg tasks" (these tasks include humanitarian and rescue missions, peacekeeping and crisis management, and peacemaking).[16]

Most European countries today face the difficult task of fundamental military reform. Many journalists commentating on the reform issue have concentrated on the need to buy high-tech equipment to match U.S. capabilities or on the need for command, control, computing, communications, and intelligence (C4I) systems. Indeed, new technology promises to offer the potential to achieve military objectives in different ways and be able to do completely new things on the battlefield. At the same time, the increasing development pace of civil technologies has led to rapid obsolescence of many components and subsystems already embedded in defense systems. As a result, a continuous upgrade and increases in technology insertion in the armed forces seem inevitable.

The governments have also recognized the military advantage that rests with those who most effectively identify and exploit technology. Many ministries of defense, including the British Ministry of Defence, have decided to invest heavily on emergent and new military and civil technologies.[17]

Trying to meet the ambitious objectives of ESDP, Europeans have decided to strengthen their capabilities in many areas. According to the Western European Union's (WEU) "Audit of Assets and Capabilities for European Crisis Management Operations" in November 1999, Europeans, in principle, have the available forces and resources needed to prepare and implement military operations over the whole range of Petersberg tasks. Nevertheless, the audit identified a number of gaps so considerable efforts are needed to strengthen the European capabilities.[18]

With regard to collective capabilities, these areas include strategic intelligence and strategic planning, and with regard to forces and operational capabilities these areas include:

- availability, deployability, strategic mobility, sustainability, survivability, interoperability, and operational effectiveness;
- multinational, joint Operation and Force Headquarters with particular reference to command control, and communications (C3) capabilities, and the deployability of the Force HQ.

As François Heisbourg points out, regarding collective capabilities, the following areas need improvement or could be developed:[19]

- *Strategic intelligence and information pooling.* The Europeans in general should have better access to commercial and dedicated high-resolution satellite imagery.
- *Deployability and mobility.* The EU could prepare the establishment of a European Transport Command capability (Euro lift). This could generate greater efficiency and considerable life cycle cost savings through the pooling of logistical, maintenance, and training assets.
- *Sustainability and logistics.* This area includes increased standardization of materials and procedures and the implementation of common standards for increased interoperability with special emphasis on medical interoperability. Logistic support capability requirements also include shore-based facilities, to sustain armed forces effectively.
- *Command, control, and communications (C3).* The NATO summit[20] has decided to develop a C3 system architecture to form a basis for an integrated Alliance core capability allowing interoperability with national systems. The EU countries will

harmonize their efforts in this field, so that it is assured that this C3 system is compatible with, or can also be used for, EU operations or Force HQs.

- *Combat search and rescue (CSAR).* The EU should establish a European CSAR capability.
- *Air-to-air refueling.* Europe has very limited air-to-air refueling capability. One option that has been discussed is to develop a European tanker fleet, preferably under a single organization.

However, as Donnelly suggests, it would be unwise to assume that technological superiority will in all cases translate into military superiority.[21] Indeed, the rapid proliferation of technology means that even small developing countries can acquire weapons and delivery means that can pose a real threat to major powers. Even worse, when this is coupled with fanaticism the threat is even more evident. Moreover, the extended use and reliance on technology in the battlefield can add significantly to the so-called fog and friction of the war.

Furthermore, the nature of modern weaponry means that, unless the technology gap is truly enormous, a competent defender today could make a "forced entry" too costly for any country to contemplate.[22] Forces that can be projected and maintained overseas are a lot more expensive than conscript forces for national defense. For example, Israel with conscript forces has been able to maintain a much greater force than Canada with roughly the same defense expenditure.[23]

Therefore the issue of reform is not only an issue of technical capabilities but also a question of the capabilities of the important human element. A closer examination reveals that the often-mentioned "capability gap" between Europeans and the United States is not only in weaponry but also in personnel availability. This is most evident in the capability to deploy, employ, and maintain forces for ready peacekeeping, monitoring, and policing capacity.

Armies today may have to be deployed in support of domestic police operations. In addition they may have to go out to deal with the threat in the countries from which it is generated. Forces today must expect to be projected—that is, sent abroad, sustained there (perhaps over long periods), and used. This will not be passive peacekeeping or, as in the Cold War, deterrence by simply waiting. Troops must expect to fight. "We have gone in a very short time from Cold War to Hot Peace."[24]

According to Donnelly, if a modern army is to be sustained in operations, a simple rule can be applied: Land forces need to have at least three times the manpower of the actual battalions making up the force structure deployed.[25] On top of this, a large number is needed to staff the infrastructure to support the whole. But the quality of European forces is more important than the quantity. For that reason it is necessary to identify European deficiencies. As it cannot be predicted where and in what circumstances a European force will be deployed, the crisis-response task (as set in Petersberg) requires a power projection capability or an expeditionary force. Flexibility through availability, modularity, interoperability, sustainability, strategic and tactical mobility, and firepower are key characteristics of such a force.[26]

Consequently, with regard to European deficiencies, the highest priority is to meet capability goals according to the Petersberg tasks and capabilities that have to be fully congruent with the overall capabilities included in the headline goal.[27]

Therefore a European capability for autonomous action requires enhanced capabilities in the field of:

- *suppression of enemy air defenses* (SEAD) and support jamming;
- *all-weather precision guided munitions* (PGMs) and *nonlethal weapons* to reduce collateral damage and risks to own troops;
- *stand-off weaponry*, such as cruise missiles;
- *composition of forces*: such as engineer units and deployable medical units;
- *readiness and availability*: EU countries have more than 1.8 million men and women under arms but are unable to sustain an operation involving a force of more than 40,000 over a period of years.[28]

As Heisbourg points out, these deficiencies should be dealt with through the national planning processes.[29] However, the really tricky question is who is going to enforce this. EU does not have the power to do this yet, and although NATO has the clauses to pressure its member states to comply it cannot enforce this.

There is a tendency today to see the issue of military reforms and capabilities enhancement as one of aggregate defense spending. It is assumed that to acquire the advanced weapons, command and control systems, and airlift necessary to develop mobile, autonomous military capacities, European spending must increase.

However, the solution to the capability gap does not exclusively lie on an increased defense spending, but rather on a more efficient one. Financial pressures point to a shift in European policies toward more competitive and common European defense markets and away from the costly acquisitions of the U.S. technology. In this context the European defense industry restructuring is being analyzed in the next sections.

The European Defense Industry Restructuring

The environment in which European defense industries operate has changed radically during the last decade. The convergence of several economic, financial, technological, and political factors led these particular companies to restructure themselves. A large merger/acquisition wave movement that has taken place in Europe since the late 1990s is one indication of this restructuring.[30]

Economic Realities

Since the end of the Cold War, European countries have reduced their defense budgets considerably. Between the years 1992 and 1998 the defense budgets of the 15 EU member states, with the United Kingdom, France, and

Germany in the forefront, saw a drop of more than 20% (see Table 4.1).[31] With the exception of the United Kingdom these reductions have affected equipment budgets (procurement as well as R&D), which directly concern defense industries (see Table 4.2) for 1995 to 1999.

Table 4.1
Defense, Acquisition and R&D Budgets of the Six Letter of Intent (LoI) Countries and the United States, 1995-1999 (in millions of constant $US, 1997)

	Defense Budget				
	1995	1996	1997	1998	1999
Germany	34,625	32,745	26,641	26,002	23,790
France	42,240	37,861	32,711	30,703	28,353
Spain	7,243	7,014	5,942	5,888	5,464
United Kingdom	35,725	34,196	35,736	36,111	33,254
Italy	16,619	20,680	18,237	17,495	15,609
Sweden	6,290	6,253	5,021	5,241	4,350
Total	142,742	138,749	124,288	121,440	111,820
United States	274,624	271,739	257,975	253,423	252,379

	Acquisition				
	1995	1996	1997	1998	1999
Germany	3,969	3,705	2,956	3,455	3,715
France	7,952	7,588	6,465	5,620	5,242
Spain	9,98	1,243	1,012	781	744
United Kingdom	7,334	8,189	8,466	9,354	8,263
Italy	1,642	2,026	2,100	2,394	1,905
Sweden	2,485	1,943	1,671	1,895	2,205
Total	24,380	24,694	22,670	23,499	22,074
United States	46,251	43,332	42,930	43,887	47,052

	R&D				
	1995	1996	1997	1998	1999
Germany	1,981	1,850	1,487	1,410	1,262
France	5,525	4,932	3,821	3,254	3,148
Spain	299	282	242	198	170
United Kingdom	3,408	3,422	3,491	3,785	3,909
Italy	579	756	751	533	298
Sweden	163	160	158	160	95
Total	11,955	11,402	9,950	9,340	8,882
United States	36,597	35,722	36,404	36,469	35,324

Source: *The Military Balance 1999/2000* (London: IISS 1999).

Table 4.2
R&D and Equipment Budgets of the six LoI Countries and the United States, 1995-1999 (in millions of constant $US, 1997)

	Percent of equipment in defense budget					*Percent of R&D in equipment budget*				
	1995	1996	1997	1998	1999	1995	1996	1997	1998	1999
Germany	17%	17%	17%	19%	21%	33%	33%	33%	29%	25%
France	32%	33%	31%	29%	30%	41%	39%	37%	37%	38%
Spain	18%	22%	21%	17%	17%	23%	18%	19%	20%	19%
UK	30%	34%	33%	36%	37%	32%	29%	29%	29%	32%
Italy	13%	13%	16%	17%	14%	26%	27%	26%	18%	14%
Sweden	42%	34%	36%	39%	53%	6%	8%	9%	8%	4%
USA	30%	29%	31%	32%	33%	13%	13%	14%	14%	14%

Source: *The Military Balance 1999/2000* (London: IISS 1999).

The decline in defense budgets is in sharp contrast with the rise in the development costs of equipment, which becomes more and more sophisticated and complex. Studies of the evolution of the cost of equipment show that in France, for example, the overall cost of the *Mirage III* (entry into service 1960) program was FF7.74 billion (at 1992 prices), that of the *Mirage F-1* (entry into service 1973) FF26.7 billion, *Mirage 2000* (entry into service 1983) FF104.5 billion, and that of *Rafale* is put at over FF202 billion.[32]

Falling equipment budgets and rising development costs have led inevitably to a reduction in the number of defense programs. In Europe it is considered that several combat aircraft programs (Rafale, Eurofighter) are not going to coexist in the future. For European companies this development will have serious consequences. Nonparticipation in a major program may even oblige a company to leave the sector. For example the German company IWKA was obliged to sell its defense division to Rheinmetall because it was not a member of the consortium that won the contract to produce the MRAV equipment for the German Army.[33]

Technological Trends

As already explained in earlier sections, the radical review of what constitutes security and defense strategies relied heavily on an analysis of modern threats. However, this review relied also on the progress being made in defense technology. From this point of view, strategic thinking is today largely dominated by the revolution in military affairs (RMA). This U.S. concept envisages the integration of new intelligence, surveillance, and reconnaissance (ISR), and command, control, communications, and computing systems (C4) and long-range precision weapons into what is often called a single "system of systems" that gives complete dominance of the battlefield.[34] The key RMA technologies are digitization, data processing, and global positioning.

Consequently, space and cyberspace are becoming dimensions in the conduct of war in the same way as land, sea, and air.[35]

RMA-related systems are based on the combination of electronics, information, and telecommunications. One of the characteristic features of these technologies is their commercial origin: To a large extent they have not been developed by defense companies but by civilian firms. One thus sees an important flow of technology from the civil to the military sectors.[36] The growing role of civil technologies in the RMA represents one of the most fundamental changes that the defense industrial base has ever experienced.

Last but not least, it is becoming increasingly difficult to define defense industries. The most innovative contributions come from sectors on the periphery of the traditional defense industry, such as telecommunications, electronics, optronics, and aerospace. It is the latter that have become the true strategic sectors and the heart of the modern armaments industry.[37]

New Relationships Between Government and Industry

The declining defense budgets on the one hand and the increasing equipment costs on the other have made governments redefine their relationships with the defense industry.

Price is becoming a major criterion for decision making compared with technological performance. The new rules of the game are: competition between manufacturers, participation by industry in the funding of R&D, a demand for gains in productivity similar to those in the civilian sector, and a responsibility of industry to ensure quality as well as low manufacturing costs.[38]

In this respect, one notes a change of governments' procurement policies. Faced with budgetary constraints, the defense sector is moving from a "regulatory mode" to a mode that is more industry-oriented and more concerned with economic considerations.[39]

This reorientation has led to new program management methods, the reorganization of procurement systems, and hence new forms of cooperation between the government and the defense industry.

The New Industry Model

Based on the previous paragraphs it is fair to say that political, economic, financial, and technological challenges of the post-Cold War era have demolished the old industry model. The drive for greater productivity and broader market access has moved away from the traditional model of a defense company, whose sole aim was to meet the requirements of the armed forces of the nation, whatever the cost of the equipment. The symbiotic relationship between states and defense industries has been gradually replaced by new forms of partnership that (more) clearly distinguish between government and business. As B. Schmitt has emphasized, as long as the government tends to behave increasingly like "real" customer, the business has been obliged to embark upon a wide-scale process of concentration, portfolio reshaping, rationalization, and internationalization.[40]

- *Concentration:* This is a means of reducing duplication, pooling resources devoted to R&D, and increasing market shares.
- *Portfolio reshaping*: This is a way for companies to redefine the range of their activities. Some have left the business by selling off their defense divisions while others have increased their presence through new acquisitions. In addition, among the groups remaining in the sector, new tasks are taken over from the armed forces in the field of maintenance and logistics.
- *Rationalization:* This involves a thorough overhaul of operating procedures and strategic cost management. Modern techniques such as concurrent engineering, modern simulation, and modeling methods can deliver greater efficiency, especially in the development and manufacturing phases.
- *Internationalization*: It is true that national markets are too small even for a consolidated industrial base. Internationalization is therefore essential but it is progressing at different speeds from country to country and sector to sector.

Whereas this process has hardly begun in naval shipbuilding, it is far advanced in aerospace and defense electronics. In these high-technology businesses, the industrial landscape changed radically over the 1998-2000 period.

Due to political obstacles, internationalization in Europe has long been limited to cooperation on specific programs. Some of these projects have led to lasting alliances, which have gradually been transformed into common structures. Under the pressure of new financial and economic constraints, these structures have in the last few years been turning into true transnational joint ventures. This phenomenon of globalization has affected Europe particularly, given its historically small and fragmented national markets.[41]

Moreover, the major companies are trying to penetrate new export markets by buying into local firms. This is another innovation in an industry that has traditionally been organized on a national basis.

The speed of the change is also remarkable since transnational restructuring has taken place before the setting up of an appropriate political and regulatory framework. Indeed, there is neither a European company status, nor common fiscal or social law. According to Schmitt, the fact that companies have nevertheless ventured into Europeanization shows just how powerful the new economic and financial constraints are.[42]

OBSTACLES AND WAYS FORWARD

Defense Spending: A "Deadly Cost Spiral"?

As most modern threats call for nontraditional military responses and require investment in paramilitary forces, police, ministries of interior, border and customs forces, crisis management facilities, and so on, it is fairly safe to conclude that as investment in internal security goes up, the pressure on defense budgets to go down is likely to increase even further.[43]

EU member states collectively spend only some 60% of what Washington allocates to its armed forces. Furthermore, due to poor coordination and basically continuing Cold War force structures, Europeans get a disproportional low return on their budgets in key areas such as procurement and R&D (see Table 4.2).

As weapons and equipment have improved in the last years, their cost has risen much faster than the rate of inflation. According to Donnelly,[44] as forces modernize, if they retain the same size of force structure, the cost of equipment procurement as a percentage of the overall budget will double in real terms every few years. If the percentage of gross national product (GNP) allocated to defense is constant and if GNP does not grow annually by a considerable amount, then the costs of procurement will lead inevitably to a reduction in the size of the force structure.

As forces need to become more flexible, versatile, and capable of being sustained abroad, their cost will increase and the size of force that can be afforded will drop.

Concluding, as defense budgets reduce dramatically and at the same time the technical complexity of new equipment and the associated costs keep their upward trend, the restructured and smaller armed forces will require equipment in smaller numbers. With fewer numbers, economies of scale in production could less and less be realized, driving the costs up. The armed forces will then have to pay a higher price for their equipment, under declining budgets. As they try to maintain a force structure incorporating all relevant and needed capabilities, a further reduction in numbers of armed forces will be necessary. Hence, a "deadly cost spiral" has been created.

More than anything else this is what drives countries to conduct defense reviews. The politician who promises that "leaner will be meaner" and "smaller equals better" is in fact making virtue out of necessity.

The U.S. Competition

Since defense equipment costs are rising and European members are facing stable or declining defense budgets, the pressure on European governments and industry to increase their competitiveness increases. Since the U.S. government has decided to shift a large amount of money (the U.S. defense budget is expected to increase faster than inflation reaching US$408 billion by 2007),[45] to improve U.S. defense capabilities in fighting the "war on terrorism" this pressure is going to increase even further.

To the extent that ESDP implies an autonomous military capability,[46] the Europeans cannot afford to be solely, or even principally, dependent on the U.S. defense industry, whose primary loyalty (in terms of product definition and overall service) is naturally to its main customer, the U.S. government. At the same time, it would be unreasonable, given current budget constraints, to exclude U.S. defense equipment from the European marketplace. The Europeans simply cannot afford to pay a significant and systematic premium for a product simply because it is non-American (even if such a premium may be called for in a small number of hypersensitive areas, which the Europeans—like the Americans—do not wish to expose to foreign eyes); nor should they deprive themselves of the pressure that American competition puts on the prices and performance of products offered by European industry.[47]

In other words, the Europeans have to pursue two goals simultaneously:[48]

- Extraction of the best value for money in terms of military equipment. For this purpose, a thorough overhaul of the procurement process—the demand side of the equation—is in order; as is a comprehensive reordering of the defense industry—the supply side of the equation.
- The capability to compete and cooperate with the American defense industry in both the U.S. and European marketplaces. Here, the reform of the supply side is paramount.

Today, many legal and political questions arise with the restructuring of the European defense industry that are directly linked to the internationalization of industries. National regulations regarding armaments are particularly complex in Europe, and for historical and cultural reasons they lack homogeneity. They make the operation of transnational companies very complicated and therefore present a major obstacle for the transformation of the industry.

Political Initiatives to Overcome Obstacles

Given the central role of states in the field of armaments, governmental support has been an essential condition for the transnational consolidation of defense companies. Governments have intervened more or less actively in the process in accordance with their influence and political will.

While the supply side has reorganized under the leadership of industry, it is up to governments to reform both the market's regulatory framework and the functioning of the demand side. It is a matter, on the one hand, of creating the appropriate conditions for transnational companies to operate in an efficient way and, on the other, of safeguarding states' interests vis-à-vis an increasingly transnational defense industrial and technological base.[49]

To gain full advantage from a consolidated industrial base, governments must change their mode of cooperation throughout the procurement process and redefine their role of customer, sponsor, and regulator. It was against this background that two governments' initiatives were put forward: First is Organisme Conjoint de Cooperation en matiere d'Armements (OCCAR), created by a four-way treaty (Britain, France, Germany, and Italy) and signed in September 1998.[50] The OCCAR treaty contains two major innovations: the first is that the principle of *juste retour*[51] (fair return) is no longer required at the level of specific armaments programs, but only at an overall level, where it is to be measured in net financial terms rather than in work-sharing terms. This should help to make it possible to give priority to economic efficiency versus a "social security," entitlement-based approach to arms procurement.

OCCAR is also empowered to run programs, including the exercise of contractual responsibilities. Hence, OCCAR has, in fact, supranational powers but these powers flow from the member states' initiative and they are not imposed from the outside, as was the case with previous initiatives by the European Commission. OCCAR's primary mission is to run cooperative programs but nothing prevents member states from handing over to it the responsibility of running the tendering process for national programs. In this context, it seems that OCCAR has substantial potential.

The other process is the Letter of Intent (LoI) signed in July 1998 by the defense ministers of the six major European armaments-producing countries (France, Germany, Italy, Spain, Sweden, and the United Kingdom) with a view to harmonize procedures and policies.[52] In other words, it is an initiative to offer a politically permissive environment for defense industries to restructure. Six working groups were set up, dealing with security of supply, export procedures, security of information, research and development, harmonization of military requirements, and treatment of technical information.

Harmonizing Procedures and Policies Across Borders

It is worth having a brief glance in the six areas mentioned above to illustrate the complexity of financial, political, technical, and legal issues involved. Although most of the technical issues have been dealt with successfully, the political issues are slowing down the overall restructuring progress.

Security of Supply

From the governments' point of view the Europeanization of industry could lead to an unbalanced geographical distribution of capabilities among the states. Since a country participating in a given program would like an appropriate share of it to take place on its territory, there is a great danger for transnational firms to distribute workload in an inefficient way according to a political rather than an economic logic.

This area also is concerned with the ownership of these companies and their shareholder structures and whether the firms' capital should be completely open to foreign investors.

Security of supply does not concern just governments, but also companies. Defense companies must be sure of obtaining a component or subsystem produced in another country without difficulty. In the absence of a common armaments market, this type of transfer requires firms in Europe to go through long, nonharmonized export procedures.

In this area, LoI countries have agreed on several general guidelines but have not reached agreement on common regulations. The six countries are in agreement that transnational companies should be free to distribute industrial capabilities according to economic logic and their own commercial judgment, but reserve the right to keep certain key strategic capabilities on their respective territory.[53]

Export Procedures

The area of exports is particularly complex because it covers various types of export such as:

- transfers of components and subsystems as part of international cooperative projects;
- the export of an item produced through international cooperation to a third, European or non-European country;
- export of a national product to European or non-European countries.[54]

Among the various types of export and transfer procedures the nonhomogeneity of many national regulations and the absence of common export policy hampers industrial cooperation and the functioning of transnational companies.[55]

A global project license has been the result of these negotiations, specifically authorizing the transfer of components and subsystems. The same procedures may be applied at the request of the companies concerned for industrial cooperation not conducted pursuant to an intergovernmental program.[56] For industrial cooperation outside the framework of an intergovernmental or approved industrial program, the six countries undertake to simplify transfer procedures. Participants must all agree on a list of permitted destinations before exporting a system produced in a cooperative program. If consensus cannot be reached regarding a specific country on the above list, then it can be removed on request.

Security of Information

Harmonizing security of information regulations raises many technical problems: the security clearance of personnel and sites, access to classified information, protection and transmission of data, and so on.

In this area the challenge is to ensure that appropriate security provisions for the protection of classified information are enforced within transnational defense companies without putting unnecessary restrictions on the free movement of personnel, information, and materiel between the different countries.

In the area of security of information, progress has been made on the security clearance issue. Security clearance given by a country will now, for a given program, be recognized *ipso facto* by the other participating countries. In the same way, personal security clearance will allow an agent to carry classified documents from one country to another permanently (formerly, authority was required for each mission). However, adjustment of existing regulations will require further time and energy.[57]

Research and Development (R&D)

Research and Development is the basis of a competitive industry. However, luck of coordination of R&D policies between the countries, luck of common vision for future technological requirements, and systematic exchange of information on R&D programs results in much duplication. The harmonization of procurement processes seems the only way to avoid this duplication. There have also been attempts to set up a European R&D system including an organization of managing and contracting R&D programs.[58] The challenge here includes issues such as the appropriate method of awarding contracts (competitively or not) and application of the principle of *juste retour*.[59]

However, so far there has been an agreement between the six LoI countries to keep each other informed of policies, strategies, and programs in this field, and to coordinate their respective relationships with transnational defense companies.

Harmonisation of Military Requirements

Harmonization of military requirements is essential for both industry and governments. Industry will be able to improve competitiveness through rationalization of manufacturing methods, while governments have a lot to gain from a combined purchasing power and interoperable armed forces.

For the last few years there have been a growing number of initiatives in this area: from NATO, the Western European Armaments Group (WEAG), and the Letter of Intent. The results of these have, however, been modest. Geostrategic orientation creates conflicting priorities between European countries. Hence, different military doctrines cannot lead to same specifications, even if there are the same requirements. A common defense policy appears to be the only promising way to solve such problems.[60]

In the field of harmonization of requirements, the treaty outlines a program of future work. The six countries are willing to produce a common military concept, harmonized acquisition planning, a common profile of future investment, and common user requirements.[61]

Treatment of Technical Information

In some countries intellectual property rights belong the companies while in others they belong to the government. Hence, practical problems in the treatment of technical information arise in case of cross-border mergers.

A common framework will be needed that allows a deregulation of the industry in a way that, on the one hand, provide a guarantee to governments that the creation of a transnational enterprise will not affect their rights concerning technical information, and, on the other, assure industry that governments will not interfere in the running of the company if it is not necessary.[62]

Progress on Harmonizing Procedures and Policies

The LoI agreement has been a powerful initiative toward agreement on harmonization of policies. However, most areas need further discussions. In view of the complexity and the politically sensitive nature of certain subjects, this is hardly surprising. In general results have been more concrete in technical areas than in political areas.[63]

It seems that there are many more problems to be solved and the negotiations will not end soon. The interesting thing to see in the future will be to invite these two agreements (OCCAR, LoI) to more European countries and reach agreement at the EU level.

However, at the EU level, nothing has been achieved beyond the Amsterdam Treaty's rhetoric on a common European Security and Defence Policy and on cooperation in the field of armaments declared in a statement of intent in the Amsterdam Treaty.[64]

The reasons seem to be that the nation-states' interests differ a lot, it is difficult to accept any supranational approach, and any attempt to establish a European armaments agency (as an all-EU-member-states enterprise) is not very likely given the deep differences of interest between those countries that have a defense industry and those that do not (or those that have a nascent defense industry that they feel requires protection).

In this context tighter processes (like the OCCAR and LOI agreements) that focus the energies of like-minded states with converging interests hold more promise of progress. However, both OCCAR and LOI start from a fairly narrow base in terms of member states. Therefore the only way forward is toward greater inclusiveness.

CONCLUSIONS

We have witnessed a revolutionary change in the European security environment. The end of the Cold War and the rapid advances of technology have been the real agents of this change. As a result, we are faced today with the urgent need to reassess what constitutes security, what are the 21st-century threats, and what should be the responses.

Military strength is the most fundamental element of power. On one hand it is too easy to think that military force is the only constituent of national security, while on the other hand it is equally possible to underestimate the role of military power in achieving national security. Although security today still contains military elements, it is surely a much broader issue.

As a result, European armed forces and their defense technological and industrial bases have to adapt to a dramatically changing environment. Fundamental military reform is an exceptionally difficult and urgent task. This reform does not only include technical capabilities but also the capabilities of the important human element. It includes changes in institutions, in force structures, in doctrines, changes in quantity as well as quality, and simultaneously it requires the engagement of many elements of society.

The reform of the defense industry, regulatory, and procurement systems is all the more urgent due to budgetary problems in Europe, which seem more likely to continue; with the exception of the United Kingdom a substantial increase in defense budgets does not seem probable. The "lean-and-mean" force projection capabilities—decided by the Europeans within the European security and defence policy ambition—imply an emphasis on procurement and on the permanent availability of mission-effective materiel, which can only be delivered by a competitive, competent defense industry.

While most European members are facing stable, or slightly declining, defense budgets, defense equipment costs are rising. In this environment, it becomes impractical for individual nations to develop and produce independently the technologies and the weapon systems needed to keep pace with the concept of the revolution in military affairs. No one, nowadays, even the United States, can hope to be totally independent. Collaboration and cooperation are necessary for success.

The European defense technological and industrial base has the capabilities to deliver whatever the equipment demand of European armed forces may be. However, in some areas it will be much cheaper to buy U.S. equipment instead of developing and producing a European system. But buying equipment "off the shelf" in the United States is not always an acceptable solution to close the various capability gaps. The development of a true European approach to force capability and armament planning is the only way to set free a considerable amount of resources that are needed to improve military capabilities in due time in order to allow European states to benefit from an ESDP ambition.

A redefinition of relations with industry is also essential. This is, however, a major challenge because it covers political, strategic, military, financial, and industrial questions. As these factors can diverge, tensions emerge that slow down progress and lead to contradictions. As customers, on one hand, governments increasingly treat defense industries as "normal" industries. As regulators, on the other hand, they insist on their prerogatives regarding exports, security of information, and the like. In an area that lies between two very different worlds—defense and economics—such contradictions are inevitable.

Whereas technological, financial, and economic considerations drive defense structures to cooperation and internationalization, defense is still, through the eyes of European nation-states, a national matter.

The governments have recognized the importance of engaging in the restructuring of the defense industries. The LoI initiative signed in July 1998 by the governments of Europe's transnational defense industries provides a common legal framework to harmonize certain procedures in order to ease the day-to-day operations of transnational companies and has paved the way to overcome the obstacles to industrial transformation.

Defense procurement is one area in which maximum efficiency truly is of the essence. European defense ambitions will not be achieved if its members spend substantially less than the United States, in terms of defense capital spending, while at the same time allocating its scarce funding in a grossly inefficient manner. Common procurement structures and managing a number of cooperative programs as proposed in the OCCAR initiative (September 1998) can lead to greater efficiency in managing programmes.

There is no doubt that both initiatives, the OCCAR treaty and the LOI agreement, represent an important step in the right direction. However, this is only the beginning of a more competitive European defense industry that is much needed for the development of the type of military forces the new European security environment requires.

Nevertheless, most of the areas the reform covers will require a long-term effort and European countries should take this into account in continuing their work in detail, in the future.

The development of ESDP can help in finding solutions to many of these problems but it requires great political will and political leadership to overcome the obstacles. If European countries continue to focus their defense policies on an exclusively national level, there is little room to move forward with transformations and allow the ESDP to practically meet its goals.

However, since almost all European Union member states are facing stable or declining defense budgets while defense equipment costs keep their upward trend, it becomes impossible for individual nations to keep pace with the requirements of the new security environment. It remains to be seen whether cross-border political obstacles will be broken down by economic realities. It seems that the most viable scenario for the future is an increased cooperation between the nation states; a real multidimensional cooperation that will have to include the political, military, and industrial arenas.

ACKNOWLEDGMENTS

From March 2003 to April 2003 I was a Research Fellow at the Private Office of the Secretary General at NATO headquarters in Brussels. During this period I was glad to work in high-level policy issues under the supervision of Mr. Christopher Donnelly, Special Advisor to the Secretary General for Central & Eastern European Affairs. This chapter draws on ideas developed during that period and is an account of my research there. In this context, I would like to thank Mr. Donnelly for his important input and ideas on this work and for stimulating my thinking during our conversations at NATO.

I would also like to thank Dr. W. Nuttall, director of the MPhil in Technology Policy program at Cambridge University for his encouragement to write this chapter and for his support and guidance during my work.

NOTES

1. Thomas S. Kuhn, *The Structure of Scientific Revolutions*, 2nd ed. Chicago: University of Chicago Press, 1970, p. 10.
2. C. Donnelly, "Security in the 21st Century: New Challenges and New Responses," private paper, 2002.

3. NATO: Speech by the Secretary General at the Economist Conference, Athens, April 17, 2002, "A New Security Network for the 21st Century," available online at http://www.nato.int/docu/speech/2002/s020417a.htm
4. Ibid., 3.
5. Speech by J. Solana, EU high representative for the CFSP, to the Institute of European Affairs, Dublin, May 21, 2003, available online at http://ue.eu.int/pressdata/ EN/discours/75856.pdf; NATO fact sheets, "Strengthening European Security and Defence Capabilities," December 2000, NATO online library http://www.nato.int/ docu/facts/2000/dev-esdi.htm; Charles Grant, "A European View of ESDP," IISS/CEPS European Security Forum, September 2001, available online at http://www.eusec.org/ grant.htm
6. Solana speech, 5.
7. Donnelly, "Security in the 21st Century."
8. North Atlantic Council, "The Alliance's Strategic Concept," Washington, D.C., 1999, NATO Press release NAC-S(99)65, April 24, 1999, available online at http://www.nato.int/docu/pr/1999/p99-065e.htm
9. Donnelly, "Security in the 21st Century"; Michael Mihalka, "Political Economy, *Weltanschauung* and the Transformation of European Security," *European Security Journal*, 10 (3), (2002); North Atlantic Council, "The Alliance's Strategic Concept"; Darwin College Lectures, "The Changing World." Cambridge: Cambridge University Press, 1996, p. 129; Institute of Peace and Conflict Studies, online database in nonmilitary threats at http://www.ipcs.org/ipcs/ipcsArchives.jsp
10. Donnelly, "Security in the 21st Century."
11. UNCTAG, World Investment Report, 2000.
12. C. Donnelly, "Rethinking Security," private paper, 2000.
13. Ibid., 12.
14. C. Donnelly, "Reshaping European Armed Forces for the 21st Century," private paper, 2000; Adam Smith, "Technology Is a Proven Force Multiplier," The Hill Newspaper, special edition: Homeland Security, February 2003, available online at http://www.hillnews.com/news/022603/ss_technology.aspx; Robert Riscassi, "Principles for Coalition Warfare", *Military Review*, 73 (6), June 1993, available online at http://www.dtic.mil/doctrine/jel/jfq_pubs/jfq0901.pdf
15. NATO website, NATO's role in Kosovo, http://www.nato.int/kosovo/all-frce.htm; Jonathan M. White, "European Defense: Not More Tanks More Cops," November 1999, European Security online, Basic publications, http://www.basicint.org/ europe/ESDP/main.htm
16. NATO handbook, "Implementation of the Petersberg Tasks," NATO Office of Information and Press, Brussels, 2001, p. 364; "The European Union's Headline Goal," Centre for Defence Information, May 2002, http://www.cdi.org/mrp/eu.cfm
17. Ministry of Defence, United Kingdom, http://www.mod.uk/issues/science.htm
18. WEU Council of Ministers, "Audit of Assets and Capabilities for European Crisis Management Operations: Recommendations for Strengthening European Capabilities for Crisis Management," Luxembourg, November 23, 1999, available online at http://www.cesd.org/eu/weulux.htm
19. François Heisbourg, "European Defence: Making It Work," ISS Western European Union, Chaillot paper no. 42, September 2000.
20. The Prague Summit, November 21-22, available online at NATO Web site at http://www.nato.int/docu/comm/2002/0211-prague/index.htm
21. Donnelly, "Security in the 21st Century."
22. Ibid., 21.

23. Israel Ministry of Defense, Draft Budget for Fiscal Year 1998, available online at http://www.mof.gov.il/budget98_e/benefits.htm
24. Ibid., 21.
25. Donnelly, "Reshaping European Armed Forces for the 21st Century."
26. "The European Union's Headline Goal," Centre for Defence Information, May 2002, http://www.cdi.org/mrp/eu.cfm
27. Ibid., 26.
28. WEU Council of Ministers, "Audit of Assets and Capabilities for European Crisis Management Operations"; Heisbourg, "European Defence."
29. Heisbourg, ibid.
30. The European Aerospace Industry (EAI), "Facts and Figures 2001," AECMA Publication RP137, Brussels, October 2002.
31. *SIPRI Yearbook 2000*. Oxford: Oxford University Press for SIPRI, 2000, p. 298.
32. Paul Quilès and Guy-Michel Chauveau, "L'industrie de défense: quel avenir?" Report 203, Defense Committee, National Assembly, Paris, 1997, p. 43.
33. Burkard Schmitt, "From Cooperation to Integration: Defence and Aerospace Industries in Europe," ISS Western European Union, Chaillot paper no. 40, July 2000.
34. Robert Grant, "The Revolution in Military Affairs and European Defence Cooperation," Konrad-Adenauer-Stiftung working paper, St-Augustin, June 1998; "The Revolution in Strategic Affairs," Adelphi Paper 318, April 1998.
35. Charles Grant, "Transatlantic Alliances and the Revolution in Military Affairs," in *Europe's Defence Industry: A Transatlantic Future?* London: Centre for European Reform, 1999, p. 67.
36. Frédérique Sachwald, "Defence Industry Restructuring: The End of an Economic Exception," Les notes de l'IFRI, 15bis, Paris, September 1999, p. 17; Schmitt, "From Cooperation to Integration."
37. Schmitt, ibid.; Jacques Gansler, "The Changing Face of Arms Production and Cooperation–Technological Trends," ESAN Projekt: Arms Production and Cooperation, Projektpapier 5, SWP-AP 3002, Ebenhausen, January 1997.
38. Schmitt, "From Cooperation to Integration"; Robert Pandraud, "L'Europe et son industrie aérospatiale," report 3219, European Union delegation, National Assembly, Paris, 1996.
39. Schmitt, ibid., 33.
40. Ibid.
41. Schmitt, "From Cooperation to Integration"; Jolyon Howorth, "European Integration and Defence," ISS Western European Union, Chaillot paper no. 43, November 2000.
42. Schmitt, ibid., 33.
43. Donnelly, "Security in the 21st Century"; Donnelly, "Reshaping European Armed Forces for the 21st Century."
44. Donnelly, "Reshaping European Armed Forces for the 21st Century."
45. "From Lean Manufacturing to Systems Integration," in *Aerospace Europe—21st Century Powerhouse, Aviation Week & Space Technology*, October 5, 1998, supplement, pp. 22-30.
46. NATO fact sheets, "Strengthening European Security and Defence Capabilities."
47. Schmitt, "From Cooperation to Integration"; Andrew James, "Medium Sized Defence Electronics Companies and US Industry Restructuring," Report to FOA, Stockholm, February 2000.
48. Jens van Scherpenberg, "Transatlantic Competition and the European Defence Industries—A New Look at the Trade-Defence Linkage," ESAN-Projekt: Arms

Production and Cooperation, Projektpapier 1, SWP-AP 2992, Ebenhausen, December 1996.

49. Schmitt, ibid., 33.
50. OCCAR Convention, available online at http://www.occar-ea.org
51. The United Kingdom Parliament, House of Commons, Select Committee on Trade and Industry, "Juste Retour," Tenth Report, Section IV-22.
52. "Letter of Intent Between the Six Defence Ministers to facilitate the Restructuring of European Defence Industry," Stockholm International Peace Research Institute (SIPRI), available online at http://projects.sipri.se/expcon/loi/loisecone.htm
53. Ibid., 19.
54. Ibid., 52.
55. Jeremy Clayton, "Export Controls must meet tomorrow's challenges, not yesterday's," IBM Conference on "The Globalisation of Export Controls and Sanctions," London, November 11, 2002.
56. Schmitt, ibid., 33.
57. Ibid., 19.
58. Ibid., 19.
59. Ibid., 51.
60. Burkard Schmitt, "European Armaments Cooperation," Institute for Security Studies, Chaillot paper no. 59, Paris, April 2003.
61. Ibid., 19.
62. Ibid., 33.
63. SIPRI, July 1998, "Measures to facilitate the Restructuring of the European Defence Industry," available at http://projects.sipri.se/expcon/loi/lointent.htm; SIPRI, July 2000, "Agreement Concerning Measures to facilitate the Restructuring of the European Defence Industry", http://projects.sipri.se/expcon/loi/lointent.htm
64. The European Union Online, "The Amsterdam Treaty," http://europa.eu.int/eur-lex/en/treaties/selected/livre546.html

LIST OF ABBREVIATIONS

C2	Command and Control
C3	Command, Control, and Communications
C4	Command, Control, Communications, and Computing
C4I	Command, Control, Communications, Computing, and Intelligence
CFSP	Common Foreign and Security Policy
CSAR	Combat Search and Rescue
ESDP	European Security and Defence Policy
EU	European Union
EW	electronic warfare
FDI	foreign direct investment
GNP	Gross National Product
ISR	Intelligence, Surveillance, and Reconnaissance
LoI	Letter of Intent
NATO	North Atlantic Treaty Organisation
OCCAR	Organisme Conjoint de Cooperation en matiere d'Armements
OEW	Offensive Electronic Warfare
PGM	Precision Guided Munitions
R&D	Research and Development
RMA	Revolution in Military Affairs
SEAD	Suppression of Enemy Air Defenses
SIPRI	Stockholm International Peace Research Institute
WEAG	Western European Armaments Group
WEU	Western European Union

PART II:
BUILDING CAPACITY

5

Emerging Technologies for Change: Mobilizing Invention and Entrepreneurship for People in Poor Places

Philip E. Auerswald

The challenge is tremendous: to turn today's technological transformations to the goals of human development. The genius of what can be done through technology is astounding. But the collective failure to turn that genius to the technology needed for development is indefensible. As the potential of what can be done continues to unfold, will innovations in science and technology be matched by innovations in policy to turn global technological advance into a tool for development? This will be the ultimate test of public policy in the new technology era.

—United Nations Development Program, *Human Development Report 2001*

INTRODUCTION

On a global scale, the positive long-term impact of technology-based innovations on human welfare is evident. However, the benefits of technological change are unequally shared. Over 1.2 billion people live on less than US$1.00 a day; 2.8 billion people—ten times the population of the United States, or 45% of the people in the world—live on less than $2/day. Access to clean water, vital medicines, and basic education is still out of reach for much of the world's population. For those in the world's persistently poor regions and urban neighborhoods, the defining features of the 21st century—an efficient and interconnected global economy and rapid technological advances—are remote at best, threatening at worst. Yet in poor places no less than rich ones, technology is critical to human development.[1] To enable the inventions and innovations that will permit the benefits of technological change to be more broadly shared—in particular, improving the lives of the majority of people in the world who live in persistently poor regions and neighborhoods—is a paramount human imperative.

The agenda is as vast as it is urgent. A common presumption is that the ability to address this agenda depends critically on the policies that governments are able to implement. However, as GrameenPhone founder Iqbal Quadir observes: "In many persistently poor places, however, governments neither develop proper policies, nor implement them effectively. In social and political environments where national governments are weak or unreliable partners, efforts to envision 'public policies' to be implemented by the State may be largely wasted."[2] Even in the more democratic developing countries, elites that control inefficient monopolies in sectors critical to technology-based growth—particularly energy and telecommunications—have managed to forestall the enactment of competition-enhancing regulations.[3] Surely there is some benefit to continued efforts to induce or pressure existing governments to implement policies directed toward national goals rather than private interests. Yet, if history is any indication, waiting for governments to change is an inadequate approach to addressing the challenge of persistent poverty in poor regions.

This chapter sketches a strategy for mobilizing the energies of people in poor places. The core of the strategy is the support of invention, innovation, and social entrepreneurship by the creation of direct, personal connections between technology/institutional entrepreneurs[4] and the resources provided by international or transnational nongovernmental organizations—where resources include financial assistance, mentoring, technical assistance, nonfinancial material aid, and information.

THE ECONOMIC SITUATION FOR MANY POOR PEOPLE

The promise of technology-based economic growth has not been fulfilled for most people. The reasons are many, varied, and related in complex ways. We know that human development is undermined by disease, drought, floods, environmental degradation, discriminatory trade practices,[5] warfare, the absence of rudimentary infrastructure, and the disempowerment of women. Valid concerns exist in developing countries regarding trade and intellectual property regimes. Poverty and environmental degradation reinforce each other in persistently poor places.[6] In some parts of the world, seemingly unending warfare has itself become a thriving industry that links violence with the exploitation of ethnic divisions and natural resources.[7] Mobilizing the energies of people in poor places means working to mitigate the adverse impact of global climate change, seeking to resolve violent conflicts (particularly in failed states), ending discriminatory trading practices, managing local resources sustainably, improving the prevention and treatment of diseases prevalent in poor places, and empowering women.

As Robert Frosch observes, while these are wholesale problems, they require retail solutions.[8] Yet, almost by definition, "retail solutions"—ones based upon invention, innovation, and social entrepreneurship in particular localities—have the potential to threaten and destabilize existing patterns of social interaction. Schumpeter (1942: p. 132) summed up the disruptive effect of entrepreneurship in general:

> The function of entrepreneurs is to reform or revolutionize the pattern of production by exploiting an invention, or more generally, an untried technological possibility for producing a new commodity or producing an old one in a new way, by opening up a new source of supply of materials or a new outlet for products, by reorganizing an industry, and so on.... To undertake such new things is difficult and constitutes a distinct economic function, first because they lie outside of the routine tasks which everybody understands and secondly, because the environment resists in many ways that vary, according to social conditions, from simple refusal either to finance or buy a new thing, to physical attack on the man who tries to produce it.

Technology itself can be similarly destabilizing. The destabilizing potential of technology—for "better" or "worse," however these terms are defined—is greatest when new technologies are imported rather than endogenously created.[9] Furthermore, new technologies can be as easily employed to restrict freedoms as they can to expanding services. Idealized promises of technology often are not realized. The impacts of technology are occasionally devastating, and the world's weak and disenfranchised are frequently the ones at greatest risk and vulnerability.[10] Under what conditions, then, can technologies and entrepreneurship be most effectively and appropriately harnessed meet the needs of people in poor places?

For a generation, scholars of science and technology have rightly emphasized the importance of culture, institutions, and particular policy contexts in the shaping of technological trajectories and outcomes. Yet human inventiveness and entrepreneurship represent a common denominator that links disparate cultures and transcends geographical boundaries. The most powerful inventions and innovations do not wait for the right policies, but rather help create the context within which the right policies can be formulated and implemented. When local entrepreneurs are linked to global resources, weak and unreliable states may be pressured to transform governance.[11] Human inventiveness and its institutional context reciprocally create each other.

Inventiveness in most contexts is the rearrangement and imaginative use of existing technologies—in Schumpeter's terms, the creation of new combinations—rather than the creation of "new to the world" knowledge. Figure 5.1, taken from the United Nations Development Program (UNDP) *Human Development Report 2001*, illustrates that developing regions of the world continue to lag OECD countries dramatically in rates of adoption of such long-established technologies as the telephone, the tractor, and electric power. Even more alarmingly, in some cases rates of adoption of these technologies are slower in developing regions than in the OECD countries. These data suggest that addressing the challenge of persistent poverty may have at least as much to do with employing inventive approaches toward the adoption and adaptation of existing technologies as with the development of new technologies at the global frontier.[12]

Figure 5.1
The Digital Divide is Nothing New. Diffusion of Decades-old Inventions Has Slowed

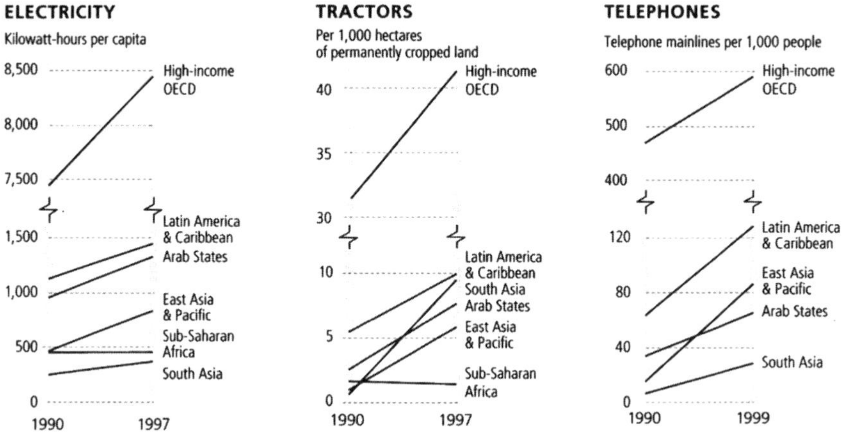

Source: UNDP, *Human Development Report 2001* (figure 2.1).

A century ago Bengali poet Rabindranath Tagore celebrated the inventive power of nontechnically trained people in a poem titled "The Invention of Shoes." The poem describes the elaborate schemes envisioned by courtiers and ministers to prevent dirt from touching the feet of the king. The schemes range from elimination of dirt, to a policy of nationwide carpeting, to wrapping the earth in leather. The poem has resonance today. In many settings—particularly those outside of health care—"low technology" adaptations of existing technology and simple inventions may offer greater opportunities to improve lives of people in persistently poor places than elaborate, remote schemes conceived at the technological frontier. This theme has been at the basis of the efforts of the "Honey Bee Network," organized by Anil Gupta at the Indian Institute of Technology.[13]

Inventors, innovators, and social entrepreneurs need many types of support. However, for the very reason that the actions of such innovators are inherently (sometimes deliberately) disruptive, local governments and institutions are as likely to suppress as to support their efforts. For this reason, the combination of effort and initiative on the part of institutional entrepreneurs working at a local level and supporting resources provided by international or transnational organizations can be powerful one. Such linkages include, but are not restricted to:

- direct support for the inventor/innovator
    - skill-development and mentoring (including peer-mentoring);
    - financial support;
    - recognition for outstanding efforts.
- dissemination of information about the invention/innovation understood in context
    - conventional publicity and press documentation;
    - case studies for both policy and teaching.
- creation of financial incentives and support networks
    - prizes, awards;
    - contracts;
    - communities of practice.

The next section sketches two examples among a set that will be the subject of case studies by the Innovations in Technology and Governance Project over the coming year.

A SKETCH OF TWO EXAMPLES: GRAMEENPHONE AND THE BAREFOOT COLLEGE OF TILONIA

The information that follows regarding GrameenPhone and the Barefoot College of Tilonia is self-reported. To develop analytically grounded case studies of such projects is the primary goal of the Innovations in Technology and Governance Project, which was launched in October 2003 by the Ash Institute for Democratic Governance and Innovation at the Kennedy School of Government at Harvard, jointly with the School of Public Policy at George Mason University, whose support is gratefully acknowledged.

GrameenPhone (www.grameenphone.com)

The Innovation

GrameenPhone provides affordable cell phone access. At the same time, it creates "new opportunities for business development, social mobilization, and political participation." The financial model for GrameenPhone is essentially the same as that for GrameenBank. A GrameenBank customer typically borrows $100, without collateral, to purchase a productive asset such as a loom or a cow. In the case of GrameenPhone's Village Phone, the productive asset is a cell phone. A woman in a village may borrow $350 to purchase a cell phone, then rent the phone to neighbors for their use. The income earned allows the woman to repay the loan, and earn a profit. In this way the phone creates entrepreneurial possibilities.

Links to International Networks

Shareholders of GrameenPhone (GP) are an internationally diverse group, including major shareholders from Norway, Japan, and the United States.

International expertise and financing was critical in building the infrastructure needed to support this innovation.

Barefoot College of Tilonia (www.barefootcollege.com)

The Barefoot College in Tilonia was founded in 1972. The college is located in a poor, arid region in Rajasthan, India. It serves a population of over 125,000 people in the immediate region, as well as some at a distance. It teaches "basic skills needed to provide critical services like rain water irrigation, solar energy provision, and wasteland reclamation to their villages." The College "encourages practical knowledge and skills rather than paper qualifications through learning by doing process of education."

The College was "entirely built by local people." The 80,000-square-foot campus includes residences, a guest house, a library, a dining room, meeting halls, an open-air theatre, an administrative block, a ten-bed referral base hospital, a pathological laboratory, a teachers' training unit, a water testing laboratory, a post office, an STD/ISD call booth, an Internet daba (cafe), a puppet workshop, an audio visual unit, a screen printing press, a dormitory for residential trainees, and a 700,000 liter rainwater harvesting tank. The energy for the College is entirely solar.

Links to International Networks

The creation of the college was "supported with funds from foreign sources and the Central Government Departments, India."

Achievements

The college self-reports the following achievements:

- provision of quality water for humans and livestock—1,317 hand pumps, 184 underground storage tanks, piped water in six districts, deepening of 175 ponds;
- effective training, installation, and use of solar energy units for night schools and for lighting homes;
- promotion of literacy—establishment of 83 night schools and 48 day schools;
- establishment of Village Education Committees; over 3,000 schoolchildren exposed to environmental awareness through the school curriculum;
- poverty alleviation—job creation for close to 7,000 people including youth, women, technicians, and artisans;
- provision of new markets for rural women and artisans;
- Income generated from sales of fodder and fuelwood from reclaimed wastelands.

NOTES

1. See Figure 5.1.
2. Personal communication. See also Quadir (2002).

3. TelMex, the dominant telecommunications company in Mexico, is a case in point. (Roger Noll, personal communication.)

4. Here we use the term "entrepreneur" broadly to refer to an institutional innovator, encompassing political, military, or social transformations as well as economic ones.

5. On this point, Stiglitz (2002) is particularly compelling, speaking from his experience as a chairman of the Council of Economic Advisors in the Clinton White House and director of research at the World Bank.

6. Stanford economist Partha Dasgupta has written extensively and persuasively on this topic (see, for example, Dasgupta, 1995).

7. For example, the Congo—a country with 150 college graduates—and Chechnya. However, it is an instructive irony that four of the countries that have most successfully implemented development strategies based on invention and innovation—Israel, South Korea, Ireland, and India—have, at the same time, been engaged in sustained conflicts with neighbors. (Ken Morse, personal communication.)

8. Personal communication.

9. Jasanoff (2002) notes that "the transnational movement of science and the artifacts that embody scientific knowledge gives rise to distinctive social and political problems, especially as societies that played no part in the original design or construction of new technologies are forced to engage with technology's widening reach." Lansing (1991) provides an interesting case study of disruption to Balinese rice production caused by the introduction of high-yield rice varieties and related cultivation practices. See also Bijker et al. (1987) and Jasanoff (1994).

10. The 1984 chemical accident at a Union Carbide plant in Bhopal, India, is one frequently cited in this context.

11. See Quadir (2002).

12. Evenson and Ranis (1990) observed that "[t]he human capacity in the developing countries to choose wisely from among technologies available outside and, more importantly, to make wise adaptations in consonance with the local environment is the single most important dimension differentiating success from failure." The UNDP *Human Development Report 2001* echoes the view: "Not all countries need to be on the cutting edge of global technological advance. But in the network age every country needs the capacity to understand and adapt global technologies for local needs."

13. See http://www.sristi.org/honeybee.html

REFERENCES

Bijker, Wiebe, Hughes, Thomas, and Pinch, Trevor (eds.) (1987). *The Social Construction of Technological Systems: New Directions in the Sociology and History of Technology*. Cambridge, MA: MIT Press.

Dasgupta, Partha (1995). "Population, Poverty, and the Local Environment," *Scientific American*, 272 (2): 40-45

Evenson, Robert and Gustav Ranis (1990). *Science and Technology: Lessons for Development Policy*. Boulder, CO: Westview Press.

Jasanoff, Sheila (1994). *Learning From Disaster: Risk Management After Bhopal*. Philadelphia: University of Pennsylvania Press.

Jasanoff, Sheila (2002). "New Modernities: Reimagining Science, Technology, and Development." mimeo.

Lansing, J. Stephen (1991). *Priests and Programmers: Technologies of Power in the Engineered Landscape of Bali*. Princeton, NJ: Princeton University Press.

Quadir, Iqbal (2002). "The Bottleneck Is at the Top of the Bottle," *The Fletcher Forum of World Affairs*, Summer/Fall: 69-89.

Schumpeter, Joseph A. (1942). *Capitalism, Socialism, and Democracy*. New York: Harper.

Stiglitz, Joseph (2002). *Globalization and its Discontents*. New York: W.W. Norton.

United Nations Development Program (2001). *Human Development Report 2001*. New York: United Nations. http://hdr.undp.org/reports/global/2001/en/

6

Industrial Modernization in Transformation Economies: Analysis of Factors and Strategies

Tobias Schauf

INTRODUCTION

Modernization of industry must play an essential part in the process of economic development in the countries under transition. Modernization requires internal strategies of enterprises as well as active support in regional and national policies. Therefore, on the one hand at the enterprise level the managers have to realize that adaptation and modernization are essential, and they should be familiar with efficient management and planning tools. On the other hand, a second prerequisite for sustainable modernization activities is the existence of a reliable macroeconomic framework and in particular a minimum standard of the national and regional technological infrastructure. Both the development of strategies and behaviors of enterprises and the important integrals of the regional and national technological infrastructures are covered in this analysis.

Since innovation processes are influenced by many factors, firms almost never innovate in isolation. Even in the case of imitating technology and for diffusion processes, it is necessary to interact with other organizations to gain and exchange various kinds of knowledge and information. Other organizations of relevance are other firms, universities, research institutes, ministries, local and regional development agencies, finance institutes, consulting firms, patent offices, self-organizations of the industry like Chambers of Industry and Commerce, and the like. Of relevance for the behavior and performance of firms are also coordination within the market, regimes to rule and incentives to create innovations by systems of industrial property rights, the existence or nonexistence of technical standards, cultural norms with consequences on business customs, and such. Several of these in different regions contain elements that constitute constraints for innovations and hinder the growth and development of firms and by this the regional and national development.

The different economic development of regions in part may be traced back to the differences in these systems and not least to the tradition, education, and attitudes of the labor force and managers as part of the systems. Therefore, a tentative benchmarking of firms' or managers' behavior and strategies and the shape of different innovation systems is suggested and may shed light on some shortcomings of the systems and lead to suggestions in respect to institution building and modification. We do not have the illusion to transfer experiences from one region to another. Nevertheless, a comparison of regions under transition with each other and with advanced regions in the West may highlight critical aspects and serve as a basis for further considerations.

An initial step to get some understanding of the potential of a region is to identify the different participants of the regional systems, these being companies and also governmental and private institutions. In federal systems like Germany these institutions can be found at the regional, national, and European Community levels. Second, the different and often overlapping activities and responsibilities between institutions have to be outlined. Third, the existing relationships between specific institutions, the potential importance for the development process, and actual use of these institutions by enterprises are outlined. To shed some more light on this picture, information available from different sources was used, including results from an expert survey and face-to-face interviews with entrepreneurs from the region, which offered valuable additional insight. To pursue this study, the Institute for World Economics and International Management (IWIM) contacted experts from different agencies and also conducted enterprise surveys and face-to-face interviews with management of the respective firms. IWIM was also engaged in enterprise surveys in Voronezh (Russia) and visits to institutions in the Voronezh district. The information and comparison of findings from Lugansk (Ukraine) and Voronezh research teams form the basis for the analysis of firms' behavior in differing environments and their relationships with different institutions of regional innovation systems.

INFRASTRUCTURE REVIEW

National Innovation Systems

For a long time technological progress was regarded by economists merely as a component of the residual factor in economic growth models. Today there seems to be a consensus within the community of academic research that the development of know-how and innovations, rather than the mere accumulation of capital, is the driving force behind economic growth, industrial change, and international competitiveness. Therefore investments in research and development and education should be seen as the basis for increasing the productive capacity of the other factors of production as well as to transform them into new products and processes. For the maximum return on these investments into economic and social benefits, the fostering of a strong knowledge base, a strengthening of the innovative capacity, and behavior of firms as well as the fostering of the conditions for diffusion and adoption of technology throughout the economy are indispensable.

Within this context, strategies directed to enhance the competitiveness under conditions of ever-intensifying global economic rivalry—or in more simple words, the modernization of industries—will be analyzed more and more under conditions of a systemic approach called the "national innovation system" (NIS).

A national innovation system may be defined as follows: "the network of institutions in the public and private sectors whose activities and interactions initiate, import, modify and diffuse new technologies" (Freeman), or "the elements and relationships which interact in the production, diffusion and use of new, and economically useful, knowledge ... and are either located within or rooted inside the borders of a nation state" (Lundvall).

There are different definitions, but there seems to be an understanding that the concept of NIS refers to structures of economic organization, whose scope is more comprehensive than the quantifiable aspects of the R&D performance financed by private corporations and the state. These structures of economic organization contribute to an increase of the production efficiency through the generation and diffusion of innovations.[1]

The NIS approach places innovations at the very center of its focus. Innovations have various tasks. They act as a driving force and point companies toward ambitious long-term objectives. They also lead to a renewal of industrial structures and are responsible for the emergence of new sectors of economic activities.

The determinants of enterprise success—and of national economies as a whole—will become more reliant on their effectiveness in gathering and utilizing knowledge. Since firms are the main carriers of innovation, their capacity to innovate is jointly determined by their own capacity as well as by their capacity to adopt and apply external sources of knowledge.

As economic activities become more knowledge intensive, a large and growing number of actors with a wide variety of specialized expertise are now involved in the production and diffusion of knowledge (other firms, public and private research institutes, universities, or transfer institutions—either regional, national or international). These rise in complexity, in addition to higher costs and risks in innovation enhanced the value of networking and collaboration.[2] Within a network or collaboration, enterprises get access to external sources of knowledge. Additionally, they can pool their technical resources and achieve economies of scale. Therefore the economy will become more and more a network society, where the opportunity and capability to get access to and join knowledge- and learning-intensive relations determines the socioeconomic position of individuals and firms.[3]

The innovation process should not be mistaken as a simple technology transfer in which knowledge production, application, or utilization can be separated. Product or process development projects designed to culminate in mass production and market success are dependent on overlaps and feedback from the research and invention activities inside and outside the particular enterprise. For this reason, technology transfer is the result of a complex process of interaction, with intensive feedback loops on various levels.

As networks become more and more important, it must be realized that networking represents a unique innovation technique. This supposition is supported by empirical studies that confirm that collaborating firms are more innovative than noncollaborating firms.[4] Furthermore, it has been argued that networking must now be considered on an equal footing with hierarchy and the market as instrument of coordination.[5]

Regional Innovation Systems

As mentioned above, the concept of a national innovation system doesn't refer necessarily to a nation. Regions also can be analyzed meaningfully with this conceptual framework. Within this context, however, the different levels of competence between regions and the federal state have to be taken into account.

The spatial concentration of knowledge from research and science, enterprises, and public institutions stimulates the growth of regional locations for innovations.[6] A reason for this development is that tendentious spillovers from the field of research are bound to the location of the knowledge emergence. New knowledge is often very complex and unstructured and for this reason it can be made more easily utilizable in interpersonal communication structures within easily comprehensible regions.

These regional networks aren't successful in single industries but in regional clusters—this means in a concentration of branches that are connected by vertical and horizontal relations with each other. The competition advantages of these regional clusters arise from the interdependent interaction of the following key factors:[7]

- innovative environment with considerable competition pressure by competitive enterprises;
- customers who press the enterprises for innovations;
- access to networks of specialised local suppliers;
- local availability of technologies and highly qualified employees.

The intention of the "innovation system" approach is to evaluate and compare the main channels of knowledge flows to identify bottlenecks and suggest policies to improve the fluidity of knowledge. Therefore attention is given to address areas such as functional mismatches, incentive conflicts, weaknesses, and asymmetries existing in the system. For policymakers this implies that an understanding of the innovation system can help find leverage points for enhancing the innovative performance and competitiveness of a region.

Western Regional Innovation Systems: The Example of Bremen, Germany

The example of the Bremen region may be used to give some insight into the Western regional innovation systems, where not untypically a close network of institutions responsible for the promotion of innovation activities exists. Bremen is host to a large university and several polytechnics as well as a number of research foundations and institutions in the field of material research,[8] which are heavily involved in transfer activities between research and applied technology. This is also reflected in the technology park surrounding the university, which hosts several small- and medium-sized technology-oriented firms. It exists in a mix of industrial and other commercial firms and networks of cooperation between these firms. Linkages also exist to Hamburg, Hannover, Oldenburg, and other neighboring regions with good traffic connections (highways, harbors, railways, airports). Since it is not possible to copy such an infrastructure, emphasis here will be on institutions in the promotion of managerial processes and business activities. According to information collected from the expert survey and inquiries made at the firm level, several institutions appear essential and may be considered a prerequisite for effective modernization processes.

Sources of Information: According to the information of the field study, the most important sources of information are suppliers and customers, as well as private contacts. However the Chambers of Commerce from the institutions active in this field are judged as most important. Additionally, state-organized promotion-agencies like the Bremen Investment Cooperation (BIG) seem to fulfil important functions.

Chamber of Commerce: The Chamber of Commerce in Bremen is the legally established self-governmental organization of the manufacturing trade in Bremen. Some 27,000 enterprises are members of the Chamber of Commerce. The federal government has transferred some duties to the chamber. It acts independently from the public administration and independently within its domain. The Chamber of Commerce is the "city hall of the economy." It

- promotes the economy in Bremen,
- represents the interests of enterprises from commerce and industry,
- advises and informs enterprises in all economy related relevant questions.

Selected tasks are

- technology consultation,
- consultation of inventors,
- industrial property consultations,

- consultation in questions relevant for public tendering,
- business start-up consultation,
- environmental protection consultation,
- information about foreign economic promotional programs,
- supervision and promotion of vocational education and reeducation.

Bremen Investment Cooperation (BIG): BIG is the central service unit for the development and business promotion of the state of Bremen. BIG covers, with its affiliated cooperations (the Industrial Development Cooperation, the Bank for Reconstruction, and the Bremen Innovation Agency) the whole palette of industrial development services. The group of cooperations formed by BIG is therefore the central address for all enterprises and start-up firms located in the city of Bremen or that want to settle down in Bremen.

The results of the written inquiry from Bremen firms showed that more than two-thirds of all enterprises found the information and additional support by these agencies very helpful.[9] The results of face-to-face interviews conducted for this project were not as promising, because conflicting competencies between different branches of the agencies were criticized.[10] The support of the Chamber of Commerce as a self-organized agency of trade and industry in Bremen is considered most useful in comparison with the other institutions. Institutions are important, but the quality of other location factors should not be neglected. Results of such inquiries often are visualized in two-dimensional figures and this approach also could be helpful in determining strengths and weaknesses of regions under transition like Lugansk and Voronezh. However, this requires that enterprises and their managers be less state guided, and regional planners take the needs and the knowledge at the firms' level about bottlenecks and weaknesses of the existing structures more seriously than they seem to do today.

Eastern Regional Innovation Systems: The Examples of Voronezh (Russia) and Lugansk (Ukraine)

It is impossible to make a direct comparison between East and West, not only because of the differences in the levels of income. Also the state's resources due to different tax income levels, the existing technological infrastructure, and the stages of industrial development are very different. Information about the regional systems of Voronezh and Lugansk are also not as easily to obtain as in Western regions. Field research for the project should be mainly conducted by local research teams. A first important finding of the Bremen research group was that modernization of industry in the opinion of Russian and Ukrainian economists may be discussed for separate firms, and mainly has to do with bad and old equipment, lack of financial resources, and other more technical bottlenecks. Following this view, problems have to be solved at the firms' level in isolation by finding some channels for infusion of technology and other resources, and perhaps in the hope of a better macroeconomic business climate (less uncertainty in the legislative base, tax incentives for innovations, financial support by the administration, etc.) in the future (see Table 6.1).

Table 6.1
Data for the Analysis of the Scientific and Technical Process in Enterprises in Lugansk (n=30)

Factors that prevent development of R&D processes	Quantity of enterprises, %	Measures needed for innovation activity support	Quantity of enterprises, %
Insufficient finance of R&D	87.32	Increasing of effective connections of enterprises and administration and organizations, supporting their development	34.58
Shortage of high-class specialists	35.03	Perfection of tax legislation for innovation stimulation	58.65
Using R&D by production	0	Making easy procedures of licensing and certification	12.36
Absence of sufficient technical and experimental base	54.31	Development of cooperation with R&D organizations	39.65
Disconnection of scientific branches	12.34	Investment attraction, enterprise supply	74.36
Estrangement from fundamental scientific research	0	Development of innovation potential of enterprises	23.12
High costs for innovation activity	66.45	Other	0
Long period of springing	34.58		
Infringement of copyright	0		
Absence of modern information about R&D projects	35.58		
Absence of will to innovate	35.58		
Problems in management	24.62		
Use of old technologies	54.31		
Uncertainty of legislative base	66.45		
Other	0		

Source: Lugansk Working Paper (2000a), Table 14.

The formation of cooperation between firms plays a minor role and the idea of institutions that form a network that supports and creates an evolutionary process seems strange to them. Therefore, the information about the regional innovation systems is by far not complete and deserves further research. The implied results mainly from the authors' field research in Russia and interviews with managers and at institutions like the local Chamber of Commerce. Voronezh and Lugansk both host universities and higher education institutions, and several institutions like the Chamber of Commerce also play a specific role in these regions. Also, for example in Lugansk, state support is given for innovative activities (Lugansk Working Paper, 2000b). The field research also gave evidence that the Chambers of Commerce in Voronezh and the neighboring district of Belgorod have only a limited scope over regional enterprises, and face-to-face interviews showed a very different degree of acceptance. Several interviewees expressed limited support by these agencies and were disappointed by the few resources available. Other interviewees gave a more positive response, because they received support in marketing activities, access to information, and so on.

The examples from the Voronezh and Lugansk regions show that some systematic promotional activities for the modernization process do exist. These approaches seem to be based mainly on financial support. Equally as important, however, is the improvement of access to information and support for the introduction of modern management systems. The state uses promotional activities more as an initiating and controlling agency. More stimulation for the self-organization of industry and commerce is needed as well.

MODERNIZATION PROCESSES AND PLANNING AT THE ENTERPRISE LEVEL

General Remarks

The performance of national economies relies on the innovativeness of companies. Science and technology establish the central base for growth and economic wealth. A permanent technical change is one of the driving forces behind innovation. This is especially true for highly developed countries where natural resources are scarce, but also applies generally.

In the framework of a market economy it is the first and foremost the task of private investors to implement new technological knowledge. Different types of innovations (adaptive or strategic, radical or incremental, processes or products) are developed and diffused by different kinds of firms in different business environments. Because firms vary in size, competitive strategies, and organizational competencies, they tend to invest in and implement a variety of innovations.

In an increasingly international environment, countries and companies develop a specific pattern of specialization with respect to technological expertise and potential for generating technological knowledge. Technological specialization follows strategic decisions of internationally operating companies, as well as national innovation policies; it reflects demand for specialized knowledge and the availability of relevant resources.[11]

Several empirical studies discuss the success factors in different enterprises, varying in size, number of branches, and so on. The Hannover Success Factor Project (HEFAP) identified ten main factors of success (success defined as the return on investment (ROI) and cash flow) for German enterprises, which are headed by innovation management, a factor considered of growing importance.[12]

To get a basic understanding about all facets of innovation processes (innovation management of a firm within the framework of a national innovation system) the OECD tried to specify the nature of the innovation process:[13]

- Innovation seldom depends on technological know-how alone. It becomes more difficult to derive a competitive advantage merely from technology; a firm needs the complementary assets (distribution networks, manufacturing capability, supporting technology, etc.) that enable it to harvest the fruits of technological innovation.
- Innovation is interactive and multidisciplinary. A firm needs to build up a network of relations with a wide variety of local and other partners, for example, consumers and users, suppliers of equipment and technology, distributors, and so on. The reliability and effectiveness of these partnerships can be obtained only through what can be called "relational" investment.
- Innovation is localized. It requires the internationalization of those "informational externalities" that pervade a location, region, or country.
- Innovation is a process of integration. The need for companies to cultivate their external relations obliges them to adopt new forms of internal organization, with such major functions as production, research, marketing, and financial planning tending to become more integrated ("organizational investment," "info-structure").
- Innovation is a learning process. A firm that innovates is characterized as much by its learning capacity as by its efficiency in producing goods, services, or knowledge ("human resource investment").
- Innovation has a social dimension. Innovation disrupts not only consumption and production patterns but also their associated power structures, whose roots are deeply embedded in the socioeconomic fabric ("public relations").
- Innovation is a process of "creative destruction." Not only must a firm "learn how to learn," it must also "know how to unlearn."
- Innovation has cultural origins in history. The concept of "path dependency" draws attention to the fact that both resources and the demand for innovation are to a certain extent subservient to chronology.
- Innovation is both expensive and risky. An innovative firm, caught between rising sunk-cost for innovation, R&D in particular, and shorter product life, has no other choice than to share the effort and risks with partners and/or broaden its depreciation base by diversifying its portfolio of products that incorporate the same technological know-how.

Innovation and Planning in German Firms

Innovation in German Firms

The innovative behavior of firms is different in the respective industries. Some industries offer a lot of opportunities due to new paradigms like microtechnologies and biotechnology. In these industries product innovation dominates process innovation. In industries with less innovative opportunities and old technological paradigms like the mineral oil processing, metal, paper, printing, and wood industries, process innovation dominates and innovations are often incorporated in machinery and equipment. An innovator in this survey is defined as an enterprise that introduced product or process innovation in the respective year. The innovation may range from marginal improvements to fundamental new products or processes.

The data presented in the following investigation were collected by IFO, Munich and covers 2,900 (IFO business cycle test) and 1,400 (IFO innovation test) firms from East and West Germany respectively.[14]

It can be seen that around three-quarters of the responding firms reported innovative activities in the years 1996, 1997, and 1998 (see Table 6.2). The highest percentage of innovative firms was given in the automotive and automotive equipment industries, office equipment, data processing equipment, equipment for electricity production and distribution, in audio, video and telecommunication technologies, medical equipment, and in chemicals and mechanical engineering. Smaller firms report discontinuous investment behavior because they have less opportunities, the reasons being smaller production programs and the fact that often these firms specialize on innovations where the market knowledge is more important than high investment expenditures.

Table 6.2
Innovators in the Manufacturing Industry in Germany (in %)

Industry[a]	**1996**	**1997**	**1998**
food/tobacco	69.9	75.2	72.8
textile	65.4	66.8	70.1
clothing	67.1	65.2	73.8
leather	71.8	59.5	73.1
wood except furniture	51.1	53.7	60.5
paper	48.7	55.4	50.1
printing	48.3	60.9	50.1
mineral processing	68.8	▭	54.1
chemicals	82.1	88.7	85.2
production of gum- and plastic products	71.9	76.7	69.8
glass, ceramic, processing of stones and earth	65.8	71.9	74.6
metal production and processing[b]	69.3	76.2	65.0
metal products	64.7	62.4	62.5
mechanical engineering	78.1	79.8	79.2
bureau and data processing equipment	88.9	100.0	100.0
electricity production and distribution equipment	81.4	84.9	86.4
radio-, television and communication technologies	84.3	87.9	77.6
medicine-, measuring-, steering and control technologies, optic	83.8	83.4	84.7
automotive and automotive equipment industry	91.0	90.4	84.8
other vehicle industry[c]	69.5	63.3	▭
furniture, jewellery, music instruments etc.	77.3	76.5	75.9
Total	74.5	77.2	75.4

a) Classification 1993 (WZ93); b) Except casting, iron, steel, iron alloy; c) Except air and space industry; ▭ Held secret, but included in the total.

Several of the surveyed firms were hampered by financial, personnel, technical, or administrative obstacles in order to fully extend the innovative potential. In 5% of the firms, financial barriers made innovative activities almost impossible. For those firms with innovative activities, the rate of return was often considered too low because of high R&D costs and the long time of amortization. Additional problems such as personnel barriers resulting from a shortage of qualified employees and a lack of innovative motivation of the workforce, works committees, or management are less important. Overall there seems to be no dominating obstacle for innovations, and overall criticism appears minimal on the state of technological infrastructure that could be influenced by economic policy measures.

The largest part of the R&D budget is spent on market- (product) related innovative activities, that is, the creation of a new product through R&D, costs of patenting, licensing, and realization by investment to schedule production and distribution.

The analysis shows that a large majority of all enterprises in Germany are involved in innovative activities. Overall, 75% of the reporting enterprises had innovative activities in the respective year. For most firms innovation is not a means to gain advantage over competitors or to achieve high growth rates, but a question of survival in the marketplace. There is no overriding obstacle for innovative activities if we overlook the well-known problem of a restrictive legislation in some industries and the complaints about the time-consuming administrative procedures. Problems arising from low profitability of product innovation indicate strong competition in the market. Thus the general technological infrastructure seems to be valid from the point of view of most reporting firms.

Field Study for Bremen: according to the expert interviews carried out in Bremen (for more details see Birkemeyer, 2000b, IWIM Working Paper No. 3), the frequently mentioned economic factors (lack of equity capital or outside finance) are of great importance. However, additional innovation obstacles include:

- recruitment of capable R&D staff,
- the problem of a restrictive legislation in some industries, and
- the problems with time-consuming administrative procedures.

In the opinion of the experts, problems such as low rate of return due to high innovation expenditures, long amortization period, and an easy imitation of products seem less important.

For the preservation and improvement of competitiveness, R&D activities become more and more important for regional firms. Each of the nine interviewed enterprises could refer to their own R&D activities, whose main emphasis lies in the development of new products and processes as well as in the improvement of existing products and processes. Two enterprises were even committed in the field of basic research. Apart from an operational suggestion system, further education/technological training of the employees is important for the enterprises interviewed. In two of these companies there is no operational suggestion system in place, since the individual willingness of the employees for creative collaboration is part of the enterprise philosophy.

The enterprise survey clarified that cooperation occurs frequently in the R&D area. Common R&D activities with competitors were not seen, but public and private research units, as well as universities, are frequently chosen as cooperation partners. Complete allocation of R&D projects to other enterprises/institutes is rare. Only three enterprises occasionally allocate complete R&D projects.

Planning in Western Enterprises

Medium- and long-term planning in Western enterprises is the foremost task of management. Most if not all of the medium- and large-sized enterprises have some kind of revolving financial planning that covers the medium term for the next three to five years. Planning also plays an important role in the budgeting of the different departments and in the process of ex post controlling. Budgeting and ex post controlling are often presented in some kind of ratio system like the Pyramid Structure of Ratio System or the DuPont Ratio System. Fundamental for this kind of planning is the existence of a workable accounting system. This system has to deliver information on certain input requirements, on the development of demand in different segments, and so on. Strategic planning is normally carried out informally and rarely routinely. It is often pursued only in extreme situations where a reorientation is urgently required because of a changing environment, the increase of international competition, or the emergence of alternative technology or products.

The majority of the (small and medium) companies interviewed in Bremen have established medium-term enterprise planning (medium-term time horizon = approximately three years; liquidity planning, projected balance sheets, performance figures, etc.). Only two companies had not set up these instruments at the time of writing. For the appraisal of the future market trend, the following forecast proceedings were primarily chosen market observations and performance figures. The estimation of future technology trends is carried out by internal and external discussions (4 companies), observation of market shares (3 companies), scenarios (1 company), and experience (1 company).

All of the enterprises questioned carried out business statistics and also had an operational cost-accounting system. In the cost-accounting process, four enterprises used external services. The use of the Electronic Data Processing (EDP) system for the different function ranges (production, sales, R&D, etc.) was carried out by all companies over a computer network. The interviews showed that the majority of the enterprises used a Management Information System (MIS). The complete range of the application possibilities of these MIS was at the enterprise's disposal. Besides the standard programs, four enterprises worked with their own EDP applications.

Even in dynamic markets or where there is great uncertainty about the future, firms have no doubt about the necessity of a medium-term plan and budgeting procedures based on this plan. All the interviewed enterprises had business statistics, and also had an operational cost-accounting system that can be used as a scheme and data source for the forecast. The use of the EDP system is considered a great help in the planning procedure. The development of strategies is a creative act and often follows intuition. However, several general principles may also be of help. A systematic approach for strategic planning was outlined in IWIM Working paper no. 4 (Sell, 2000) in English and Russian languages and served as the basis for a discussion on the need and applicability in enterprises of the Voronezh region.

Modernization and Planning in Economies under Transition

General Remarks

In an informative survey Inzelt gives a brief introduction into the socialist past and its heritage (a concentrated listing of main weaknesses within the socialist system) and some elements of transition.[15] These are recent developments in science, technology and innovation on the one hand, and those changes that have had a strong influence on commercialization, R&D results, innovation processes, and competitiveness on the other hand. To quote Inzelt:

> In order to transform the R&D resources of former socialist countries into an economic success, these countries need to redefine their institutional and behavioural operations. The socialist system was very weak in diffusing knowledge and commercialisation and command economies were clearly in a technological deadlock. To emerge from this deadlock the proper economic environment for commercialisation and innovation needs to be created.

The question arises as to how these countries in transition can move from a one-way linear model of innovation toward the feedback-loops model. In this summary I focus—on the basis of Inzelt's work—on: (1) the main weaknesses of the socialist innovation system, (2) results concerning organizational changes, and (3) results concerning commercialization of R&D results.

1. Main weaknesses of the socialist innovation system

- one-way linear innovation model instead of feedback loop (improvements in capital goods and related technologies were created outside the enterprises in the relevant, sectorwide research and design bureaus; users' experiences were neither fed back to design bureaus/institutes, nor diffused to other similar users)
- restricted international networks in science and technology (limited exchanges in higher education, lack of mobility for researchers, low level of technology transfer outside networks and alliances)
- the tendency to reproduce or duplicate foreign research in an autarchic or semiautarchic economic environment
- institutionally granted political and ideological state and party control over higher education and research
- the bureaucratic approach to research programming, noncoherent educational, research and economic policy
- basic research and doctoral training performed separately (academies and universities respectively)
- lack of competition between institutions and research teams based on merits and competence; selection and criteria evaluation omitted
- planning and financing of research in firms by central administration
- budgetary funding of profit-oriented applied research and its implementation
- maintaining incompetent institutions and pseudoscientific work
- "phoney" product innovations to allow for higher prices without extra benefit to consumers

- business enterprises and production areas isolated administratively and geographically from the domestic, organizationally separate R&D sector
- a gap between operation of technology used and state-of-the-art technology
- weak incentives for commercialization and innovation (e.g., inventions were given to firms in monopoly positions with little incentive to develop and exploit them)
- inadequate development and testing activities
- unsophisticated users in local market and Council for Mutual Economic Assistance (CMEA) market
- weak user-producer linkages
- remote from the major international clusters of innovative suppliers and users
- poor knowledge of Western languages preventing flow of information and exchange between Eastern and Western academic communities
- the remarkable bridging of the technological gap within the military industry failed to transfer knowledge, and consequently the gap in core civilian technologies remained

2. Results concerning organizational changes. In the original Soviet model, the science and technology system was divided into three sections: (1) higher education institutions, (2) Academy of Sciences, (3) sector institutes, laboratories, and design offices. The role of in-house departments of enterprises was marginal. This fragmented R&D organizational structure was the main cause for the research-push model and non-demand-pull model.

3. Results concerning commercialization of R&D results. The major weaknesses of commercialization during the transition period are:

- financing sources (lack of capital, proper credit system, venture capital, etc.)
- activity from applied research to technology dissemination
- critical mass of capabilities (innovation potential too small)
- managerial skills (in project management, R&D management)
- knowledge on process engineering, quality control, cost control
- marketing skills
- linkages between related sectors
- integration with foreign counterparts
- strategic partnerships, connection with networks
- new markets

The most important changes within privatized companies with foreign owners in Central and East European countries can show the way to solve some of these problems in the transition period:

- spending cuts by new owners (expenditure, personnel)
- change in the research portfolio
- a shift away from the broad learning strategy toward a more focused approach
- closing down research fields and concentrating on fewer areas

- cutbacks in the portfolio of interests of privatized companies in advanced technologies to concentrate narrowly on the development of less advanced products with market potential
- cuts in R&D expenditure (for implementation of new product ideas from outside in order to better adapt them)
- increase in collaborations
- introducing new equipment and new R&D methods
- innovation-related activities (total quality management, just-in-time delivery and modernization of machinery, on-the-job training)

Modernization of Equipment and Planning in Voronezh (Russia) and Lugansk (Ukraine)

From the field studies of the teams from Voronezh and Lugansk and also the insight gained by several firm visits, the need for modernization of equipment becomes quite clear. According to results taken from the investigation in Lugansk, around 60% of the firms purchased capital equipment more than 15 years ago (see Table 6.3).

However, modern equipment is often essential to meet quality standards and also diversification into new product lines, and it is in many cases impossible with inflexible equipment. Modernization of equipment is financially restrained, and in several cases information and visions about potential markets do not exist. Cost reductions through energy savings and improved efficiency by introducing new equipment are also underestimated. Technological backwardness, therefore, is not primarily a technical problem, but a general management problem. Modernization in many cases would be more effective by using international channels of technology transfer and the potential of international technological cooperations.

Table 6.3
Age Structure of Equipment and Technology in Lugansk Region (n=30)

Index	Enterprise quantity (%) in which age of equipment and technologies is:				
	1 year	2 years	5 years	10 years	More than 15 years
Basic technology age			19.87	10.82	60.31
Average equipment age				28.21	71.79
Time of the last equipment purchase	18.65	11.20	10.40		59.75
Time of the last alteration made in basic technology	10.33	18.01	41.12	9.83	20.71

Source: See Lugansk Working Paper (2000a), Table 10.

The field studies and the analysis of the experts from Lugansk and Voronezh gave evidence that systematic planning is performed only in a minority of firms. This lack of planning may be attributed to the fact that the firm's strategy is directed more to guaranteeing the survival in the short term, and also because of an uncertainty about future markets, the availability of raw material, and the like. A prerequisite of systematic planning is the existence of a reliable accounting system and enterprise statistics on sales, costs, and other figures, divided for different products or product groups. Both modern accounting systems and reliable statistics in most enterprises do not exist and thus cannot support the planner. Thus, on the one hand there is a limited awareness for the need of systematic planning and also a basic lack of knowledge about appropriate planning instruments, and on the other hand a lack of reliable internal and external data to serve as a basis for the planning process. Therefore, planning and especially strategic planning is done more or less on an ad hoc basis, and not in a systematic way.

Germany versus Voronezh (Russia) and Lugansk (Ukraine)

Modernization Attitudes

Empirical evidence shows that for German and other Western enterprises, modernization and innovation are considered musts and not exceptions. The above-mentioned IFO study demonstrated that about three-quarters of the responding firms reported innovative activities in the previous business year. The percentage of innovative firms varies from industry to industry, and in particular smaller firms show discontinuous investment behavior. However, a great awareness for modernization and innovation in small enterprises was also shown in the interviews of firms in Bremen. The analysis carried out in Lugansk and Voronezh and also the field studies in Voronezh showed that in these regions other problems dominate the modernization processes. In many cases the need to adapt and to improve the quality of products to meet international standards has yet to be met (see also Voronezh Working Paper, 2000). The survey of Lugansk firms showed that the struggle for survival seems to paralyze management, and severe problems of raw material shortages and complicated payment systems absorb most of managements' capabilities (see Table 6.4).

This is in contrast to the situation of Western firms. Several firms of the Voronezh region feel that they have a monopoly in the Russian or even the Commonwealth of Independent States (CIS) market, and that pricing and not quality competition is essential. This attitude neglects the fact that in the long- and medium-term, Russian firms must also compete with foreign products in their markets, and this also holds true for third markets like the CIS states and especially for Western markets. It also ignores the fact that diversification is necessary to outplace products in the final stage of the product cycle and in some cases it is overlooked or even disregarded that several more demanding segments of their markets are still occupied by foreign suppliers.

Table 6.4
Resource Provision Estimation of Enterprises in Lugansk Region

Name of enterprise	Raw materials	Completing parts	Fuel	Energy
Closed JSC "Lugansk Crankshaft making plant"	ss	ss	ss	ss
JSC "Donez"	ps	ip	p	p
JSC "Lugansk Carassambly plant"	ps	ps	ss	ss
Lugansk "Machine tool making plant- corporation"	ps	ps	ss	ss
Closed JSC "Lugansk foundry-mechanical plant"	ss	ps	ss	ss
Closed JSC Lugansk machine building plant Parchomenko	ss	ip	ps	ss
JSC "Elecro-apperatus making plant"	ip	ip	ps	ps
Closed JSC "Lugansk pipe making plant"	ss	ps	ps	ip
JSC SPF "Lugansk-Accumulator"	ss	ss	ip	ip
Lutuginsk state amalgamation for shaft production	ss	p	ps	ss
Krasnorechenk Machine tool making plant Frunze	p	p	ps	ip
JSC "Sverdlovsk machine building plant"	ss	ip	ss	ps
Closed JSC "Stachanov machine building plant"	ss	ss	ss	ip

ss –sharp shortage; ps – provided, shortage; p – provided; ip – insufficiently provided
Source: See Lugansk Working Paper (2000a), Table 12.

Planning Attitudes

As shown above, Western medium and large enterprises are well aware of the need to forecast their business activities and apply more or less systematic approaches, often integrated in MIS and supported by EDP. Most firms can base their planning on a reliable accounting system, internal business statistics, and systematic external information. As discussed above, the field studies and the analysis of the experts from Lugansk and Voronezh gave evidence that systematic planning is performed only in a minority of firms. A firm's strategy is directed more to guaranteeing the survival in the short term. Uncertainty about future markets, shortage of raw material, and disruptions in the production process are some of the reasons not to do formal and systematic planning. The obvious underdeveloped accounting systems and a lack of enterprise statistics certainly make planning more difficult. Several firms are aware of the shortcomings of their accounting system and business statistics, but the planning departments seem not to be prepared to change the situation soon. Compared to Western firms there also exists a basic lack of knowledge about appropriate planning instruments, which in part also may be traced back to limited access to consultancy in this field as one sign of the underdeveloped regional innovation system.

CONCLUSIONS

The analysis of the regional innovation system in Voronezh and Lugansk shows severe deficits in respect to the existence and the functioning of institutions that are considered essential for the development process. Because firms are no longer able to innovate and modernize in isolation, and there is a minimum network of institutions offering access to information and potential cooperation partners in the region, national or international cooperation is required. In Western countries this function is often fulfilled by the Chambers of Commerce, which identify themselves as the representatives of the business community in the region, independent from the state and self-governed by the enterprises. The field research gave evidence that the majority of the firms underestimate the importance placed on self-organization in industry, and they do not identify themselves with their organization, which should represent the interests of industry, and they also trade against government interference and guidance. Because there is a severe lack of resources in these institutions, external support is needed to strengthen the capabilities of these and similar institutions.

An important topic of future research could be a more detailed analysis of the technological infrastructure in the regions under transition, beginning with an in-depth study of the existing institutions. Topics of such studies should take into account the reasons for the limited acceptance of some of these institutions by the local enterprises. Forms of organizations have to be found by which the different kinds of firms (small, medium or large; service, trade, or industry; firms under different ownership; modern or traditional sector) feel represented and may express their needs. As a result, a forum for an exchange of information and the basis for the creation of regional networks may emerge. Support should also be given to mobilize technical assistance by the international community and/or to initiate bilateral cooperation. As is shown in practice by several examples and was also indicated by the expert interviews in Western institutions,[16] a knowledge transfer from some institutions is feasible and effective.

On the enterprise level, stress should be laid on the improvement of organizational and management tools. The awareness for systematic economic planning is underdeveloped in many firms—or heavily dominated by the daily struggle for survival. However, in the medium term a positive development is possible only if the firms get a clear understanding of their strengths and weaknesses, systematically analyze the national and international markets and their competitors, and use effective tools to find and evaluate market opportunities. One requirement for more effective enterprise planning is the availability of reliable internal data. In many cases these data are not available due to organizational deficits, outdated accounting systems, and the like.

Great efforts are required to convey the message that a permanent change for successful enterprises will be the normal case, and that modernization and innovation are not the exception, but will become a routine for which information systems and planning procedures have to be developed. Thus there is a great need for education and further education not only at the middle- but also at the top-management level.

NOTES

1. See David and Foray (1994), p. 9.
2. See OECD (1998), p. 58.
3. See OECD (1996), p. 14.
4. See OECD (1998), p. 59.
5. See Powell (1990), pp. 295-336.
6. See Gutowski (1999), p. 26.
7. See Porter (1990).
8. See Birkemeyer (2000a).
9. Sell (1997).
10. See Birkemeyer (2000b).
11. Whitley (1998); Klodt (1996); OECD (1995).
12. See Steinle (1996), pp. 14-23.
13. Excerpt from OECD (1995).
14. Schmalholz and Penzkofer (1999), pp. 3-11.
15. See Inzelt (1999), pp. 163-193. The general framework of this study is an empirical analysis of the present situation in six Central and East European Countries (Lithuania, Bulgaria, Hungary, Czech Republic, Poland and Ukraine).
16. See Birkemeyer (2000b).

REFERENCES

Birkemeyer, H. (2000a). IWIM Working Paper No. 2: "Regional Innovation Systems with Special Reference to Bremen."

Birkemeyer, H. (2000b). IWIM Working Paper No. 3: "Expert and Enterprise Survey for the Bremen Region."

David, P. and Foray, D. (1994). "Accessing and Expanding the Science and Technology Knowledge Base." In OECD (ed.), DSTI/STP/TIP(94)4. Paris.

Freeman, C. (1987). *Technology and Economic Performance: Lessons from Japan*. London: Pinter.

Gutowski, A. (1999). "Innovationen als Schlüsselfaktor eines erfolgreichen Wirtschaftsstandortes" (Innovation as a Key-Factor for Successful Economic Outcomes.) In *Materialien des Universitätsschwerpunktes Internationale Wirtschaftsbeziehungen und Internationales Management*, Vol. 17, Bremen.

Inzelt, A. (1999). "Science, Technology and Innovation: Institutional and Behavioural Conditions for Innovative Industrial Development." In B. Widmaier and W. Potratz (eds.), *Frameworks for Industrial Policy in Central and Eastern Europe*. Aldershot: Ashgate, pp. 163-193.

Klodt, H. (1996). "The German Innovation System: Conceptions, Institutions and Economic Efficiency," Kiel Working Paper No. 775.

Lugansk Working Paper (2000a). "Analysis of Modernisation Processes on Engineering Companies," unpublished.

Lugansk Working Paper (2000b). "State Support of Innovative Activity of Industrial Companies in Ukraine," unpublished.

Lundvall, B.-Å. (ed.) (1992). *National Innovation Systems: Towards a Theory of Innovation and Interactive Learning*. London: Pinter.

OECD (ed.) (1995). *National Systems for Financing Innovation*. Paris: OECD.

OECD (ed.) (1996). *The Knowledge-Based Economy*, OECD/GD(96)102. Paris: OECD.

OECD (ed.) (1998). *Technology, Productivity and Job Creation: Best Policy Practices*. Paris: OECD.

Porter, M. (1990). *The Comparative Advantage of Nations*. New York: The Free Press.
Powell, W.W. (1990). "Neither Market Nor Hierarchy: Network Forms of Organisation," *Research on Organisational Behaviour*, 12: 295-336.
Schmalholz, H. and Penzkofer, H. (1999). "Innovation in Deutschland. Ergebnisse des ifo Innovationstests nach der neuen Klassifikation der Wirtschaftszweige" (Innovation in Germany. Results of the IFO Innovation Test According to the New Classification of Industries), *ifo* schnelldienst 5/99: 3-11.
Sell, A. (1997). "Nationale Wirtschaftspolitik und Regionalpolitik im Zeichen der Globalisierung" (National Knowledge Policy and Regional Policy as Indicators for Globalization,) Report of the World Economics Colloquium of the University of Bremen, No. 54.
Sell, A. (2000). IWIM Working Paper No. 4: "Stepwise Approach for Strategic Planning." Bremen.
Steinle, C. (1996). "Erfolgsfaktoren und ihre Gestaltung in der betrieblichen Praxis" (Success Factors and their Arrangement in Workplace Practic,) *Aus Politik und Zeitgeschichte*, B 23/96: 14-23.
Voronezh Working Paper (2000). "Expert Reports Review."
Whitley, R. (1998). "Innovation Strategies, Business Systems and the Organisation of Research" (prepared for the Sociology of Sciences Yearbook meeting held at Krusenberg, Sweden, September).

7

Human Capital, Technology Adoption, and Export Performance in Mexico's Manufacturing Industry (1989-1999)

Cristina Casanueva

INTRODUCTION

The main objective of this chapter is to identify trends in the practices introduced by manufacturing companies within the context of the totally new conditions arising from the opening up of the Mexican economy. The chapter specifically examines the educational attainment of the workforce, the training of personnel (human capital), and technological innovation in Mexican companies belonging to 52 manufacturing-industry branches during the 1990s. The decade of 1989-1999 can be seen as a critical period for manufacturing companies in Mexico because of the pressure on them to increase efficiency and productivity in order to face the challenge created by the opening up of markets.

A second aim this chapter is to analyze the existence of possible associations between the above-mentioned variables, examining whether education, company training, and technological innovation have any effect on productivity and export performance. The research also has a dynamic component, because it compares the behavior of the aforementioned variables at the beginning and end of the 1990s.

THE CONTEXT: THE OPENING OF THE MEXICAN ECONOMY

In the mid-1980s, the Mexican government began to implement a set of policies that resulted in the Mexican economy's becoming more export oriented. This liberalization policy began in 1983 and was ratified when Mexico joined the General Agreement on Tariffs and Trade (GATT) in 1986. The country has also been involved in an intense program of negotiation of free-trade agreements with other countries,[1] thus gradually evolving from an industrial import-substitution economy to an export-oriented one. The opening up of markets resulted in an unprecedented increase in trade flows. Indeed, the combined importation and exportation of commodities went from 24% of the GNP in 1985 to 70% in 2000, while total exports increased by 93% between 1994 and 2000 (Villarreal and Ramos, 2001: 772).[2]

Along with the impressive increase in trade, there has also been a major change in the overall composition of Mexico's exports. By the end of the decade, manufactured goods had replaced petroleum as the major export commodity. In the early 1980s (1981-1983), crude oil and natural gas made up 72% of Mexico's total exports, while manufactured goods comprised only 19%. In 1998, manufactured goods replaced petroleum as the dominant element in the export mix. Exports from the manufacturing sector represented 90%, while petroleum constituted only 6.1% of total exports (see Loria, 1999).[3] The exportation of manufactured goods became the engine of the Mexican economy, growing by almost 20% between 1991 and 2000 (León and Dussel, 2001).

Trade Deficit and Limited Growth of Local Supply Chains

In spite of Mexico's impressive export performance, the growth in exports was associated with an increase in the importation of both input materials for production and finished goods, resulting in a structural trade deficit. Between 1991 and 1994, this deficit increased from $7.3 to $18.5 billion. In the middle of the decade, trade leveled out due to a fall in imports stemming from the 1994 liquidity crisis and the subsequent recession. However, a trade deficit emerged again with the recovery of the economy in 1997, and by 2001 there was a $10 billion trade deficit.

Implicit in the trade deficit is the problem of structuring exports around finished goods that are dependent on export input for their production. There were no policies, at the time, to sustain and improve existing suppliers or to develop new ones so as to build mutual links, both among the said suppliers and with the end-product exporter (León and Dussel, 2001), with the limited integration of value chains into the production of final goods threatening to make pressures on the trade balance an endemic problem.

Limited Diversification of Exports

An additional problem derived from the export structure, which was largely concentrated in a few industries: the terminal automobile sector (28.3% of exports and 13.8% of imports), the electric and electronic equipment sector, which produces audio and video equipment (radio and television receivers, and audio equipment) (19.8% and 24.7% of imports respectively), and the sector producing office, data-, and information-processing equipment, which includes manufactures of computer equipment and peripherals (15.6% of exports and 16.6% of imports respectively) (Secretaría de Economía, 2002).[4]

Export activity was also concentrated in the larger companies, which were responsible for 51.8% of total exports. In contrast, medium, small, and micro-sized companies jointly exported only 6.6% of their production in 1999 (Secretaría de Economía, 2002) (see Table 7.1).

Table 7.1
Share of Exports by Size/Type of Company, 1999

Size/type	Share of Exports
Large	51.80%
Maquiladora	41.50%
MicroSME*	6.60%

*Micro-, Small-, and Medium-sized Enterprises
Source: Secretaría de Economía (2000).

The other sector constituting a major exporter of manufactured goods was that of the *maquiladora* or in-bond companies, which had a 41.5% share of total exports in 1999, and a 50.7% share for the whole decade. The North American Free Trade Agreement (NAFTA) extended *maquiladora*-like privileges to all producers located in Mexico, regardless of their country of origin, as long as they complied with "Rules of Origin."[5] However, in spite of the more flexible rules favoring the local integration of parts and components, this has not occurred. At the end of 2001, domestic input still represented only 3.4% of finished products, the other 96.6% of input being imported (Secretaría de Economía, 2002).

The trade deficit, the limited creation of links between exporters and local suppliers, and the excessive concentration of export activity in a few industrial branches and in *maquiladoras* and large companies have widened the differences between the so-called export sector and the rest of the manufacturing industry.

The export sector, encompassing *maquiladoras*, large companies, and subsidiaries of foreign companies, has achieved a sustained rate of growth and stable levels of employment. In contrast, the rest of the manufacturing sector, comprising the bulk of employment-generating activities,[6] achieved irregular growth and found it more difficult to introduce more advanced technology and to create an innovative environment in companies. This sector has also experienced more difficulty in making the transition from a mass-production organization to a modern, more efficient work-organization scheme. Similarly, companies in this sector find it harder to require, and create, more sophisticated skills in their workers and staff in general. Above all, this sector has had greater difficulty in increasing the exported proportion of its total production, being threatened, in local markets, by more competitive imports.

The following section contains a review of the literature and a conceptual discussion on the relationship between educational level, in-firm training (human capital), technological innovation, the restructuring of production processes, and companies' productivity and export performance.

REVIEW OF THE LITERATURE

Human Capital, Technology and Trade

Since World War II, economic thinking about growth has been based on what has been referred to as the growth model, originally developed by Solow (1956, 1957). This model assumed that both capital and labor were homogenous, thus eliminating any consideration of differences in the quality of labor (and capital) that might arise from education, technology, or other factors. The assumptions of homogenous labor excluded any consideration of the differential effects of education on labor. This model also assumed that technological innovation was exogenous (Bailey and Eicher, 1995: 106).

Empirical tests of the Solow model (Solow, 1957; Abramowitz, 1956; Denison, 1962) suggested that most output growth could not be explained by population growth and the accumulation of capital. Abramowitz labeled this enigma the "unexplained residual" or the "economic index of ignorance" (later known as the "Solow residual"). Economists turned their attention to exploring the components of the Solow residual by dropping the assumption of homogeneous capital (Solow, 1960; Kaldor and Mirrlees, 1962), recognizing that physical and human capital should be included in the growth models. Regarding physical capital, these new versions took into account the newness or oldness of equipment, assuming that newer equipment embodied more advanced technology (Bailey and Eicher, 1995: 107).

Research into the importance of investment in human capital was pioneered by Schultz (1960). His work led to a large body of literature on human capital and on-the-job training (Becker, 1964; Mincer, 1974; Schultz, 1960, 1961). By the late 1980s, studies had shown that educational level (Manwik, Romer, and Weil, 1992), the size of the educated work force (Romer, 1989; Rivera-Batiz and Romer, 1991), the number of patents issued, and the volume of research expenditure (both privately and publicly funded) were associated a country's income growth and the makeup and volume of its trade (Grossman and Helpman, 1991).

Economic theory concerning trade followed the same lines as the theoretical propositions presented earlier, at first also assuming that technology and education played no role in the creation of comparative advantages in developing countries, the main determinants of such advantages being the relative endowments of labor and capital, and emphasizing the importance of free trade. In the theories of Heckschler-Ohlin, technology and skills were not discussed at all. Production functions were assumed to be identical, with technology fully diffused across countries. Firms automatically selected techniques suited to their factor (i.e., capital: labor) prices and, once they had made the right choice—that is, labor intensive techniques in the case of developing countries—they used technologies efficiently, without a need for learning or effort. Because labor was taken to be homogenous and technology users as automatically reaching optimal levels, there was inefficiency only if governments intervened to distort factor prices or to prevent free trade (Lall, 2000a).

According to the above-mentioned models, countries optimized their competitiveness by facilitating technology inflows and opening up their economies to trade, technology licensing, and direct foreign investment. Comparative advantages then depended entirely on factor endowment and any attempt to change this—apart from providing the appropriate conditions for the faster accumulation of factors—was assumed to be inefficient.

A later version of the economic theory (Wood, 1995) assumed capital to be fully mobile and held comparative advantage to be dependent on two factors: skills and natural resources. Skills were treated as a generic resource, created by the education system and generally measured in terms of levels of school attainment. The notion that the efficient use of a given technology required skills and knowledge specifically related to it, and that such skills and knowledge were acquired by means of prolonged experience and problem-solving via, the said technology, was ignored.

With the emergence of new growth theories in the 1980s, factors such as human capital, learning by doing, technology transfer, and endogenous technological innovation were taken in account. Bailey and Eicher (1995) grouped the different approaches regarding how human capital contributed to economic growth into three rough categories: the first of these categories pertained to "education as a separate factor of production," the second to "learning by doing," and the last to an approach stressing "the mutual interaction of technology, human capital, and economic conditions."

Approaches Regarding Human Capital and Productivity

Human Capital as a Separate Factor of Production

Representative of this last approach is the early work of Romer (1989) and the work of Lucas (1988), in which it was suggested that human capital, just like physical capital, can be viewed as a production input which can be accumulated. However, the authors did not specify how the human and physical capital and technological change were interrelated (Bailey and Eicher, 1995). In their analysis, human capital represented the average level of technological knowledge of an economy, without any implied relation to current levels of technological development. The policy implication was that a country's economic growth rate was closely related to its level of educational attainment. However, since the human capital in these models was included in a highly aggregate form, these studies were unable to generate insights into relative investment in primary, secondary, and tertiary education; into how such education should relate to the rate of technological change; or into the appropriate role of government in subsidizing on-the-job training.

"Learning by Doing"

According to this approach the benefits of "learning by doing" were seen to be twofold (Rosenberg, 1982; Mowery and Rosenberg, 1989; Lucas, 1988; Boldrin and Scheinkman, 1988; Forbes and Wield, 2002): the first benefit pertained to the traditional notion (Arrow, 1962) that the more of a particular good produced, the further labor moved down the learning curve and the more efficiency and productivity increased. Second, the more of a particular good (or service) produced, the greater the generation of skills pertaining to the related technology, making it easier to learn about new, relatively similar, production processes. Increased output therefore led to lower unitary costs and to important knowledge spillovers, which in turn facilitated the adoption of new technology. Research along these lines suggested that prior education influenced the effectiveness of learning by doing, and that schooling and learning on the job were complementary. Thus learning by doing would be more effective if built on a minimal foundation of schooling (see Bailey and Eicher, 1995; Mincer, 1974, 1989, 1991).

Interaction of Technology, Human Capital, and Economic Conditions

The third category of study, rather than viewing education as a simple input into the production process or considering its influence on the process of learning by doing, proposed that the invention, adoption, and assimilation of technology, the accumulation of human capital, and economic conditions were all interdependent and endogenous to the model (Nelson and Phelps, 1966; Romer, 1989; Grossman and Helpman, 1991; Lall, 2003).

Nelson and Phelps (1966) suggested that the introduction of new technologies radically transformed the production environment, and that skilled workers differ from unskilled ones in their ability to function in this new environment, since skills enhance their ability to handle new demands created by new technologies. Nelson and Phelps (1966) went on to rank jobs according to the extent to which they require change from unskilled (highly routinized) abilities to highly skilled ones involving the necessity "to learn to follow and understand new technological developments" (Bailey and Eicher, 1995).

The proposition that education promotes both the adoption and creation of new technology has strong empirical support. Benhabib and Spiegel (1992) have shown that human capital better explains economic growth when modeled to facilitate the adoption of new technologies, as opposed to being just another input into the production function. Other empirical work, by Bartel and Lichtenberg (1987), Mincer (1989, 1991), has shown a large degree of complementariness and reciprocity between technological change and human capital. These studies find that a higher rate of technological innovation and adoption increases the demand for skilled as opposed to unskilled labor.

Skills, Technological Capabilities, and Competitive Advantage

The ability to compete in the developing world is changing from a traditional base of primary resources and cheap unskilled labor to one of manufactured products incorporating higher levels of skill and technological input (Lall, 2000b). As a result, in both developed and developing countries the employment of skilled workers has been on the rise. In advanced countries, job growth in the period 1981-1996 has usually been highest for professionals and technicians. This has also been the case in developing countries, but in contrast, in the latter, job growth in the production-and-related-workers category—which contains some skilled workers and craftsmen, but mainly unskilled and semiskilled ones—has been very low and often negative (International Labour Organisation [ILO], 1988).

Even for activities where low wages still constitute an important competitive advantage, technical change and new patterns of demand lead impose strict skill needs. In addition, the pervasive use of new information and communications technologies in all activities means that all countries, regardless of their level of development, must have the ability to use such technologies. Traditional modes of competition, based on low costs and prices, are being replaced by the "new competitiveness," driven by quality, flexibility, design, reliability, and networking (Best, 1990).

New forms of management and organization create different skill needs, and new skills are entailed in the setting up of new production systems with different hierarchies, information flows, and responsibilities. These have to be complemented with new attitudes toward work, as well as new working relationships and management systems (Lall, 2000b). The ILO (1999) identified four features of both these practices: (a) greater use of work teams, involving greater group responsibility, higher skill levels, and more frequent job rotation; (b) involvement in off-line activities such as problem solving, quality improvement, health, and safety; (c) flattening of organizational hierarchies, with greater responsibility being assumed by shop-floor workers and more intense information exchange; (d) tighter links to other resource policies, with training and remuneration systems adapted to prepare and reward employees for new responsibilities. To sum up, the determinants of competitive advantage (for export-oriented and other activities) go far beyond primary resources or cheap unskilled labor; indeed, technological competence, skills, on-the-job discipline, and trainability—largely dependent on the availability of a well-educated labor force—are required.

Technological Structure: Industry Classification According to Technological Complexity

Creating and sustaining export growth in a world of ever more intense competition and increasingly fast technical change require technological depth (Lall, 2000a). In this context, the concept of *technological structure*, proposed by Lall (2002) and based on technological complexity, is very helpful, because it facilitates the classification and identification of segments and subsegments of industries with similar technological activities, functions, and processes.[7] These categories, based on different levels of technological complexity, make it possible to distinguish between *resource-based*, *low-technology*, *medium-technology*, and *high-technology manufactures*, which are defined as follows:

Resource-based (RB) products tend to be simple and labor intensive.[8] Competitive advantage in these products arises generally from the local availability of natural resources. A distinction is made between agriculture-based (*RB1*) and other natural resources such as natural rubber, construction materials, base metals, and foundry activities (*RB2*).

Low-technology (LT) products tend to involve stable, well-diffused technologies. Such technologies are primarily embodied in capital and equipment; the low end of the range has relatively simple skill requirements. Many traded products are undifferentiated and compete on price: thus labor costs tend to be a major element of cost competitiveness. Economies of scale and barriers to entry are generally low. However, there are exceptions to these features, because there are particular low-technology products in high-quality segments where brand names, skills, design, and technological sophistication are very important, even if the technology does not reach the levels of intensity of other categories. A distinction is made between *LT1*, consisting of textiles, garments, and footwear, and *LT2*, consisting of other low-technology products.[9]

The *Medium-technology (MT)* products category is larger and comprises skill- and scale-intensive technologies for the production of capital goods and intermediate products. These products tend to involve complex technologies, with high levels of R&D, advanced-skill needs, and lengthy learning periods. *MT* products are divided into three groups: *automotive products (MT1)*, *process industries (MT2)*, and *engineering products (MT3)*.

MT1 are parts and components that are "linkage intensive" and need considerable interaction between firms to reach "best-practice" technical efficiency. They are produced for both terminal automobile companies and for their suppliers in the transportation-equipment industry.

MT2 have stable and undifferentiated products, often with large-scale facilities and considerable technological effort aimed at improving equipment and optimizing complex processes. *MT2* comprise chemical-product industries, including those producing synthetic fibers.

MT3 emphasize product design and development. Many of them have mass assembly or production plants and extensive supplier networks. Products are heavy and need advanced capacities to reach global standards. This subsegment includes machinery and equipment for specific or general purposes, audio and video equipment, household appliances, and instrumental equipment.

High-technology (HT) products have advanced, fast-changing technologies, with high R&D investment and a primary emphasis on product design. The most advanced technologies require sophisticated technological infrastructures, high levels of specialized technical skills, and close intercompany interaction and interaction between firms and universities or research institutions. However, some products (electronic ones) require labor-intensive final assembly, and their high value-to-weight ratios make it economical to place this stage in low-wage areas. These products lead in new, international, integrated-production systems where different processes are separated and located by multinational corporations (MNCs) according to differences in production costs.

The theoretical discussion section that follows constitutes the basis for identifying the main variables and formulating research questions, both of which activities orient the empirical analysis. The section on methodology describes the characteristics of the sample firms and the data-collection techniques, and also discusses the statistical tools used for data analysis.

DEFINITION OF THE STUDY VARIABLES

This study is of an exploratory nature and sets out to identify and describe the behavior of the operational indicators of two groups of variables: the first group comprises variables, such as education and training (human capital), technological innovation, and the restructuring of work processes, that underlie productivity and competitiveness (independent variables); the second group is related to the expected outcomes—namely, productivity and export performance (dependent variables in the stepwise regression analysis).

Human Capital

Education

This variable involves two indicators:

- The educational attainment of the labor force (workers and personnel), in each industrial segment or in specific industrial branches, and the evolution (or changes) observed during the period from 1992 to 1999.
- The percentage of total personnel in each company that has reached a university level of educational attainment, and the evolution (or changes) observed during the period from 1992 to 1999.

In-firm Training

The in-firm training variable consists of the range of activities carried out by a firm in order to increase the amount and quality of worker skills and to enhance competencies in workers and staff. These training activities can be organized in the form of courses or on-the-job training processes. For this study, the training indicators are:

- Percentage of companies providing training to their labor force and the evolution (or changes) observed during the period from 1989 to 1998.
- The average of hours devoted to training, by industrial segment and by specific industry, and the evolution (or changes) observed during the period from 1989 to 1998.

Technological Innovation

Technological innovation refers to company endeavors related to the following three independent but related technological activities: (1) the acquisition or licensing of technology readily available on the technology market; (2) research and development activities directly undertaken by the companies; (3) the acquisition of new equipment, assuming that such equipment embodies new technology. The specific indicators for this variable are as follows:

- Percentage of income from sales invested in the acquisition of technology licenses, by industrial segment, and by specific industry during the period from 1989 to 1998.
- R&D expenditure as a percentage of the company income from sales, by each of the industrial segment, and by specific industry during the period between 1989 and 1998.
- Percentage of companies introducing new equipment by each of the industrial segment, and by specific industry, and its changes during the period between 1989 and 1998.
- Type of equipment introduced by companies, and its evolution (or changes) during the period between 1989 and 1998. Assuming that equipment embodies technology with different levels of flexibility and automation.

Productivity

The proxy for productivity used in this study is the ratio between the total volume of production and the number of workers. This ratio is estimated by industrial segment and by specific industry, and the evolution (or changes) during the period from 1991 to 1999.

Exports

The amount of exports is estimated based on the portion of production that is sold on foreign markets, by industrial segment, and specific industry (in pesos at 1989 value), and its evolution (or changes) during the period from 1991 to 1999.

RESEARCH QUESTIONS

Based on the above discussion and on the conceptual framework, the research questions can be stated as follows:

Is it possible to identify trends of change—in educational attainment of personnel, in-firm training (human capital), and technological innovation, as well as in exportation, productivity, and employment—in the different industrial segments grouped in accordance with their technological complexity (i.e. resource-based, low-technology, medium-technology, and high-technology) during the 1990s?

Next, are there statistically significant links between educational attainment of personnel, workplace training, technological innovation (*independent variables*), and exportation and productivity (*dependent variables*) in 1989 and 1999?[10]

METHODOLOGY

Nature of the Data

The information analyzed herein was gathered using a nationwide survey based on a stratified, representative sample of the manufacturing sector. Both the survey and sample were designed by the Mexican National Institute for Statistics Geography and Information (Instituto Nacional de Estadística Geografía e Informática, INEGI), which also carried out the data collection.[11]

The sample of firms belonging to the manufacturing sector includes 8 large industrial-sector divisions and 52 manufacturing-activity subsectors, and is stratified according to company size, as defined by the number of employees working in a company, as follows: large companies are those with 251 or more employees; medium-sized companies are those with 101 to 250 employees; small companies are those with 16 to 100 employees; micro-companies are ones with 15 employees.

In the case of large- and medium-sized firms, the sample attempted to include all the companies in the manufacturing sector (based on the 1984 industrial census). For the selection of small and micro-sized firms, a random sampling was employed, which defines the size and distribution of the sample per stratum. In 1989 the sample comprised 138,774 companies. The final sample in 1999 comprised 308,508 firms. The data on which this study is based were collected on three different dates, the first between 1989 and 1992, the second in 1994, and the third between 1998 and 1999. However, only the data gathered at the beginning of the period (1989, 1990, and 1992, according to availability) and at the end of the period (1998 and 1999) are analyzed when examining the behavior of the different variables throughout the decade.

Descriptive Analysis (Identification of Trends)

The analysis of these data was performed in two stages. The first stage consisted of a descriptive analysis based on the comparison of all industrial segments, classified according to their technological complexity and to the other variables of the study (education, training, technological innovation, productivity, and exports) in each industrial segment. This descriptive analysis provided us with valuable information supporting a more robust, multiple-stepwise regression-based examination of the relationships among variables.

Regression Analysis

In a second stage, *stepwise regression* was applied, in order to analyze the possible impact of factors such as education and training, and technological change (*independent variables*) on exports and productivity (*dependent variables*).

So as to minimize possible distortion arising from the heterogeneous nature of the manufacturing sector as a whole, two models were defined, three subsamples have been separately analyzed; the first includes those industrial sectors exhibiting lower levels of complexity (39 low-technology industrial branches based on natural resources); the second sample comprises more technologically complex industries falling within the medium- and high-technology segment;[12] the third subsample contains industries with a low level of technological complexity (see Table 7.2). In each case, the behavior of the variables in the years 1988 and 1989 is taken into account.

Table 7.2
Stepwise Regression Models

Independent Variables	Dependent Variables
• Average number of years of schooling of personnel by industrial sector.	Model 1
• Average hours of training by industrial sector.	• Exports as a percentage of total sales.
• R&D expenditure as a percentage of total sales income by industrial sector.	Model 2
• Investment in technology (licensing) as a percentage of total sales income.	• Production per worker.
• Percentage of companies which have implemented changes in their work processes based on flexible production systems.	

DISCUSSION OF RESULTS: IDENTIFICATION OF TRENDS (DESCRIPTIVE ANALYSIS)

Human Capital

Education

School attainment. At the start of the decade, the average educational level of the company employees who comprise the sample was 8.4 years of school attainment, which corresponds to a level above that of elementary school but below that of secondary school. At the end of the 1990s, a slight increase to 9.0 years of school—equivalent to secondary level—can be noted (see Table 7.3).

Table 7.3
School Attainment of Personnel, by Technological Segment, 1992-1999

	1992 (A)	1999 (B)	1992-1999 B-A	1992-1999 Annual Average Growth %
Total Group Average	**8.4**	**9.0**	**0.6**	**1.0**
RB1: Agro-Based	8.0	8.4	0.5	0.8
RB2: Other	8.3	8.8	0.5	0.9
Low Tech. 1	8.0	8.5	0.5	0.9
Low Tech. 2	8.4	9.1	0.7	1.1
Med Tech. Process	9.7	10.4	0.7	1.1
Med Tech. Engineering	9.0	9.7	0.7	1.1
Med. Automotive	8.9	9.5	0.7	1.0
High Technology	9.4	10.9	1.5	2.2

This slight increase in the educational input received by the employees in the manufacturing sector is partially explained by the increased supply of middle-school graduates, which allowed employers to make the possession of secondary-level education a minimum hiring requirement. Secondary education increased to 12.8% between 1970 and 2000 and the vocational school went up at a rate of 32.3% during this period (see Table 7.4).

The examination measuring the average educational level of staff by industrial segment, grouped according to technological complexity, shows that the most highly educated staff work in the *high-technology* segment. In this segment, staff educational level is almost 11 years (one year below senior-high-school level). Moreover, growth in the educational levels prevailing in companies comprising the *high-technology* segment was higher during the 1990s (2.2%, which amounts to more than double the average yearly growth in the manufacturing sector as a whole).

Table 7.4
People Enrolled in the Educational System, 1970-2000 (in thousands)

Education Attainment	1970-1971	1980-1981	1990-1991	2000-2001	Annual Average Growth 1970-1980	Annual Average Growth 1980-1990	Annual Average Growth 1990-2000	Annual Average Growth 1970-2000
Preschool					16.8	15.5	2.5	25.2
Elementary	400.1	1,072.0	2,734.0	3,424.0	5.9	-0.2	0.3	2.0
Secondary	9,248.2	14,666.0	14,402.0	14,793.0	17.5	3.8	2.8	12.8
Vocational School	1,102.2	3,034.0	4,190.0	5,350.0	26.1	21.0	-0.5	32.3
High School	33.9	122.0	379.0	362.0	27.8	6.3	5.1	27.6
Teachers College	279.5	1,058.0	1,722.0	2,594.0	27.2	-10.0		13.6
Professional Teachers College	55.9	208.0	0.0	0.0	55.4	-1.2	8.4	31.8
Professional (B.A.)	19.0	125.0	109.0	201.0	22.2	3.5	5.7	19.4
Graduate	252.2	811.0	1,097.0	128,947.0	---	---	18.1	18.0
Job Training	147.8	369.3	413.6	1099.6	15.0	1.2	16.6	21.5
Total	**11,538.9**	**21,464.9**	**25,092.0**	**29,669.0**	**8.6**	**1.7**	**1.8**	**5.2**

Source: Secretary of Public Education (SEP) databases. Available at www.sep.gov.mx

The data on the average educational level of employees in specific branches of industry show that the pharmaceutical and computer (*high-technology*) industries, closely followed by the chemical and synthetic-fiber (*mid-level processing-technology*) industries and the printing and publishing (*low-technology*) industries are those that demand more highly educated staff (see Table 7.5).

Table 7.5
School Attainment of Personnel, by Industry (highest/lowest), 1992-1999

	1992 (A)	1999 (B)	B-A	1992-1999 Annual Average Growth
Total Group Average	**8.4**	**9.0**	**0.6**	**1.0**
High				
High Tech.				
Pharmaceutical	10.6	12.0	1.4	1.9
Computer	9.0	10.8	1.8	2.9
Med. Tech. Process				
Chemical	10.3	10.7	0.4	0.6
Synthetic fibers	9.3	10.4	1.1	1.7
Low Tech.				
Printing and Publishing	9.6	10.5	0.9	1.3
Low				
Low Tech.				
RB-1 and RB-2				
Hard-yarn textiles	7.8	5.3	-2.5	-4.6
Corn milling and tortillas	6.0	6.9	0.9	2.1
Wood manufactures	6.6	7.4	0.8	1.7
Low Tech.				
Clay refractory bldg.	6.7	7.3	0.6	1.3
Pottery and ceramics	6.6	6.6	0.0	0.0

In contrast, those employed by the majority of the companies comprising the *low-technology* industries have only elementary education, and no changes are observed in this variable throughout the 1990s (see Table 7.5).

Personnel with higher education, 1992-1999. Measuring the numbers of staff with higher education employed by the manufacturing-industry companies comprising this sample is an additional means of ascertaining the educational composition of the workforce, above all when one considers that higher education has been the area showing most growth over the last 30 years (see Table 7.4).

Observation of the sample of companies analyzed suggests that the number of people with higher education who were employed varied little during the 1990s, increasing by a single decimal point (from 9.1 in 1992 to 9.2 in 1999) (see Table 7.6).

Table 7.6
Percentage of Personnel with Higher Education by Technological Segment, 1992-1999

	1992	1999		1992-1999
	(A) %	(B) %	(B-A)	Annual Average Growth
Total Group Average	**9.1**	**9.2**	**0.1**	**%**
RB1: Agro-Based	11.5	8.9	-2.6	-3.2
RB2: Other	10.0	10.6	0.6	0.8
Low Tech.: 1	6.0	6.2	0.2	0.4
Low Tech.: 2	8.7	9.7	1.0	1.6
Med. Tech. Process	15.4	15.7	0.2	0.2
Med. Tech. Engineering	11.0	13.2	2.2	2.9
Med. Tech. Automotive	8.9	9.8	0.9	1.4
High Technology	14.5	23.2	8.8	8.7

A breakdown of the data by industrial segment shows that the sector whose staff have the highest educational levels, in terms both of total years of schooling and of participation in higher education, is the *high-technology* one. This segment displays the highest level of employment of people with tertiary-level education, from 14.5% in 1992 to 23.2% in 1999, with a bigger relative average yearly growth (8.7%) than the rest of the industrial segments, which suggests that the companies comprising the *high-technology* sector employed a larger share of university graduates during the decade in question (see Table 7.6).

Accordingly, an analysis of each industrial branch reveals that the pharmaceutical and computer industries employed the largest proportion of university-educated staff. At the end of the decade, one-third of those employed in the pharmaceutical industry had received higher education (an increase from 22.4% in 1992 to 33.7% in 1999). There was also a significant increase of the proportion of staff with higher education in the computer industry, from 10.9% of the total number of employees in 1992 to 21.2% in 1999 (see Table 7.7).

In contrast, during the 1990s the number of employees with higher education is considerably lower in companies belonging to those industries with the lowest level of technological complexity (*Resource-Based* and *Low-Technology segments*)—pottery, ceramics, and hard-fiber textile ones—where a decrease in the number of employees with higher education can be observed (see Table 7.7).

Table 7.7
Percentage of Personnel with Higher Education, by Industry (highest/lowest), 1992-1999

	1992	1999		1992-1999 Annual Average
	(A) %	(B) %	(B-A) %	Growth
Total Group Average	**9.1**	**9.2**	**0.1**	**%**
High				
High Tech.				
Pharmaceutical	22.4	33.7	11.3	7.2
Computer	10.9	21.2	10.3	13.5
Med. Tech.				
Chemical	20.0	22.9	2.9	2.1
Audio and video equipment	11.7	20.8	9.1	11.1
RB-2				
Charcoal fuel	17.9	19.5	1.6	1.3
Low				
RB				
Corn milling and tortillas	1.3	2.9	1.6	17.6
Hard-yarn textiles	5.9	2.6	-3.2	-8
Low Tech.				
Pottery and ceramics	4.4	2.4	-2	-6.5
Shoe	3.8	3.8	0	0
Garments, apparel	3.2	4	0.8	3.6

Based on an analysis of the above data, we can assert that, with the exception of the industries in those sectors with the highest levels of technological complexity, the educational level of the staff working in the manufacturing-sector companies that make up the sample is nine years, and it changed very little during the 1990s. Despite the increased supply of middle-high-school graduates, of senior-high-school graduates, and of tertiary-level graduates, the changes in the economic environment and the pressures to be more competitive did not translate into decisions to hire better-educated staff. Decision makers in firms did not give the impression to have taken in account that the determinants of competitive advantage go far beyond primary resources or cheap unskilled labor; indeed, technological competence, skills, on-the-job discipline, and trainability are all largely dependent on the availability of a well-educated labor force.

The above evidence paves the way for an analysis of technological change; it poses questions about the possible link between educational levels in industry and technological change (or lack of it). Furthermore, it indicates the extent to which companies seek to make up for, or supplement, the educational level of their workforce by offering training and endowing their staff with specific production-oriented skills and know-how, a phenomenon that will be analyzed in the following section.

In-Firm Training

Personnel who received in-firm training. There was a substantial increase in the training offered by manufacturing-sector companies overall during the 1990s. While 35.2% of staff received training in 1989, this figure grew at a rate of 8.5 per annum between 1989 and 1998, reaching 62.0% at the end of the decade (see Table 7.8).

A breakdown of the information by industrial sector, grouping the industries in accordance with their levels of technological complexity, shows that it was the companies belonging to the industries comprising the *medium-technology* sector—both the *process* and the *engineering* and *automotive* industries—that offered the most training to their employees. Outstanding among the *medium-technology-process* industries is the industry devoted to other chemical substances, where trained personnel increased from 39.2% to 269.0% between 1989 and 1998. Noteworthy among the *medium-technology engineering* industries is the electrical-equipment industry, which shows the biggest increase in trained employees, from 60.1% in 1989 to 175.5% in 1998. Outstanding among the companies comprising the *medium-technology-automotive-industry* subsector is the terminal auto industry, where the proportion of trained employees increased from 44.4% in 1989 to 104.7% in 1998 (See Table 7.9).

Table 7.8
Percentage of Personnel Who Received In-Firm Training, by Technological Segment, 1989-1998

	1989 (A)	1998 (B)	B-A	1989-1998 Annual Average Growth
Total Group Average	**35.2**	**62.0**	**26.8**	**8.5**
RB1: Agro-Based	31.0	43.9	12.9	4.6
RB2: Other	35.4	58.9	23.5	7.4
Low Tech.: 1	30.4	46.1	15.8	5.8
Low Tech.: 2	36.3	50.9	14.6	4.5
Med. Tech. Process	57.3	141.4	84.1	16.3
Med. Tech. Engineering	41.3	101.1	59.8	16.1
Med. Tech. Automotive	33.7	83.6	50.0	16.5
High Technology	40.7	83.6	42.9	11.7

A breakdown of the information for each industrial branch shows that the companies comprising the chemical-substances industry were the ones that gave the most training to their staff (with a rate of increase of 65.1% per year), followed by the instrumental-equipment industry and the home appliance and computer industries (see Table 7.9).

In contrast, in the industries belonging to the sector with least technological complexity—e.g. the corn-processing-and-tortillas, pottery and ceramics, and hard-yarn textile industries, among others—interest in training staff decreased (see Table 7.9).

Table 7.9
Percentage of Personnel Who Received In-Firm Training by Industry (highest/lowest), 1989-1998

	1989 (A)	1998* (B)	B-A	1989-1998 Annual Average Growth
Total	**35.2**	**62**	**26.8**	**8.5**
High Tech.				
Computer	30.1	101.3	71.2	26.3
Med. Tech.				
Other chemical substance	39.2	269.0	229.8	65.1
Instrumental equipment	60.1	175.5	115.4	21.3
Household appliances	31.6	105.5	73.9	25.9
Automobile industry	44.4	104.7	60.3	15.1
Low Tech.				
Metal structure	20.8	25.4	4.6	2.4
Bread	30.1	20.3	-9.8	-3.6
Hard-yarn textiles	44.7	19.2	-25.4	-6.3
Pottery and ceramics	19	17.5	-1.5	-0.9
Corn milling and tortillas	4.4	6.8	2.4	6.1

*In 1998 the percentage of personnel receiving training is higher than 100 per cent due to the fact that these members of personnel were recipients of more than one training event during the year.

Companies offering in-firm training. An additional indicator of the intense growth in training activities during the decade is the number of companies, in diverse industrial sectors, that offered training to their employees. The data show a shrinking, rather than growing, number of participants in training, indicating, in most segments, a decrease in the number of companies that offered training to their staff. This implies that, although less companies offered training, those firms that did offer it did so more intensively (see Table 7.10).

An analysis of each industrial branch shows that the companies that offered the most training were those belonging to the industries with higher levels of technological complexity, as in the case of the audio and video and electronic-equipment industries (see Table 7.11). It is worth emphasizing that, though more people in the pharmaceutical and computer industries received training, fewer companies offered training. It is probable, as was pointed out earlier, that, faced with either hiring more-qualified staff or offering more training, most firms opted for hiring staff with more schooling.

Table 7.10
Percentage of Companies Offering In-Firm Training by Technological Segment, 1989-1998

	1989 (A)	1998 (B)	(B)-(A)	1989-1998 Annual Average Growth
Total Group Average	**34.4**	**29.8**	**-4.6**	**-1.5**
RB1: Agro-Based	25.9	21.2	-4.8	-2.0
RB2: Other	29.7	32.5	2.9	1.1
Low Tech.: 1	26.4	20.9	-5.0	-2.3
Low Tech.: 2	26.3	25.4	-0.9	-0.4
Med. Tech. Process	76.3	48.6	-27.6	-4.0
Med. Tech. Engineering	43.9	42.4	-1.5	-0.4
Med. Tech. Automotive	69.5	64.1	-5.4	-0.9
High Technology	69.5	64.1	-5.4	-0.9

Table 7.11
Percentage of Companies Offering In-Firm Training by Industry (highest/lowest), 1989-1998

	1989 (A)	1998 (B)	(B)-(A)	1989-1998 Annual Average Growth
Total Average	**34.38**	**29.77**	**-4.62**	**-1.0**
High				
Audio and video equipment	71	78.1	7.1	1.0
Pharmaceutical	88.6	76.2	-12.4	-2.0
Computer	72	62.3	-9.7	-1.0
Synthetic fibers	55.9	58.2	7.2	2.0
Electronic equipment	48.1	53.9	5.9	1.0
Low				
Hard-yarn textiles	14.9	1.5	-13.4	-10.0
Clay refractory bldg.	1.2	1.8	0.6	6.0
Corn milling and tortillas	2.7	4.3	1.6	7.0
Containers wood and cork	7.7	5.2	-2.5	-4.0
Wood furniture	10	7.7	-2.3	-3.0

Finally, education and training can be seen to complement each other (Mincer, 1974), with the highest levels of these two variables occurring in the chemical industry and the industry devoted to other chemical substances, and in the electronic equipment industries (devoted to instruments, audio and video equipment, and home appliances).

Technological Innovation

Technology Licensing

The adoption and application of technical know-how available on the international technology markets has been the most important source of technological change in Mexico (see Casanueva, 2001). In effect, of the various sources of technological change, that most frequently occurring in the companies comprising the sample is the acquisition of technology that is available in the marketplace via licensing. Investment in the acquisition of technology doubled during the 1990s (from 2.7% in 1991 to 5.6% in 1998).

The biggest investment in the acquisition of technology licenses was made by the companies comprising the *low-technology* and *medium-process-technology* segments, which were the ones investing most in the acquisition of technology licenses (see Table 7.12).

Table 7.12
Investment in Technology Licensing, by Technological Segment, 1989-1998

	1989 (A)	1998 (B)	B-A	1989-1998 Annual Average Growth
Total Average	**2.7**	**5.6**	**2.9**	**12.1**
RB1: Agro-Based	2	3.8	1.7	9.8
RB2: Other	2.7	4.9	2.2	8.9
Low Tech.: 1	3.4	6.6	3.4	10.8
Low Tech.: 2	3.7	4.5	0.8	2.4
Med. Tech. Process	1.6	6.9	5.3	36
Med. Tech. Engineering	2.7	5.4	2.7	11
Med. Tech. Automotive	2.2	6.1	3.9	20.4
High Technology	2.3	5.1	2.8	12.6

This is especially clear in the case of some industrial branches, above all those devoted to wooden furniture, publishing, and synthetic fibers (see Table 7.13). In the case of the first two, investment in technology licenses is very possibly due to the increasing adoption of design-oriented information technologies.

Table 7.13
Investment in Technology Licensing, by Industry (highest/lowest), 1989-1998

	1989 (A)	1998 (B)	B-A	1989-1998 Annual Average Growth
High				
Printing and publishing	4.2	16.5	12.3	33.0
Synthetic fibers	0.9	10.6	9.7	118.0
Plastic resins	5.2	9.8	4.6	9.8
Clay refractory bldg.	4.4	9.4	5.1	12.8
Cocoa and chocolate	3.7	1.6	-2.1	-6.3
Cereal processing	1.3	1.5	0.2	1.8
Fats and oils	0.8	1.3	0.4	4.8
Food for animals	0.2	1	0.8	37.7
Transportation equipment	7.9	0.5	-7.4	-10.4

Surprisingly, investment in technology licenses in the auto-parts (transportation-equipment) industry decreased in the decade in question, from 7.9% in 1989 to 0.5% in 1998 (see Table 7.13). This is probably because technological change in this industry occurred in the 1980s, in obedience to worldwide restructuring and innovatory tendencies in the automotive industry during that period.

The data also suggest that those industries with the least technological complexity, above all the food industry, were the ones that invested less in technology licenses (see Table 7.13).

Research and Development

Investment in R&D by the manufacturing-industry companies included in the sample was very low and, indeed, tended to decrease during the 1990s (see Table 7.14). The slight increase observed during that period (from 0.64% in 1989 to 0.71% in 1998, see Table 7.15) is due to the behavior of a very select group of industrial branches, and, for the most part, as is more fully analyzed below, to possible confusion between R&D activities and new-product-development activities where the emphasis is on design.

In the pharmaceutical industry spending on R&D increased more than twofold, as was the case in the wooden furniture industry, the soft-yarn textile industry, the pottery and ceramics industry, the knitted fabrics industry, and the printing and publishing industry (see Table 7.15). However, in the case of the *low-technology* industries, the data regarding investment in R&D reflect possible confusion between R&D activities and Development (D) activities, where the emphasis, as is characteristic of these industries, is on design processes and new-product development.

Table 7.14
R&D Expenditure as a Percentage of Total Sales, by Technological Segment, 1989-1998

	1989 (A)	1998 (B)	B-A	1989-1998 Annual Average Growth
Total Average	**2.7**	**5.6**	**2.1**	**12.1**
RB1: Agro-Based	0.5	0.7	0.2	3.5
RB2: Other	0.7	0.6	-0.1	-1.5
Low Tech.: 1	0.5	0.1	0.5	11.5
Low Tech.: 2	0.7	0.4	-0.3	-5.3
Med. Tech Process	0.6	0.4	-0.2	-3
Med. Tech Engineering	0.9	0.8	-0.2	-2.1
Med. Tech Automotive	1.2	0.7	-0.5	-4.5
High Technology	0.8	1.2	0.3	4.5

Table 7.15
R&D Expenditure as a Percentage of Total Sales, by Industry, 1989-1998 (thousands of pesos = 1989)

	1989 (A)	1998 (B)	B-A	1989-1998 Annual Average Growth
Total	**0.64**	**0.71**	**0.1**	**1.3**
High				
RB				
Wood manufactures	0.18	3.55	3.4	211
Soft-yarn textiles	0.62	3.07	2.4	43.5
Low Tech.				
Pottery and ceramics	0.22	2.03	1.8	89.2
Printing and publishing	0.63	1.9	1.3	22.1
High Tech.				
Pharmaceutical	0.83	2.24	1.4	19
Low				
Corn dough and tortillas	0.11	0	-0.1	-11.1
Bread	0.15	0.1	-0.1	-4

In contrast, the pharmaceutical industry is a research-intensive one, which is why the observed increase in R&D (0.83% in 1989 to 2.24% in 1998) can be associated with R&D activities.

The literature on technological innovation acknowledges the growing, interfirm cooperation at the international level, above all in the case of multinational companies with plants located both in developed countries and in developing ones such as Mexico.

The sharing of R&D activities allows multinational corporations to take advantage of qualified human resources that are internationally distributed in a global innovatory endeavor. On the one hand, multinational companies benefit from the human capital available in developing countries. On the other hand, such relationships between companies on a global scale may lead to a scattering of technology, which also benefits developing countries. In India and Singapore, technological development has been documented in the biopharmaceutical, software, and telecommunications industries (see Reddy, 2000).

Moreover, the trends observed with regard to variations in the R&D investment made by the companies comprising the sample coincide with those observed with regard to levels of staff education. As previously observed, of the set of industries that makes up the manufacturing sector, it is the pharmaceutical industry that has the highest levels of education among its employees. Besides making the adoption of existing technologies easier, this concentration of human capital is a factor that can favor the carrying out of R&D activities in countries such as Mexico.

Technology Embodied in Equipment

In 1998, more than the half of the companies comprising the sample (51.2%) acquired new equipment, which suggests that there was some innovation based on the technology embodied in the purchased equipment (in 1989 the above figure went down to 42.1%).

However, the depth of technological change brought about by the new equipment acquired, during the 1990s, by the manufacturing companies comprising the sample largely depends on how complex and flexible the said equipment is.

In 63% of cases, the companies reported that they acquired only manual equipment and other tools, which suggests that, rather than technological change, what was registered in these cases was upkeep of the operations in question. Below, we look at the remaining 37% of the companies using automated equipment[13] and numeric-control equipment[14] (see Table 7.16).

Automated Equipment. An overall decrease can be observed in the incorporation of automation processes in the companies comprising the sample (from 33.2% of the companies in 1989 to 29.6% in 1998), above all in the *medium-processing-technology* and *medium-automotive-technology* sectors (with a marked decrease being observed in the latter) (see Table 7.17).

There was also a decrease in automation in the *natural-resource-based* and *low-technology-based* segments, as in the case of the food industry (with the exception of the fats and oils and animal food industries, where a slight increase can be observed).

Table 7.16
Percentage of Companies Acquiring Manual Equipment and other Tools, 1989-1998

	1989 (A)	1998 (B)	B - A
Companies Acquiring New Equipment	**41.9**	**51.2**	**9.3**
Manual Equipment and Other Tools	**54.3**	**63.1**	**8.9**
BR 1 Agro-Based	50.8	60.8	10.0
RB2 Other	60.2	69.4	9.2
Low Tech. 1	50.9	63.3	12.4
Low Tech. 2	75.6	73.8	-1.8
Med. Tech. Process	35.8	51.4	15.5
Med. Tech. Engineering	59.4	67.7	8.3
Med. Tech. Automotive	54.8	71.2	16.4
High Technology	42.0	38.8	-3.2

Table 7.17
Percentage of Companies Introducing Automated Equipment, by Technological Segment, 1989-1998

	1989 (A)	1998 (B)	B - A
Percentage of Companies Acquiring New Equipment	**41.9**	**51.2**	**9.3**
Automated Equipment Total Average	**33.2**	**29.6**	**-3.7**
BR 1 Agro-Based	38.2	34.2	-4.0
BR 2 Other	57.5	49.4	-8.1
Low Tech. 1	29.6	25.1	-4.5
Low Tech. 2	19.1	20.7	1.7
Med. Tech. Process	50.4	38.4	-12.0
Med. Tech. Engineering	26.9	21.1	-5.8
Med. Tech. Automotive	36.5	13.8	-22.7
High Technology	32.4	40.7	8.3

Although the overall trend was toward less automation of industrial processes (as a result of the introduction of new equipment), there were branches in the food and tobacco industries which intensify their automation by introducing new equipment. In addition to these industries, an increase in automation was also observed in synthetic fibers, printing and publishing industries, and pharmaceutical sectors (see Table 7.18).

Numeric-control equipment. The technological complexity of numeric-control equipment[15] resides in its potential for programming, which gives the production process a flexibility that enables the industry to adopt to trends and fluctuations in market demand. Staff possessing complex skills and know-how, and, hence, higher levels of education, along with intensive training input by the company, are necessary in order to operate numeric-control equipment.

Table 7.18
Percentage of Companies Introducing Automated Equipment, by Industry (highest/lowest), 1989-1998

	1989 (A)	1998 (B)	B - A	Annual Growth 1989-1998
Companies Acquiring New Equipment	**41.9**	**51.2**		**9.3**
Automatic Equipment Total Average	**33.2**	**29.6**	**-3.7**	**-1.2**
High				
RB-1				
Fats and oils	44.4	62.1	5.3	4.4
Tobacco	55.6	60.9	-0.2	1.1
Sugar processing	57.1	56.9	0.0	0.0
Med. Tech. Process				
Synthetic fibers	42.9	62.5	0.0	5.1
High Tech.				
Pharmaceutical	53.6	60.2	0.0	1.4
Low				
Metal structure	11.3	6.3	-7.0	-5.0
Wood furniture	11.6	4.6	-10.0	-6.7
Metallic furniture	14.4	4.4	1.2	-7.7
Machinery for general purposes	2.3	3.5	-13.1	5.6
Clay refractory bldg.	14.0	0.8	0.0	-10.5

In the set of manufacturing-sector companies that comprises the sample, the proportion of companies using numeric-control equipment decreased, in all the segments, regardless of their level of technological complexity, with the exception of the automotive-industry segment, from an average of 12.5% in 1989 to an average of 7.3% in 1998 (see Table 7.19).

Productivity

During the 1990s an increase in productivity can be observed in the companies comprising the manufacturing-industry sample. Average productivity, expressed as a ratio between total production volume and number of workers, increased by 13.1% (from 103.1% in 1989 to 116.2% in 1998), at an average annual growth rate of 1.4% (see Table 7.20).

An analysis of specific industries shows that the printing and publishing industry was the one that most increased its productivity during the decade. This leap in productivity is due to the replacement of linotype technologies by information technologies. In the 1990s, the application of hardware and software to the design, printing, and production of printed materials had a decisive impact on productivity in this industry.

Table 7.19
Percentage of Companies Introducing Numeric-Control Equipment by Technological Segment, 1989-1998

	Total			**Numeric Control**		
	(A) 1989	**(B) 1998**	**B - A**	**(A) 1989**	**(B) 1998**	**B - A**
Total Average	**41.9**	**51.2**	**9.3**	**12.5**	**7.3**	**-5.2**
BR 1 Agro-Based	34.7	39.1	4.4	11.1	5.1	-6.0
RB2 Other	35.5	53.8	18.3	10.3	5.5	-4.8
Low Tech. 1	28.5	45.8	17.3	19.6	4.6	-15.0
Low Tech. 2	51.6	64.7	13.1	5.3	5.5	0.1
Med. Tech. Process	72.3	57.7	-14.6	13.8	10.2	-3.6
Med. Tech. Engineering	52.9	65.1	12.2	13.7	11.2	-2.5
Med. Tech. Automotive	53.6	70.7	17.1	8.7	15.0	6.3
High Technology	65.1	69.8	4.7	25.6	20.4	-5.2

As well as those in the printing and publishing industry, important productivity increases were observed in the *high-technology* segment, specifically in the computer, electronics, and pharmaceutical industries (from 95.8% in 1989 to 150.1% in 1989, at an annual growth rate of 6.3%).

The *medium-technology* industries that displayed the greatest increases in productivity were the home-appliance, audio-and-video-equipment and automotive ones. Outstanding among the industrial branches with least technological complexity are the *natural-resource-based* ones, in which production-process automation increased, as is the case with the tobacco industry, some branches of the food industry, and the iron and steel industry (see Table 7.20).

Those companies that experienced negative growth in their productivity during the decade are located in the least technologically complex segments. It is in the companies composing these segments that lower levels of human capital (education and training) and of technological and organizational innovation are observed (see Table 7.21).

An analysis of productivity trends in manufacturing-sector companies during the 1990s shows possible linkage between the human capital, technological-innovation, and productivity variables. These possible links will be verified using regression analysis in the corresponding section of this chapter (see Discussion of Results Regression Analysis).

Table 7.20
Productivity by Industry (highest), 1989-1998

	1989 (A)	1998 (B)	B-A	Annual Growth
Total Average	**103.1**	**116.2**	**13.1**	**1.4**
Resource Based				
Tobacco	179.5	344.2	164.7	10.2
Food for animals	253.0	267.0	14.0	0.6
Fats and oils	217.0	259.0	42.0	2.2
Basic metals	221.9	230.0	8.1	0.4
Low Technology				
Printing and publishing	65.9	265.8	199.9	33.7
Steel and iron	125.5	204.4	78.9	7.0
Medium Technology				
Automobile industry	200.9	291.7	90.8	5.0
Chemical Industry	178.9	196.9	18.0	1.1
High Technology				
Computer	115.1	209.8	94.7	9.1
Pharmaceutical	120.8	166.7	45.9	4.2

Table 7.21
Productivity by Industry (lowest), 1989-1998

	1989 (A)	1998 (B)	B-A	Annual Growth
Total Average	**103.1**	**116.2**	**13.1**	**1.4**
RB Agro-Based				
Natural rubber, gums	98.3	74.9	-23.4	-2.6
Milk products	145.4	101.5	-43.9	-3.4
Hard-yarn textiles	50.9	16.5	-34.4	-7.5
Wood furniture	73.8	22.9	-50.9	-7.7
Low Technology				
Garments, apparel	24.7	18.4	-6.3	-2.8
Metallic furniture	46.2	31.0	-15.2	-3.7
Shoes	40.6	25.9	-14.7	-4.0
Pulp and paper	156.1	97.9	-58.2	-4.1
Metallic products	123.2	75.1	-48.1	-4.3
Metal structures	178.2	27.5	-150.7	-9.4

Exports

During the period 1989-1998, the companies comprising the sample increased the volume of exports from 13.3% in 1989 to 21.3% in 1998, though the level of exports varied greatly across the different industrial sectors.

The industry with the best export performance was the computer one, pertaining to the *high-technology* segment (with 53.3% of its production being exported in 1989 and 74.0% in 1998). In second place came the automobile industry, which exported 46.4% of its production in 1989 and 51.4% in 1998 (see Table 7.22).

Table 7.22
Exports by Industry (highest), 1989-1998

	1989 (A)	1998 (B)	B-A	Annual Growth
	13.3	**21.3**	**8**	**6.6**
High & Medium Tech.				
Computer	53.3	74.0	20.7	4.3
Automobile	46.4	51.4	5	1.2
Audio and video	23	49.8	26.8	12.9
Home appliances	15.2	42.7	27.5	20.1
Machinery specific applications	8.9	40.0	31.1	38.8
Electronic equipment	18.8	30.9	12.1	7.2
Resource B. & Low Tech.				
Knitted fabrics	4.2	49.6	45.4	120.1
Tobacco	0.3	36.2	35.9	1329.6
Pottery and ceramics	2.7	34.3	31.6	130.0

Isolated industries in the various segments showed important export growth during the decade. In the *agro-based natural resources* segment, the tobacco industry increased its exports from 0.3% to 36.2% during the period in question, and knitted fabrics increased its exports from 4.2% to 49.6% over the same period. In the *low-technology* segment, pottery and ceramics increased from 2.7% in 1989 to 34.3% in 1998. In the *medium-technology engineering* segment, audio and video equipment and home appliances also increased their share of exports (the former from 23.0% in 1989 to 49.8% in 1998, and the latter from 15.2% in 1989 to 42.7% in 1998).

Some of the companies in the food-industry branches have slightly increased their export percentages, though their production is basically aimed at the domestic market. This is the case for such industrial branches as the pulp-and-paper and charcoal-fuel ones.

With the exception of the printing and publishing industry in the *low-tech segment* and the pharmaceutical industry in the *high-tech segment*, the rest of the industrial branches comprising the industrial segments based on *natural resources* (RB1 and RB2) and the companies comprising the *low-tech segment* did not experience significant changes in the variables associated with human capital competition and production organization, despite the fact that, in some industries, technological innovations were introduced via one or other of the available sources of technology (see Table 7.23).

In contrast, as has already been mentioned, the companies comprising the printing and publishing industry experienced a radical change in the composition of their human capital and the structure of their productive organization. These changes evidently redounded in significant increases in productivity, though this did not mean an increase in exports due to the way in which the said industry's production and distribution are globally organized.

Unlike the printing and publishing industry, the tobacco industry (from the section pertaining to industries *based on natural resources*) experienced an unprecedented increase in its exports.

Table 7.23
Exports by Industry (lowest), 1989-1998

	1989 (A)	1998 (B)	B-A	Annual Growth
Total	**13.3**	**21.3**	**8**	**6.6**
Resource Based-1				
Beverage	4.0	5.7	2	4.7
Bread	0.3	4.2	4	144.4
Fats and oil	1.1	2.5	1	14.1
Food for animals	0.1	2.0	2	211.1
Meat	0.4	1.2	1	22.2
Milk products	0.2	0.6	0	22.2
Corn milling and tortillas	0.0	0.5	1	0.0
Pulp and paper	5.3	5.2	0	-0.2
Resource Based-2				
Charcoal fuel	3.9	5.1	1	3.4
Low Tech. -1				
Printing and publishing	1.4	0.8	-1	-4.8

The case of the pharmaceutical industry (belonging to the *high-tech* segment) is very different from the two aforementioned ones. This industry employed a larger amount of qualified workers (the most highly qualified ones in the manufacturing sector) than before, intensified its research activities, and also significantly increased its productivity. However, it only slightly increased its export levels during the decade (from 4.3% in 1989 to 15.5% in 1998), since 84.5% of its production continued to be aimed at the domestic market.

An analysis of developments in exportation by the companies comprising the various industrial branches during the 1990s shows a twofold tendency. On the one hand we have companies that considerably increased their exports during the 1990s, consisting of industries that export more technically complex goods such as automobiles, information-processing equipment (computers and peripherals), and electrical appliances (audio and video equipment and home appliances items); on the other hand we have a large group of industries, mostly belonging to the less technically complex segments (i.e., those based on *natural resources* and that are *low-tech*) that have marginally increased their export levels (by between 2 and 20%) and continue to aim their production at the domestic market, where, with very few exceptions, they face competition by imported products.

To sum up, while it is possible to discern associative trends between the variables of human capital, technological innovation, and, to a lesser extent, changes in productive organization and productivity, the link between the aforesaid variables and export performance is not so clear. From the late 1980s on, some industrial branches exhibited outstanding export performance, very possibly based on foreign investment and technology, while an enormous, heterogeneous group of industries introduced isolated changes, from among the set of variables herein analyzed, that had little impact on their export performance. The possible impact of variables such as human capital and technological and organizational innovation on productivity and export performance is analyzed in the following section, in which we present the results of the regression analysis.

DISCUSSION OF THE RESULTS OF THE REGRESSION ANALYSIS

Export Performance as a Dependent Variable (Model 1)

In the sample comprising all of the industrial sectors, no statistically significant associations were observed between the independent variables and companies' export performance. These results suggest that changes in *education and training,* along with *innovation* and the *reorganization of industrial work*, have been weak and had little effect on the export performance of the firms examined in the various sectors between 1989 and 1998 (a phenomenon examined in more detail in the descriptive analysis).

As in the above case, no statistically significant associations were found in the regression analyses carried out on the two subsamples of industrial sectors—the high-technical-complexity and low-technical-complexity ones. To some extent, this analysis mitigates the effects of the heterogeneity of the industrial sectors and strengthens the interpretation regarding the low impact of the education, training, technological-innovation and reorganization of industrial-work variables on the export performance (a dependent variable) of the companies examined in the various industries between 1989 and 1998.

Productivity as a Dependent Variable (Model 2)

The regression analysis finds a direct relationship between the education variable and productivity, both at the start of the decade (1989) and at the end of it (1998) (see Tables 7.24-7.28). However, it is interesting to note that, in the case of the lower-complexity, less-technological-content industries, this association between education and productivity is not in evidence at the start of the period (1989), but can be found at the end of the said period. These results suggest that educational level takes on importance for the hiring, in the 1990s, of workers and staff in the lower-complexity, less-technological-content industrial sectors.

Table 7.24
Productivity as Dependent Variable, 52 Industries, 1989

Model	Nonstandardized Coefficients		Standardized Coefficients	T	Sig.
	B	Standard error	Beta		
Constant	-154.764	63.929		-2.421	0.019
Years of schooling	33.398	7.529	0.518	4.436	0.000
Spending on acquisition of technology	-10.657	4.396	-0.283	-2.424	0.019
F	12.228				
Sig.	0.000				
R^2 adjusted	0.306				

Table 7.25
Productivity as Dependent Variable, Resource-Based and Low-Technology Industrial Segments, 1989

Model	Nonstandardized Coefficients		Standardized Coefficients	T	Sig.
	B	Standard error	Beta		
Constant	-227.875	77.749		-2.931	0.006
Years of schooling	39.514	9.565	0.562	4.131	0.000
F	17.067				
Sig.	0.000				
R^2 adjusted	0.297				

Table 7.26
Productivity as Dependent Variable: High-Technology Segment, 1989

Model	Nonstandardized Coefficients		Standardized Coefficients	T	Sig.
	B	Standard error	Beta		
Constant	158.735	27.401		5.793	0.000
Spending on acquisition of technology	-17.330	7.449	-0.574	-2.326	0.040
F	5.412				
Sig.	0.040				
R^2 adjusted	0.269				

Table 7.27
Productivity as Dependent Variable: Resource-Based, Low-Technology and High-Technology Segments, 1998

Model	Nonstandardized coefficients		Standardized Coefficients	T	Sig.
	B	Standard error	Beta		
Constant	-276.845	69.813		-3.966	0.000
Years of schooling	42.487	7.723	0.614	5.501	0.000
F	30.262				
Sig.	0.000				
R^2 adjusted	0.365				

Table 7.29
Productivity as Dependent Variable, Resource-Based and Low-Technology Segments, 1998

Model	Nonstandardized Coefficients		Standardized Coefficients	T	Sig.
	B	Standard error	Beta		
Constant	-356.897	84.128		-4.242	0.000
Years of schooling	52.775	9.731	0.666	5.424	0.000
F	29.416				
Sig.	0.000				
R^2 adjusted	0.428				

The results show that, except for the education variable, no statistically significant association is detected between the independent variables (*training, technological innovation*, and *reorganization of industrial work*) and *productivity* (a dependent variable). These observations are consistent, with regard both, to the sample containing all of the industrial sectors and to the lower-complexity, less-technological-content industrial sectors.

Stepwise Regression, Discussion of Results

The *stepwise* regression analysis shows a positive and statistically significant correlation between education and productivity throughout the full set of industries and, above all, in the less technologically complex ones, at both the beginning and the end of the decade, in support of the human-capital hypothesis.

With the exception of this correlation between education and productivity, no significant relationship is observed between *technological innovation*, changes in the *organization of production* and *training* (independent variables) and *productivity* and *development (*dependent variables). This is because the changes in the first group of variables (*personnel training* and *technological innovation*) were not intense enough in the manufacturing industry during the period analyzed.

CONCLUSION

The results found in the study described in this chapter coincide with the argument presented in the introduction of the chapter: the coexistence within the manufacturing industry, of segments comprised by companies with different levels of productivity and competitiveness. A group of companies that has successfully adjusted to the conditions of international competition coexists alongside a large group of businesses that have shown uneven growth and that are struggling to adjust to the new rules of competition in both domestic and export markets.

The sector comprising automobiles, computers, electronic equipment, including the audio and video and home appliances industries, has achieved a sustained rate of productivity and exports. In contrast, the rest of the manufacturing sector achieved irregular growth and found it more difficult to introduce more advanced technology and to create an innovative environment in companies. Similarly, companies in this sector find it harder to require, and create, more sophisticated skills in their workers and staff in general. Above all, this sector has had greater difficulty in increasing the exported proportion of its total production, being threatened, in local markets, by more competitive imports.

This explanatory study suggests that, while greater emphasis should be placed on education and training for work (human capital), it should be recognized that, to a large extent, the success of such investment is linked to investment in technological innovation and the acquisition of *endogenous technological capabilities*. These findings match Lall's (2000b: 2) contentions about the unevenness observed in developing countries with regard to improving the latter's skills base.

The lack of any significant effect of independent variables on productivity and export performance could be attributed to variables associated with financial and institutional factors that, while they could not be considered in this study, should be taken account by future research. Just a few such variables are listed below:

- Investment in human capital, stressing the quality of education and in-firm training.
- The cost of access to capital (credit on competitive terms and foreign direct investment).
- The cost of business-related transactions associated with pending government reforms, above all the high cost of energy, transport and telecommunications.
- Links between industry and both national and international centers of technological know-how (higher education and R&D centers).
- Incentives to direct foreign investment so as to complement domestic investment, transfer recently developed technology, favor the assimilation and adaptation of technology, and contribute to the development of skills and know-how among workers and staff in general.

Indeed, such investment is required to enable Mexican firms to adjust to the dynamics of international production and to forge better links with supply networks inside Mexico, with more added value and technological depth, with their consequent effect on employment and income distribution.

ACKNOWLEDGMENTS

The author would like to acknowledge the contribution of Alejandro Márquez and the valuable assistance of Ana Delgado and Paulina Ruiz Íñiguez. The views and opinions expressed in this chapter correspond to the author and do not necessarily represent those of the Universidad Iberoamericana.

NOTES

1. To date, Mexico has signed 31 trade agreements, with countries of three continents (Villarreal and Ramos, 2001: 772), the most relevant are the North American Free Trade Agreement, or NAFTA, in effect since 1994, and one with the European Union, in effect since July 2000.

2. This aggregate estimate also includes the production of the *maquiladora* or in-bond industry, which has been primarily export oriented, as discussed later in this chapter.

3. The remaining 3.7% pertained to agricultural, livestock, and natural-resource exports.

4. These figures include both *maquiladora* and non-*maquiladora* companies. http://www.economia.gob.mx/?P=36

5. NAFTA, which came into effect in 1994, establishes that *maquiladora* companies of member countries are able to sell their products as local firms on local markets, or

anywhere within the member countries' markets. Rules of origin were established for those companies whose capital investments do not come from the member countries' region. In order to take advantage of NAFTA tariff advantages, such companies must comply with Rules of Origin, integrating inputs from the member countries into their end products. NAFTA also grants tariff advantages when products are subject to major transformation within a member country to the extent that the end product qualifies for a different tariff classification.

6. Micro-, small-, and medium-sized companies employed 71.6% of the economically active population in 1993 and 67.1% in 1998 (Garza Castaño, 2000).

7. These categories were also adopted by the Economic Commission for Latin America (Comisión Económica para América Latina CEPAL) in its 2003 analysis of the incorporation of Latin America and the Caribbean into international markets.

8. However, Lall (2000b) is aware of that there are industries using capital-, scale-, and skill-intensive technologies (e.g., petroleum refining or modern processed foods).

9. According to Lall (2000a) the *LT2* or "other low-technology-products" group has undergone massive relocation from rich to poor countries, with assembly operations shifting to low-wage sites and complex design and manufacturing functions being retained in advanced countries. This relocation has been the engine of export growth in this industry. Other exports that have benefited from active relocations in this group are toys, sports and travel goods, and footwear.

10. As discussed before, the examination on the evolution of the different variables throughout the decade covers the beginning of the period (1989, 1990, 1992 according to availability) and the end of the period (1998, 1999, also according to availability).

11. Instituto Nacional de Estadística Geografía e Informática (1992, 1995, 1999), Mexican Statistics Bureau: "Encuesta Nacional de Empleo, Salarios, Tecnología y Capacitación" (National Survey on Employment, Wages, Technology and Training).

12. It should be pointed out that, although they formed part of the *medium-technology* sector (Lall, 2000a), those sectors identified in this study as exhibiting greater complexity and higher levels of technological content—such as the chemical industry, the machinery and equipment industry, and the automobile industry—were included in this sample.

13. The variable pertaining to automated equipment also includes the use of robots. The latter category was not analyzed separately, because robots amount to only 0.0062% of the equipment incorporated in 1989 and this figure goes down to 0.0048% in 1998.

14. This includes computerized numeric-control equipment.

15. Numeric-control equipment includes also computer numeric-control equipment.

REFERENCES

Abramowitz, M. (1956). "Resources and Output Trends in the U.S. since 1870," *American Economic Review*, 46: 5-23.

Arrow, K. (1962). "The Economic Implications of Learning by Doing," *Review of Economic Studies*, 29: 155-173.

Bailey, T. and Eicher, T. (1995). "Education, Technological Change and Economic Growth." In J. Puryear and J. Brunner (eds.), *Education, Equity and Economic Competitiveness in the Americas: An Inter-American Dialogue Project*. Washington, DC: Organization of American States.

Bartel, A. and Lichtenberg, F. (1987). "The Comparative Advantage of Educated Workers in Implementing New Technology," *Review of Economics and Statistics*, 69 (1): 1-11.

Becker, G. (1964). *Human Capital*. New York: Columbia University Press.

Benhabib, J. and Spiegel, M. (1992). "The Role of Human Capital in Economic Development." C.V. Starr Center for Applied Economics, New York University, Economic Research Report No. 92.

Best, M. (1990). *The New Competition: Institutions of Industrial Restructuring*. Cambridge, UK: Polity Press.

Boldrin, M. and Scheinkman, J.A. (1988). "Learning by Doing, International Trade and Growth: A Note." In P. Anderson, K. Arrow, and D. Pines (eds.), *The Economy as an Evolving Complex System*. Reading, MA: Addison Wesley.

Casanueva, C. (2001). "The Acquisition of Firm Technological Capabilities in Mexico's Open Economy: The Case of Vitro," *Technological Forecasting and Social Change*, 66 (1): 75-85.

CEPAL (2003). "Panorama de la inserción internacional de América Latina y el Caribe, 2001-2002" (International Insertion Panorama of Latin America and the Caribbean). División de Comercio Internacional e Integración.

Denison, E. (1962). *The Sources of Economic Growth in the US and the Alternatives before Us*. New York: Committee for Economic Development, Library of Congress.

Forbes, N. and Wield, D. (2002). *From Followers to Leaders, Managing Technology and Innovation*. London: Routledge.

Garza Castaño, R. (2000). "Creación de Pymes: Objetivo Emprendedor" (Creation of SMEs: Fostering Entrepreneurship,) *Ingenierías*, 3 (9), October-December.

Grossman, G. and Helpman, E. (1991). *Innovation and Growth in the Global Economy*. Cambridge, MA: MIT Press.

International Labour Organisation (1988). *Yearbook of Labour Statistics 1988*. Geneva: International Labour Office.

International Labour Organisation (1999). *World Employment Report 1998-99. Employability in The Global Economy: How Training Matters*. Geneva: International Labour Office.

Kaldor, N. and Mirrlees, J. (1962). "A New Model of Economic Growth," *Review of Economic Studies*, 29 (3): 174-192.

Lall, S. (2000a). "The Technological Structure and Performance of Developing Countries' Manufactured Exports, 1985-98," Working Paper No. 44, June. Oxford Development Studies.

Lall, S. (2000b). "Skills, Competitiveness and Policy in Developing Countries," Working Paper No. 46, June. Oxford Development Studies.

Lall, S. (2002). "The Employment Impact of Globalization in Developing Countries," ILO mimeo, October, Geneva.

Lall, S. (2003). "Industrial Success and Failure in a Globalizing World," Working Paper No. 102, February. Oxford Development Studies.

León González, A. and Dussel, E. (2001). "El comercio intraindustrial en México" (The Intraindustrial Commerce in Mexico,) *Comercio Exterior*, 51 (7), July: 652-664.

Loria, E. (1999). *Efectos de la Apertura Comercial en la Manufactura Mexicana, 1980-1998* (Effects of the Commercial Opening in Mexican Manufacturing, 1980-1998). Universidad Nacional Autónoma de México (UNAM), Facultad de Economía.

Lucas, R. (1988). "On the Mechanics of Economic Development," *Journal of Monetary Economics*, 22: 3-42.

Manwik, N., Romer, D., and Weil, D. (1992). "A Contribution to the Empirics of Economic Growth," *Quarterly Journal of Economics*, 107 (2): 407-437.

Mendiola, G. (1999). "Empresas maquiladoras de exportación en los noventa" (The Mexican In-Bond Exportation Companies in the 1990s,) Serie Reformas Económicas, No. 49, December. Santiago: CEPAL.

Mincer, J. (1974). *Schooling Experience and Earnings*. New York: Columbia University Press.

Mincer, J. (1989). *Labor Market Effects of Human Capital and of Its Adjustment to Technological Change*. New York: Institute on Education and the Economy, Teachers College, Columbia University.

Mincer, J. (1991). "Human Capital, Technology, and the Wage Structure," NBER Working Paper No. 3581. Cambridge: MA: National Bureau of Economic Research.

Mowery, D. and Rosenberg, N. (1989). *Technology and the Pursuit of Economic Growth*. Cambridge, UK: Cambridge University Press.

Nelson, R. and Phelps, E. (1966). "Investment in Humans, Technological Diffusion and Economic Growth," *American Economic Review*, 56 (2): 69-82.

Reddy, P. (2000). "Globalization of Technology: Issues in Technology Transfer and Technological Capability Building." Adobe Reader eBooks.

Rivera-Batiz, L. and Romer, P.M. (1991). "International Trade with Endogenous Technological Change," *European Economic Review*, 35 (4): 971-1001.

Romer, P. (1989). "Human Capital and Growth: Theory and Evidence," NBER Working Paper No. 3173. Cambridge: MA: National Bureau of Economic Research.

Rosenberg, N. (1982). *Inside the Black Box: Technology and Economics*. Cambridge, UK: Cambridge University Press.

Schultz, T. (1960). "Capital Formation and Education," *Journal of Political Economy*, 68: 571-583.

Schultz, T. (1961). "Investment in Human Capital," *American Economic Review*, 51: 1-17.

Secretaría de Economía, http://www.economia.gob.mx/?P=36

Solow, R. (1956). "A Contribution to the Theory of Economic Growth," *Quarterly Journal of Economics*, 70 (1): 65-94.

Solow, R. (1957). "Technological Change and the Aggregate Production Function," *Review of Economics and Statistics*, 39 (3): 312-320.

Solow, R. (1960). "Investment and Technical Progress." In K. Arrow, S. Karlin, and P. Suppes (eds.), *Mathematical Methods in the Social Sciences*. Palo Alto: CA: Stanford University Press.

Villarreal, R. and Ramos, R. (2001). "La Apertura de México y la Paradoja de la Competitividad: Hacia un Modelo de Competitividad Sistémica," *Comercio Exterior*, 51, (9), September: 772-788.

Wood, A. (1995). "How Trade Hurts Unskilled Workers?" *Journal of Economic Perspectives*, 9 (3): 57-80.

8

The Use of Nontraditional Policy Tools to Support Technological Innovation and Economic Growth: The Practice of Offsets Processes in Developed and Developing Countries

João Pedro Taborda, Pedro Conceição, and José Rui Felizardo

INTRODUCTION: OFFSETS AS A POLICY TOOL TO FOSTER INNOVATION

Practices on the demand of economic compensations associated with large public acquisitions, as a nontraditional mechanism to finance innovation, have increased during recent years, raising the need for a better understanding of the factors playing a role when these kinds of tools are associated with large public acquisitions.

Cases on the purchase of weapons systems by developed and developing countries provide data on the use of compensation programs, especially in their *offsets* form, as a policy tool to foster innovation and economic growth. Two motivations of political, social, and economic nature are considered: (1) to compensate the taxpayer for the amount of currency leaving the country (in the scenario of offsets where the system is paid in convertible currency); (2) to compensate the economic asymmetries arriving from the fact that, although of political importance, the use of this type of goods is not reflected directly in the economy.

Offset transactions include purchases of locally produced goods under the responsibility of the seller of the purchased system,[1] subcontracting of industrial capacity, transfer of new technology, foreign direct investment, credit transfer, or other considerations of added value for the receiving country. A total volume of offsets as a percentage of the price paid for the systems is agreed between the foreign seller and the buying government under the offset program. The fulfillment of this value, using a combination of the economic activities mentioned above, takes place during an extended period of time (that may exceed a decade) and considers the specific guidelines provided by the purchasing country.[2]

Evidence from international cases shows that governments in developing and developed economies have been giving priority to activities where the local industry becomes involved in long-term projects with global players, exporting new products and services, and supporting new networks for knowledge involving local and foreign specialists. This group may include not only industrial companies, but also other players from local innovation systems, including universities and laboratories, that may also be involved as partners in the offset transactions included in each offset program (Lundvall, 1992).

Research on the subject of offsets as a tool to foster innovation and knowledge flows within innovation systems has been limited. Therefore, the main aim of this chapter is to provide an exploratory discussion and a first approach toward how the subject can be structured, allowing the analysis of the more visible cases of countries using offsets to balance economic asymmetries resulting from the acquisition of weapons systems.

The theory to generate from the research will address the main questions related to the use of offsets in managing those asymmetries:

- How are offsets being used by countries with different levels of development to foster innovation, institutional development, and the growth of the local economy?
- How are offsets being used to create and consolidate systems of innovation that can maximize knowledge flows and economies of scale around core competencies at the regional level?
- How are offsets being used to develop the local defense industries, and to support the upgrade of the military capabilities of strategic importance for the country?
- Which are the social, political, and economic factors playing a role in the technological path of countries moving from direct and semidirect offsets to industrial participation as their preferred form of compensation?

Based on the research options and data currently available, the research will consider the reality in developed countries, and the nearly developed countries starting to be involved in compensations and in offset programs in particular (Hennart, 1990). The first group includes the West European countries, Canada, Australia, and Japan; and the second the countries from the Warsaw Pact like

Poland, Czech Republic, and Hungary, now adjusting their military capabilities to NATO standards at the same time as they are entering the European Union.

Analysis on data related to the practice of offsets from companies from the United States is presented (during the 1993-1999 period), followed by the presentation of a methodology for the analysis of cases of international offset programs, and tested in 17 cases. Conclusions try to combine the inferences taken from the case analysis using the methodology with the analysis of current trends.

THE USE OF OFFSETS AS A POLICY TOOL TO FOSTER INNOVATION

Economic compensations through their different forms including offsets are used by both developed and developing countries in most regions of the world. They refer to reciprocal trade agreements involving purchase of goods and services by the seller, from the purchaser of the product, or arrangements whereby the seller assists the purchaser in reducing the amount of net cost of the purchase through some form of compensation. In developed countries, compensations are mainly associated with the acquisition of weapons systems and products of high unit prices, while in developing nations it is common to identify forms of compensation also associated with other products like food and public infrastructures.

In both cases, compensations and offsets in particular are used by the local governments to sustain and develop the local economies, combing the management of the compensation programs with other policy tools. To understand the factors playing in this interaction is the main focus of the current research. Offset agreements between seller and buyer intend to seek and generate business and value over and above what would occur as the result of a cash sale (Hammond, 1990), and may take place for periods that, in some cases, may be close to 30 years.

Transactions with work-sharing like offsets are used today as a policy tool to foster innovation and economic growth, based on the regular interaction between local companies and their international partners.

Transactions with no work-sharing (barter, countertrade, and buyback) are usually related to the access to goods and services of strategic importance for the country, such as the import of food products or the export of natural goods generated from local resources. Throughout the years, these forms of compensation have been used in the acquisition of both military and civil goods by developing countries dealing with the lack of convertible currency or having to manage large debts, or nations with surplus of goods.

Compensation transactions with work-sharing (referred to since the 1970s as *direct offsets*, *semidirect offsets*, and *industrial participation*) are more common to find among developed countries when these are acquiring weapons systems. As described later, this practice started after the World War II, in the form of coproduction and licensed production agreements between U.S.

companies and the European countries that were trying to rebuild their capabilities (Martin, 1996).

Several steps have been taken up to today, including: an increased number of nations dealing with larger debts, lack of resources and surplus of goods, an increase in the number of countries requiring compensations under acquisitions; an increase in the level of requirements as a percentage of the volume of the acquisition; the emergence of indirect offsets as a tool to develop the local economies; and the increase in the number of countries able to develop their industrial capabilities to a stage where they can take part as partners in new programs for the production of highly sophisticated systems.

Developed and Developing Nations

The use of compensations and offsets in particular has been following different approaches by developing and developed countries. Organizations supervising international trade like GATT imposed that its signatories could not demand offsets in government procurement transactions, allowing developing to be the exceptions. However, it was understood that any contractor who wanted to do business with a developed country could present a substantial offset proposal on a "voluntary basis," meaning that, in practical terms, offsets continued to play an important role when companies try to export highly sophisticated goods (Egan and Shipley, 1996).

As the leading arms exporter and with higher offset commitments in different areas of the world, the United States has been a source of opinions for and against offsets. Some analysts see that military sales outside the United States are likely to increase domestic employment by somewhat more than would comparable sales without offsets—largely because offsets are a substitute for (but are less labor intensive than) the imports that would replace them to finance sales (Udis and Maskus, 1996). Others defend that, when providing offsets with transfer of technology, the U.S. companies are developing their future competitors and losing jobs, with the aerospace sector (military and commercial) more exposed (Lumpe, 1994).

Under an attempt to sustain the practice of offsets, the United States signed a memorandum of understanding (MoU) with several NATO nations, with nine containing language committing the parties to consult for the purpose of limiting the "adverse effects of offsets." When negotiating the MoU with the United States, the Netherlands added some words calling for consideration of the nonpositive impact of "other Buy National initiatives."

Other perspectives also maintain that the impact of offsets on the aerospace sector is difficult to isolate from other factors contributing to the outsourcing of non-U.S. suppliers by the American prime contractors. Data from the U.S. government states that exports involving offsets maintain 38,400 work-years, with the Presidential Commission on Offsets and International Trade finding that offset transactions displace 4,200 work-years annually (Johnson, 1999; Scott, 1999).

Small companies with a lack of strong proprietary technological capabilities have been under a higher competitive pressure in their interaction with the prime contactors, being more exposed to the type of risks raised by offsets. But, more than offsets, reductions in U.S. defense spending during the 1990s made some of these companies leave this business (Mowery, 1999).

At the same time, practice of offsets in developed countries shows that, beyond technological capabilities, suppliers tend to be asked by their prime contractors to assist in their offset obligations.

Further data on the specific impact of offsets is also difficult to gather, especially when related to indirect offsets. Generally, offsets have been a commercial argument for U.S. industries, providing opportunities for contractors to increase their economies of scale when working with suppliers owning distinctive competencies in the buying country. For the United States, exports are considered as very important to sustain the industrial base, contributing to decrease the unit cost of the military equipment, and encourage the utilization of U.S. equipment by their allies.[3]

Today, offsets are also a strong tool for contractors exporting high-value weapons systems to other developed countries like France, Israel, Russia, the United Kingdom, France, and Germany. Offset transactions may contribute to enhancing their international reach, leveraging the ability to innovate through foreign technologies, diversifying manufacturing locations, and encouraging the corporation to pursue strategic alliances with entities in the target country (Redlich and Miscavage, 1996).

Although with the same type of political orientation toward the West, the Czech Republic and Hungary are choosing another contractor, the Swedish-British consortium headed by Saab and BAE Systems, proposing the *Gripen* jet fighter. The same aircraft model was sold to South Africa, with offsets playing an important role in the purchase decision and supporting authors who position European offsets obligors as more aggressive than their American counterparts, especially when transferring technology (Cheng and Chinworth, 1996).

Figure 8.1 presents a first approach toward a taxonomy for countries approaching offsets as a mechanism to sustain or develop the economy. Two factors are considered: the level of debt and the development stage, with the diameter of each country's circle representing the volume of offset transactions between local companies and U.S. companies. Countries absorbing the bigger volumes of offset transactions from U.S. companies are developed economies with no classified debt. Among the top 15, only two are outside this group, Turkey and Malaysia, both moderately indebted and developing nations.

Figure 8.1
Clusters of Countries Using Offsets, According to Level of Development and Debt

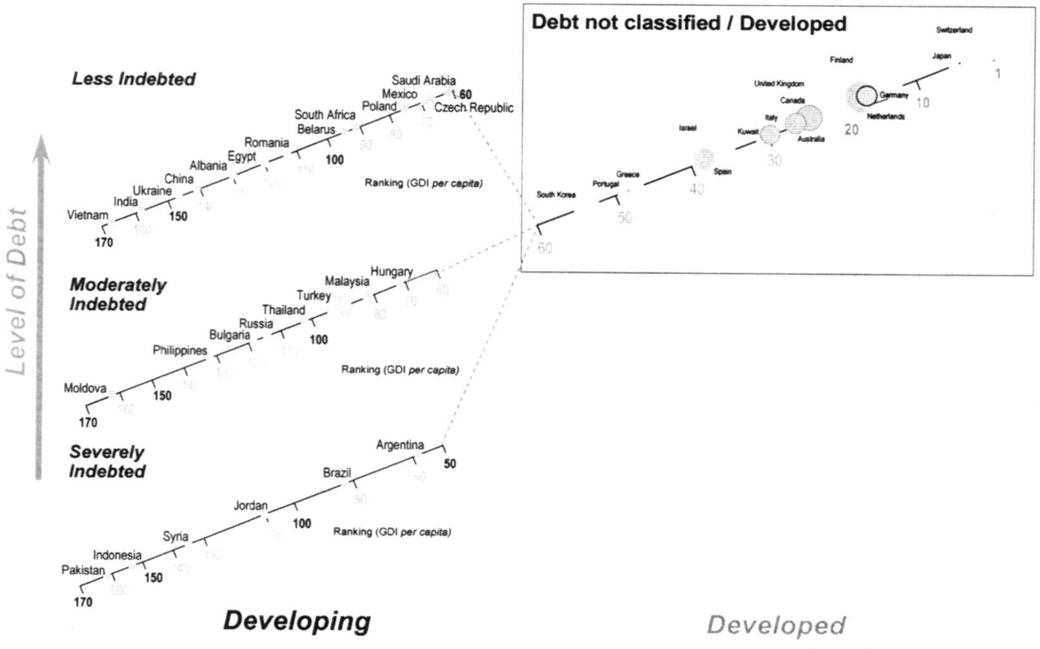

Sources: BXA and World Bank

- Less than 260 million of offsets with US companies, between 1993 and 1998
(Circle diameter grows with offsets received from US obligors between)

The case of Turkey, eighth in the list with $666 million credited in offsets to U.S. companies between 1993 and 1999, suggests the analysis of noneconomic and nonindustrial factors that should also be taken into account in the political context in each country. As the recent conflict in Iraq shows, Turkey may act as an important strategic and tactical partner to the United States, influencing the cooperative environment between the two countries. For example, Turkey had plans to spend a part of the planned $2 billion Iraq-related military grant to purchase U.S.-designed attack helicopters. The 50 units would be produced jointly by Bell Helicopter Textron from the United States and Turkish industry.[4]

Other countries in Figure 8.1 like Poland may change their position and offsets volume in the next years, based on economic and political aspects. As one of the East European and former Warsaw Pact allies nations entering the European Union and NATO (together with the Czech Republic and Hungary), Poland is adjusting its military and economic guidelines to Western standards.[5] As a result of this political reorientation following the end of the Soviet Union, Poland decided to buy North American jet fighters to replace their Soviet-built fleet.

On the economic side, by entering the European Union, Poland, Czech Republic,[6] and Hungary[7] are becoming attractive destinations as far as foreign direct investment is concerned. In line with practices in Israel and the Netherlands, these nations will coordinate the use of offsets with the priorities of their economic policy, setting as main issues the education and training of entrepreneurs as well as the upgrade of local industry capabilities (Fletcher, Barrett, and Wilkinson, 1996; Liesch and Palia, 1999).[8]

Europe versus Rest of the World

Data available concerning the period after 1980 show an increase in the number of countries demanding offsets as well as in the offset requirement as a percentage of the volume of the system being purchased (Viana and Hortinha, 1997). Especially in Europe, where the average offset required in 1998 was 82% compared with 37% cent average from non-European countries like Canada, Turkey, Israel, South Korea, Taiwan, Kuwait, and Australia.[9]

Data from Table 8.1 show that, between 1993 and 1999, Europe was the main client for U.S. companies, with $16,969 million in exports (42%) followed by Asia with $15,508 (38.5%). This close proportion in sales is not reflected in the offset agreements, where 66.8% were with European countries and only 17.5% with Asian countries.

Asian countries asked for offsets, on average, of 25% over the export value, compared with 88% in Europe. With a less-developed industrial base, most Asian nations, like some cases from the Middle East, have a minimum offset requirement close to 30% based on the expected response by the local industry (Al-Ghrair and Hooper, 1996; Welt and Wilson, 1998). However, and according to what the contenders decide to propose, this value may be higher. That was the case with three U.S. export programs to Taiwan in 1999 worth $364.2 million, under which the local government required, on average, 95% in offsets.

Table 8.1
Military Export Sales and Offset Agreements (1993-1999, millions of dollars)

Region	Export Contracts	Offset Agreement	Offset Percentage
Asia	15,508	3,904	25%
Europe	16,969	14,899	88%
Middle East	7,385	3,131	42%
North/South America	378	359	95%
TOTAL	**40,240**	**22,290**	**55%**

Source: US DOC/BIS Offsets Database.

Table 8.2 presents the major countries receiving offsets under agreements with U.S. firms, with European countries accounting for 66% of the total volume of the new offset agreements between 1993 and 1999. These agreements accounted for $22.3 billion in offsets, and were associated with export contracts totalizing $40.2 billion (55% of the value of the contracts).

Worldwide, 34 countries went into new offset agreements with U.S. companies, with transactions executed in 36 nations. For a sample of transactions involving U.S. companies in the 1993-1998 period, this value was 87 months (7.25 years), while some of the programs will have a time of implementation above 120 months (10 years).

Table 8.2
Top 15 Receiving Countries from Offset Transactions with U.S. Companies (1993-1999, millions of dollars)

Country	Actual Value	Credit Value	Average Multiplier
Finland	3,145	3,372	1.07
United Kingdom	2,819	2,839	1.01
Israel	1,206	1,263	1.05
Switzerland	1,068	1,076	1.01
Netherlands	1,017	1,305	1.41
South Korea	824	1,157	1.30
Spain	705	917	1.09
Turkey	615	666	1.08
Germany	551	551	1.00
Italy	529	529	1.00
Greece	489	782	1.60
Australia	475	501	1.06
Canada	428	432	1.01
Taiwan	383	972	2.54
France	310	552	1.78
TOTAL	**14,564**	**16,916**	**1.16**

Source: US DOC/BIS Offsets Database.

The five nations receiving more offsets from U.S. companies are European (Finland, United Kingdom, Israel, Switzerland, and Netherlands). Together, they credited 52.4% of all the offset transactions implemented between 1993 and 1999, while the value for all Europe is 69.1%.

Compared with the postwar period, the new approach is characterized by other priorities as far as the involvement of the local industry is concerned. Recent data mention cases on new offset agreements with contenders proposing more than 100 percent. Particularly some exports involving East European countries have been supporting this evidence.

Hungary agreed with Gripen International on a 110% offset commitment under the acquisition of 14 multirole combat aircraft,[10] while Poland plans to sign an offset agreement with Lockheed Martin under the purchase of 48 F-16 aircraft fighters, covering 170% of the price of the equipment ($6 billion over $3.5 paid for the aircraft).[11] The Czech Republic is also preparing the same type of acquisition, although this has been delayed due to budget constraints induced by the unexpected expenses resulting from the floods during the summer of 2002.

Based on the new approach and new data coming from cases like the Polish and the Hungarian, an impact from the current figures may be expected—especially an increase in those concerning the average offset requirement as a percentage of the export value, and the average duration of the compensation programs.

Considering the data made available by the U.S. Department of Commerce on the practice of offsets since 1993 involving American exporters, developed countries with debts not classified are currently the main players in exploring the positive incentives for compensations, in the form offsets (Table 8.2).[12]

Under the offsets programs (a group of transactions taking place to fulfill the volume associated with the offset obligation in a specific acquisition), compensation transactions are used to develop the local economy through activities with an innovative content, such as the transfer of new technologies, access to new markets, and knowledge acquisition in industrial sectors with high levels of technology intensity and risk-sharing R&D.[13] Cases following this approach are nations with higher gross national income (GNI) per capita, such as Switzerland, United States, Japan, Norway, Denmark, Finland, Sweden, Austria, United Kingdom, Israel, France, and Germany (Martin and Hertley, 1996).

Using the data available, it was possible to introduce a third column with the average *multiplier*, corresponding to the factor that the local government uses when evaluating the value of the offset transaction as far as the structural benefits for the local industry are concerned. Each country may have different guidelines in their multiplier policy.

Taiwan has the highest value among countries listed in Table 8.2, with an average of 2.54, while Italy and Germany did not give any multiplier on top of the actual transactions. Among the European countries, France is providing the higher multipliers to their American suppliers.

Finland has been a case where the local government tries to explore the positive incentives around the use of indirect offsets, setting the development of their transportation sector as one of the main priorities (shipbuilding in particular). The positioning for the United Kingdom and Israel may also reflect the strong political links between the United States and these two nations. Further developments in these relationships brought the involvement of both companies in industrial participation projects coordinated by the United States, with the Joint Strike Fighter by Lockheed Martin as the most visible case.

According to the local economic climate, governments and their suppliers (as current or potential offset obligors) should be able to work together in identifying compensation transactions that, once implemented, may enhance this type of interaction, facilitating knowledge transfer to the local companies and from their international partners.

The dynamics enabling the diffusion of these competencies through the development of local supply chains will contribute to multiply the benefits beyond those received by the partners directly involved in the transactions. The consolidation of these networks of companies should also be based on the exchange of knowledge, both in tacit and explicit forms (Nonaka, 1997).

The level of development of these nations, and the ability of their industries to absorb offset transactions, may contribute to the concentration of offset agreements and transactions. From the 307 new offset agreements celebrated by North American companies, 67% came from only 30 of those agreements.

Compared with the postwar period, relations between suppliers and local governments moved to a different stage, with this trend becoming more evident on work-sharing types of compensations involving European nations. With countries using offsets to forge long-term relationships between local companies and international players, the governments are interacting closely with the prime contractors. This may not be as evident in forms of compensation without work-sharing involved, where transactions include mainly goods as commodities, and less frequent exchange of knowledge between seller and buyer.

ANALYSIS OF THE INTERNATIONAL PRACTICE OF OFFSETS AS A POLICY TOOL

The previous sections tried to provide an overview of the main countries using offsets in terms of volume, as well as their approach during recent decades. This chapter presents an analysis of the content and priorities within the offset transactions, and how they are managed by the different nations to achieve a certain degree of efficiency when those are used to innovate, to develop the local industry, and to make the economy grow.

When asking for offsets, countries have to deal with the costs associated with their management, as well as a price increase, because the supplier will also have to spend extra resources in pursuing the compensation transactions. This effect raises the importance of the efficient use of offsets by local governments, with Switzerland and the Netherlands providing some information in this respect.

For the first case, the Swiss government states that, on average, a 10% increase in the price may be expected if offsets are required, when compared with an *off-the-shelf* purchase with no offsets involved (Udis, 1994, 1996).

In the Netherlands, the minister of economic affairs published the results of an audit that assessed the efficiency of offsets. Considering data since the 1980s, the cost of managing offsets in the Netherlands has represented 2.9% of the total volume of acquisitions in this period. Based on this figure, some specialists consider offsets as a cost-effective tool in the development of the local economy (see Table 8.3).[14]

Table 8.3
Netherlands: Benefit Analysis from the Use of Offsets (values in millions of euros)

	Offset Agreements	Offset Transactions	Gross Margin	Employment
Direct	324.5	217.5	9	1,847
Indirect	946.0	713.8	20	5,766
Total	1,270.5	931.3	29	7,613

Source: CTO/Dutch Minister of Economic Affairs.

A distinction is presented in Table 8.3 for the Dutch data, between direct and indirect offsets. Although generating a lower volume of transactions and fewer margins, direct offsets were less expensive to manage, accounting only for 0.2% of the 2.9% mentioned above. For indirect the value is 2.4%, with the rest (0.2%) related to costs related with the execution of the policy by the ministries of economic affairs and defense. These data suggest that an analysis of the international practice of offsets when used as a policy tool should also include the types of offsets, together with their categories and the priorities as far as industries are concerned.

Types and Categories of the Offsets Transactions

Between 1993 and 1999 the volume of offset transactions was $15.9 billion of which 40 percent ($6.4 billion) related to *direct offsets types*. Volumes and percentages of direct and indirect offsets depend on the volumes of the exports, the offset requirement as a percentage of the sale, and the type of country. The only year when direct transactions were above indirect was 1998 when a few large transactions took place (62% of the transactions) with countries with a developed defense industry, and preferring direct offsets (Netherlands and United Kingdom).

Except for Israel, the Middle East has been demanding mostly indirect offsets focused on technology transfer and direct investment, with several Gulf States (including the United Arab Emirates[15]) declaring little or no interest in direct opportunities.

Indirect activities also allow the contractor to benefit from the supply of innovation and technology in the target country, helping to deal with the rising concern around the potential loss of U.S. jobs resulting from direct offset activities. Meaningful indirect offsets can bring enormous benefits to procuring nations, as well as benefiting the contractor.

Contractors are also realizing the value of cooperating with third-party companies operating in their circle of influence, and providing the opportunity of all parties involved of identifying innovative projects. As shown by the figures from the Dutch audit on the efficiency of offsets, the use of indirect transactions may raise the costs of managing the offset program, increasing the flows of information and presenting new challenges to the local offset authorities when evaluating the value added of the projects and their causality.

As far as *offset categories* are concerned, purchase of goods from the local industry accounted for the biggest share of transactions (see Table 8.4), and were implemented exclusively as indirect offsets. For direct offsets, subcontracting is the main category, with the prime contractors using the local competencies in the development or production of the system being exported. This type of transaction is possible only if the local industry is well developed and the companies are competitive and able to respond.

With 89% of the exports being aerospace related, high volumes of direct offsets tend to be possible only in countries where the local players have a level of technological development that makes it possible for them to satisfy the quality standards of this particular sector.

Table 8.4
Offset Transactions by Category and Type (1993-1999, millions of dollars)

Offset Category	Offset Type	Actual Value	Credit Value
Coproduction	Direct	407	411
Credit Transfer	Direct	4	66
	Indirect	1,052	1,144
Investment	Direct	4	4
	Indirect	285	855
Licensed Production	Direct	91	109
	Indirect	4	26
Purchase	Indirect	5,181	5,574
Subcontract	Direct	4,494	4,788
Technology Transfer	Direct	797	1,131
	Indirect	1,015	1,353
Training	Direct	398	602
	Indirect	188	331
Other	Direct	189	270
	Indirect	1,224	1,483

Note: The unspecified are not included in the table.
Source: US DOC/BIS Offsets Database.

Training and technology transfer may be part of direct offset transactions, upgrading the local capabilities in order to make subcontracting possible. Credit values from transfer of technology are higher in indirect transactions, with the offset projects being used as tools to develop the technological capabilities in economic sectors not related to the imported system.

These new assets, combined with others like training and investment, may contribute to foster innovation among the local companies through the design and production of new products that might be purchased and exported to new markets. If combined with adequate policies to protect intellectual property, these initiatives may also attract private investors (Nelson, 1997).

Credit transfer and investment are the main categories used in indirect transactions behind purchasing and technology transfer, although with a different approach in terms of valorization by the governments (measured by a multiplier associated with each project).

Table 8.5 presents the most recent data covering the year 1999. Training has a very high value, which is due to a particular transaction, since the average multiplier between the 1993-1999 period is 1.60. Apart from this isolated event, investment and licensed production have the highest multipliers, suggesting a preference by the governments for these categories.

Products from these industries are characterized by a high level of integration, with the same product using more technologies in its design and manufacture. At the same time, the technologies used may be applied to other industries in the design and manufacture of an increasing number of systems, providing the conditions to diffuse new technologies among the local companies (von Tunzelmann, 1999).

Table 8.5
Offset Transactions, Credits, and Multipliers (values in millions of dollars)

	Offset Transaction	Credit Value	Multiplier
Purchasing	768.2	782.1	1.02
Subcontracting	404.7	434.3	1.07
Technology transfer	295.9	361.8	1.22
Other	249.3	358.9	1.44
Coproduction	40.5	40.5	1.00
Investment (FDI)	26.1	191.7	7.36
Credit transfers	20.0	20.0	1.00
Licensed production	3.7	26.2	7.16
Training	0.5	27.5	59.78
Total	**1,808.8**	**2,243.0**	**1.24**

Source: US DOC/BIS Offsets Database.

Using offset programs as a tool to develop its economy and innovation of its industry, Taiwan has been focused on the development of supply chains with small and medium-sized enterprise (SME) suppliers but with a capacity to design and perform prototyping, before moving to the assembly operations.

Further research on the Taiwanese case suggests that activities under offsets projects should start from production, moving to R&D, including the design of products and systems (Cheng and Chinworth, 1996).

SMEs are recognized as an important player in the innovation systems also in the United States[16] and Europe,[17] because of their role in fostering the introduction in the industry of new ideas developed in knowledge centers like universities, laboratories, and technology centers (Neto, 2002).

As Table 8.5 shows, FDI is a major priority for nations requiring offsets. However some cases may suggest that not all foreign direct investment represents a value added for the local economy. The experience between the United States and Japan suggests that the participation in the capital of local companies by foreign entities may present some disadvantages, such as the loss of currency abroad, through the sharing of profits (Chinworth and Matthews, 1996).

More than taking part as shareholders of the Japanese companies, foreign investors should help in the process of acquiring new competences and accessing new markets, especially in the civil and commercial sectors, because Japan always imposed restrictions on the export of military systems, focusing its strategy on the capacity of satisfying its needs indigenously.

This decision obliged the Japanese companies to pursue dual-use strategies in order to generate the necessary economies of scale. The country, through its main companies, became an important partner to American companies like Boeing under commercial programs like the 747, 757, 767, and the 777 model aircraft. Positioning itself ahead of the United States in this area of dual use, Japan shared the same resources (people and technology) between the military and commercial area (Chinworth and Matthews, 1996).

Priorities within Industries

Priorities on the economic sectors receiving offset transactions may be associated with areas where economic and political status are recognized. These are usually associated with the advanced technologies used in the design and production of the systems under acquisition. Examples like the South Korean case are automotive, nuclear technologies, and semiconductors, with the country giving priority to high-tech competencies as an instrument to enhance political and commercial power internationally (Cheng and Chinworth, 1996).

However, and even with a strong industrial base, countries may fail in their goal of reaching self-sufficiency, when a complete understanding of the local needs and *strategic industries* is not developed. For instance, Japan does not have a definition for *strategic industry*, declaring that it should be related with activities that include: (a) added value from production, (b) rapid growth of output, (c) innovation with intensive use of knowledge, and (d) vertical and horizontal links with other industries. Main industries in starting this dynamic were aerospace, telecommunications, electronics and computers, ceramic packaging, and automotive (bearings and equipment).

Japan's technology strategy had three main goals: (a) priority to local suppliers, (b) licensing, and (c) purchased equipment must have other applications beyond the one it has been acquired for. Dual use allows industrial synergies in other areas beyond production, such as project management and systems integration (Chinworth and Matthews, 1996).

Table 8.6 presents the top five economic sectors receiving transactions under agreements celebrated by U.S. exporters. *Transportation Equipment* accounts for 48 percent of the total offset credits generated. This sector includes components of complete systems like aircraft, missiles, railway systems, ships, vessels, frigates, and automotives, or other systems integrating different technologies. Of these transactions, 52.9% took place as direct offsets ($4,136.9 million), being by far the economic sector absorbing the higher volume of direct offset transactions. The second sector is *Electronic and Electrical Equipment* with $695.7 million in direct offsets.

Transportation Equipment absorbs the highest volume of indirect offsets ($3,299.6 million), followed also by *Electronic and Electrical Equipment* with $1,309.7. These figures show that countries with a developed industrial structure in sectors like transportation (aerospace, automotive, shipbuilding, and railways) tend to be more efficient in receiving offset transactions, in both their direct and indirect forms.

Table 8.6
Total Offset Transactions per Economic Sector (1993-1999, in millions of dollars)

Economic Sector	Actual Value	Credit Value	Direct	Indirect	Unspecified	Total Credit Percentage
Transportation Equipment	7,814.5	9,026.8	4,136.9	3,299.6	382.8	48%
Electronic/Electrical Equip.	2,018.8	2,425.6	695.7	1,309.7	12.7	12.9%
Industrial Machinery, except electrical	1,245.1	1,496.6	122.3	1,119.2	-	8.0%
Business Services	824	1,005.6	196.3	647.9	14.0	5.3%
Measuring/Analysis Instruments	653.3	812.9	555.0	98.4	-	4.3%

Source: US DOC/BIS Offsets Database.

A Methodology to Analyze Offset Programs

Based on the data and empirical evidence from the practice of offsets among developed countries, a theoretical framework and methodology to structure and analyze international cases are proposed in Figure 8.2. Both civil and military acquisitions where offsets were used as a policy tool to foster technological innovation and economic growth may be considered. Nine independent variables and one dependent were considered.

The dependent variable is the *efficiency* of the offset program. For the purpose of the research, case *efficiency* was defined as the contribution of the offset program in establishing an institutional environment that may foster the learning and absorptive capacity of the local organizations, as well as their ability to develop innovative actions (Kinder and Lancaster, 2001). Evaluation of the *efficiency* of each offset program was based on evidence of projects with an innovative content for the local industry, meaning that an *efficient* case is not a program where all the projects were innovative, but at least one was innovative (Udis, 2001).

A particular focus was established toward cases where local companies had improved their positioning in international supply chains, operating in sectors with a higher degree of technological intensity, such as automotive and aerospace.

Under the same definition, *efficiency* also considers the way the benefits from the offsets projects that took place under an offset program were spread among a extended number of local players (beyond the organizations taking part directly in the activities as project partners), and maximizing the structural and multiplying effects expected from the efficient use of public resources.

Figure 8.2
Methodology Used in Case Analysis

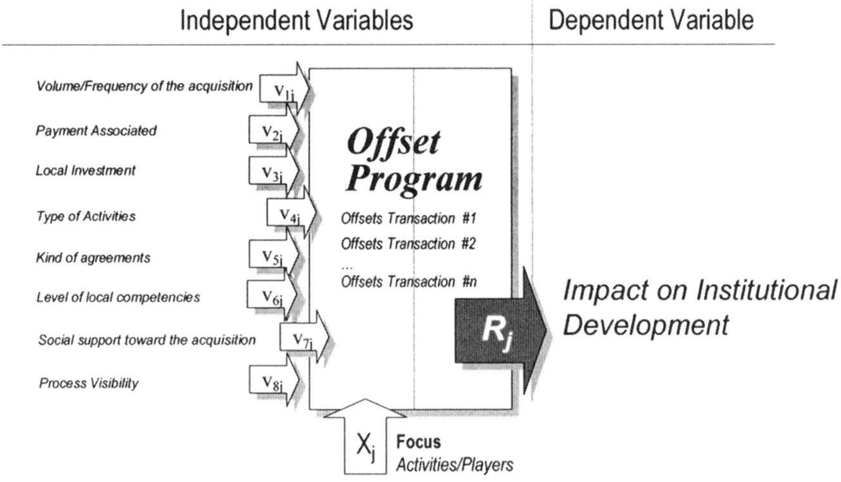

Independent variables in the methodology are the focus in the management of the offset program (activity or player-focused), the volume or frequency of acquisitions made by the buying country, the association of contact payment with the fulfillment of the offsets obligations, the kind of local investment prevailing in the offsets projects, the type of activities under the same projects, the kind of agreements, the level of local competencies in the beginning of the implementation of the offsets projects, the social support toward the acquisition, and the visibility of the acquisition and offset program associated.

Table 8.7 presents the levels for each of the variables considered in the methodology, including the dependent variable (*efficiency*). Its validation was based on a set of data and information related with 17 international cases (consolidated from an initial group of 33), related both with civil and military acquisitions. Each case regards offset processes, from both developed and developing countries. Nine cases were structured from the Portuguese experience, and were structured from primary information never made public previously, and gathered through a series of personal interviews.

Table 8.7
Levels Used for Each Variable in the Methodology

Variable	Levels considered in the analysis
X, Focus	0 = Players-focused 1 = Activity-focused
v_1, *Volume/Frequency of acquisition*	0 = Low 1 = Average 2 = High 3 = Very High
v_2, *Payment/Penalties associated*	0 = Independent or no penalties involved 1 = Payment associated or penalties
v_3, *Kind of local investment*	0 = Small or inexistent 1 = Financial support per project 2 = Participation in the capital of local companies 3 = *Joint-ventures* 4 = Greenfield 5 = Mixed, according to each project
v_4, *Type of activities*	0 = Purchase / Subcontracting / Intermediation 1 = Intermediation / Subcontracting with transfer of technology 2 = Access to licenses 3 = Subcontracting with transfer of technology, and licenses
v_5, *Kind of agreements*	0 = *Barter* 1 = *Buy-back* 2 = Direct offsets mainly 3 = Balance between direct and indirect offsets 4 = Indirect offsets mainly
v_6, *Level of local competencies*	0 = Very low or inexistent 1 = Under development 2 = Developed
v_7, *Social support for the acquisition*	0 = Not seen as a priority 1 = Important 2 = Strategic
v_8, *Process Visibility*	0 = Low 1 = Medium, some references in the press 2 = High visibility, including references from several sources
R, Process Efficiency (Dependent Variable)	-1 = Negative Impact on the existing institutions 0 = No impact or minimal 1 = Positive impact on institutions toward technological innovation, but involving a small number of players 2 = Positive impact on institutions, involving a larger number of players, with small impact on supply chain position or capability to enter new sectors with higher degree of technological intensity 3 = Creation of new institutions, in a positive direction toward technological innovation. Impact on supply chain position and/or acquisition of new capabilities to enter new sectors with higher degree of technological intensity

Analysis of Cases Using the Methodology

The analysis of the international cases using the proposed methodology was based on qualitative and quantitative methods, with the latter identifying an inference with statistical significance between one of the independent variables (*focus on activities/players*) and the dependent variable (*efficiency* of the process).

The econometric model below was considered, where R represents the dependent variable:

$$Y = R = \alpha + \beta.X + \gamma.V + \varepsilon$$

Since R is categorical, a different technique was used from the usual least squares tool. A *liaison function* was used to link the discrete values of R (-1, 0, 1, 2 or 3). The assumption was that this function could be logarithmic. A logistic linear regression was then used to compute the β coefficient associated with variable X (Gujarati, 1995). X represents the independent variable *focus on activities/players*, V the other eight independent variables, v_1 to v_8, and ε a scholastic term. Coefficients β and γ and their statistical correlation with R were measured in probabilistic terms. Data from cases used in the calculations are presented in Table 8.8 and values obtained for β and γ in Table 8.9.

Table 8.8
Application of the Methodology to 17 Cases

	X	v_1	v_2	v_3	v_4	v_5	v_6	v_7	v_8	R	Accuracy
Case #1.	1	2	-	0	0	4	2	1	0	2	3
Case #2.	1	3	-	1	3	2	1	2	1	3	2
Case #3.	1	3	-	5	3	3	1	2	2	2	3
Case #4.	1	3	-	1	3	2	2	2	1	2	2
Case #5.	1	2	-	5	3	3	1	1	1	3	2
Case #6.	1	2	-	5	3	3	1	1	1	3	3
Case #7.	1	3	-	3	3	2	0	2	2	3	2
Case #8.	1	0	0	0	2	3	0	1	0	1	2
Case #9.	1	3	-	3	1	4	0	2	2	1	3
Case #10.	1	1	1	0	0	4	1	2	1	1	2
Case #11.	1	1	-	5	3	3	0	2	2	3	2
Case #12.	0	0	-	0	2	2	0	1	0	0	2
Case #13.	0	3	0	0	2	2	0	2	2	-1	2
Case #14.	0	0	0	0	2	2	1	1	0	0	1
Case #15.	1	2	1	1	3	2	2	2	2	3	1
Case #16.	0	1	0	1	2	2	0	2	2	0	2
Case #17.	1	1	1	0	0	4	2	1	0	1	2

Note: Data Accuracy was assigned according to the number of sources used in the validation of information gathered on the cases. Full description of cases may be found in Taborda (2001).

Data for v_2 (use of penalties in offset programs) did not allow us to compute any estimate for this variable. However, empirical evidence shows that it should also be considered by governments when managing offset programs. Factors p_n measuring the accuracy of the estimates were used to obtain the statistical significance of each estimate associated to each of the independent variables. With a confidence interval of 10%, variable X is statistically relevant with a positive factor ($1- p_X = 1 - 0.085 = 0.915 > 0.9$), meaning that, when X is 1 instead of 0, R tends to be higher. Other variables close to this interval are v_3, v_4, and v_5, suggesting that statistical significance may be achieved when more data and cases are available.

The inference between R and X suggests that, when approaching an offset program as an alternative tool to foster technological innovation, nations should define *what* to do under the scope of the offset transactions, before deciding on *who* is going to work in their implementation. Implementation should be focused on activities that can maximize the multiplying benefit resulting from a process usually involving substantial spending of public resources combining two effects: a *diffusion* effect and a *specialization* effect.

The first aims to maximize the number of local entities directly or indirectly involved, together with the development and consolidation of local supply chains. The *specialization* effect should be focused on the development of the absorptive capacity of the local organizations, enabling them to analyze information coming from external sources, innovating and integrating it in new solutions with market acceptance, and targeting industries with a higher degree of technological intensity.

Table 8.9
Estimates from the Linear Regression

	Estimate, γ_n	p_n
v_1	.296	.895
v_3	13.345	.153
v_4	-40.265	.163
v_5	-64.709	.135
v_6	-5.498	.263
v_7	-23.594	.219
v_8	5.042	.379
X	**79.863**	**.085**

A first analysis under this framework suggests that the diffusion effect depends on the volume of the offset transactions implemented throughout the period of the offset process, which could be more than a decade long. A regular flow of transactions involving local supply chains will sustain the establishment of learning routines and networks that may provide new opportunities reinforcing the multiplying benefit to generate. The specialization effect may depend on the industry. Industries with a higher degree of technological intensity may use offsets to continuously improve their ability to select information from external sources such as other local entities and from the international partners taking part in the offset projects.

Significance of variable v_3, empirical evidence, and data presented above in the text also suggest that investment is playing an increasing role in the successful implementation of offset transactions. The same is true for variable v_4, where subcontracting of the industrial capacity is absorbing more volumes of offsets and consolidating partnerships between the local companies and foreign players. Although without statistical significance inside the confidence interval of 10%, v_5 suggests also that offsets in their indirect form may contribute to increase the efficiency of offsets as a policy tool to foster innovation, economic growth, and institutional development.

CONCLUSIONS: OFFSETS AS A TOOL TO FOSTER INNOVATION AND ECONOMIC GROWTH

The text provides an exploratory analysis on how offsets may be used to foster innovation and economic growth among countries and regions. Factors under and outside the control of governments were identified together with some trends that may be considered for future work. Although it was possible to validate some of these trends, the research carried out by the authors on the subject is expected to consolidate it further, based on data coming from cases where offsets will be used by countries in the development of their economies.

Data suggest that offsets are expected to see an increase in their influence in the final decision regarding the acquisition of highly sophisticated goods, while evidence from the 1990s shows that the developed countries have been their main users, involving their industrial base in the implementation of offset transactions. As a main goal, these projects try to compensate the economic asymmetries emerging from the acquisition of very expensive items, while consolidating core competencies in particular regions with the countries.

The process is supported by the establishment of knowledge flows within partnerships between local companies and foreign partners, under the responsibility of the seller of the system to the government (taking the *offset obligor* role). These flows tend to increase when focused on the development of core competencies that can support the design of innovative products to be sold in international markets. As well as for the development of the products, the offset obligor may play an active part in helping local companies to access the channels to those markets.

However, and as explored in this chapter, to be used effectively, offsets should be combined with other policy tools, including those aiming to reinforce human capital in the region (examples: R&D, education, and training), as well as the absorptive capacity of local companies in gathering and selecting information from their business environment, particularly from sources outside the region with the support of the offset obligor.

When involving these organizations in offset transactions that may take place throughout a period of eight years or more, governments may work together with the offset obligor in achieving economies of scale that enable the local industry to design, manufacture and export products with a higher level of technological integration. This dynamics may multiply the benefits by combining a *specialization* effect with a *diffusion* effect.

From the side of corporate strategies, further benefits may be achieved under corporate strategies and entrepreneurial activities, where each of the companies involved in the offset transactions may explore the economies of scale through the exploitation of other business opportunities.

Industrial sectors like aerospace and defense are the two main examples of this kind of practice. In 2000, 89% of the sales volume of military systems by U.S. exporters was related to aerospace products, with the buying countries exploring opportunities to establish links between local companies and the foreign players acting in the business sphere of the seller. The path of South Korea and its relationship with U.S. supplier Lockheed Martin shows how offsets can be used by countries to improve the inflow, absorption, and use of knowledge flows in the development of aerospace products with an increased level of technological integration, while exploring the economies of scale with other industries (electronics and automotive).

When managing offset programs, governments should then provide the most appropriate institutional framework according to local circumstances in order to make possible the generation of this kind of benefit. Offsets require an extra effort from the buyer and the supplier sides, suggesting that, if not adequately managed, offsets may become an additional burden of financial resources (Taborda, 2001).

The dominant strategy for governments and potential offset obligors may be to create long-term relationships between the local industry and foreign partners acting on behalf of the offset obligor. To act as a source of innovation, the offset programs and projects implementing this strategy should be supported by routines that may contribute to foster the absorption of knowledge, especially in industries with a high level of technological intensity, and where different technologies are produced and integrated in a final product.

Beyond the economic and industrial perspective, social and political content associated with offsets should also be considered. From the pure economic perspective, some authors support that on theoretical grounds, offset transactions raise impediments to competitive markets, therefore leading to trade distortion and welfare reduction. This suggests that a part of the justification for this evidence may be found in the influence of political aspects, of both internal and external natures, on the purchasing decision.

On the internal political side, the distortion might be introduced due to local policies in the defense and industrial fields. For example, a country's search for autosufficiency scenarios, in which the country is able to fulfill all its needs in terms of weapons systems. In pursuing this goal, data from offset practice in countries like China, Japan, Taiwan, and India suggest that some nations may be willing to invest in learning how to produce what they are buying, making the purchase more expensive than when buying off-the-shelf.

Therefore, and when addressing all these issues, governments may be more efficient in the use to foster innovation if two levels are considered:

1. the offset program itself, as well as type of activities, industries, and players involved in the offset transactions;
2. the interaction between offsets and other policy tools also used to support the industry when developing innovative products and services.

Specific conclusions on these two issues are developed below.

Figure 8.3 proposes a relationship between political, industrial, and defense policies, and the way they should be reflected in the management of offset programs and in the implementation of offset transactions, associated with large public acquisitions (civil and military), in order to maximize impact on institutional change and economic performance.

Figure 8.3
Approach to the Validation of the Hypothesis

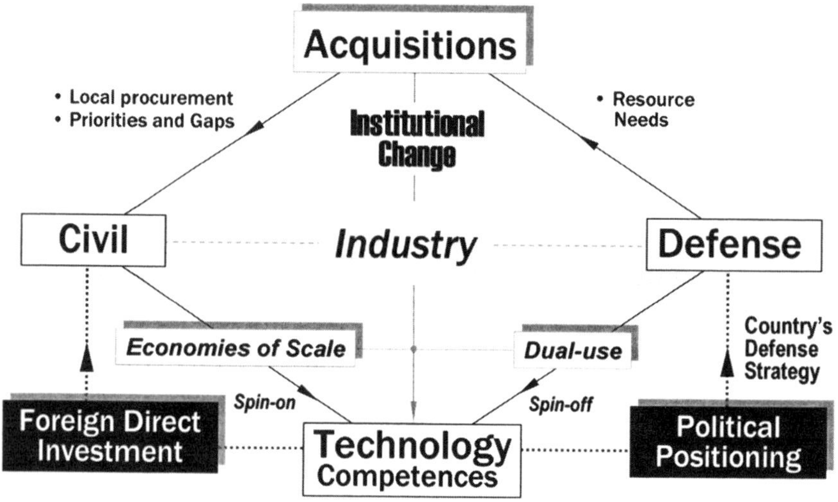

The approach in Figure 8.3 tries to illustrate the conclusions taken from the data available on some of the more efficient cases on the use of offsets. Countries in a leading group (that includes Japan, South Korea, Taiwan, United Kingdom, and Israel) seem to continuously integrate their priorities for the offset transactions with their industrial, economic, and defense policies. This is supported by the data that show the existence of large companies in these regions, able to design and manufacture final products or very complex systems for the aerospace and industries.

Political positioning of the country and its external conditions (like an imminent threat) should be considered when establishing the priorities of weapons systems in national defense plans, together with the technology and the capabilities that should be available in the local industry to support its manufacturing, operation, and maintenance.

Increased sophistication of weapons systems suggests that countries should focus on core competencies, establishing partnerships with other entities abroad in order to fulfill all their needs. Due to the strong influence of political aspects in all these dynamics, economic and industrial relationships should be based on political agreements. For instance, maintenance services that the country cannot assure should be outsourced to countries with whom the client nation has a favorable political relationship.

On the other side, and in the 1990s, full military suppliers have been facing decreasing defense budgets among their potential clients, obliging them to consider business opportunities in civil/commercial areas in order to sustain existing competencies, or to generate economies of scale in new areas. Again, areas where these dynamics may be implemented should be considered as targets for offset transactions, and should integrate the conditions of each specific program.

This practice tends to increase the efficiency of offsets as a tool to implement industrial strategies. When deciding which core competencies to stress, governments should consider dual-use approaches. Highly sophisticated goods (military and nonmilitary) integrate an increasing number of different technologies, some of them from a knowledge base that was developed under pure civil applications. That has been the case, for example, with solutions in the software and telecommunications areas.

In the interaction with other policy instruments, offset programs should combine efforts with foreign direct investment initiatives in order to consolidate competencies and economies of scale, while facilitating the entry of new players. These should not focus exclusively on the transferring of new technology for the local industrial base, but also on the sustaining of strategic competencies for their activity worldwide through the implementation of new institutions based on formal or informal networks with local organizations (companies, universities, and laboratories).

The development of these new assets under offset transactions may be possible only if education and R&D policies are able to provide the necessary human capital, which is key to assure the absorptive capacity in the region that may support the role of the local innovation systems for activities of increasing complexity and technological integration.

Together, all these aspects should be among the ones to be considered by governments when using offsets to foster innovation. Due to their long-term nature, data suggest that offsets may be a very powerful tool if combined with other policy measures. When declaring one main goal associated with offsets such as "export increase" or "attracting more foreign direct investment," governments should not isolate offsets. Better identified practices show that structural benefits for the economy may be achieved if countries and regions are able to absorb, share, and develop the knowledge based on the institutional arrangements with the foreign players provided by the offsets transactions.

In order to achieve those benefits nations should consider *what* to do under the offset programs before *who* is going to implement them. That may require long-term priorities for the development of core and distinctive competencies in the region that may sustain exports of goods and services to the international markets. Therefore, and before aligning priorities with the management of offsets, governments should also act on other policy areas that may act on the development of human capital that will facilitate the paths toward the improvement of the absorptive capacity of local firms.

NOTES

1. Offsets are usually associated with the purchase of military systems. However, recent cases of South Africa and India shows that this kind of agreement may also be used under the acquisition of goods for civil use with high unit prices such as airliners.

2. Generally, offsets may have a direct and indirect form. If the good or service provided by the local industry under the offset transaction will be incorporated in the system under acquisition, that is considered a direct offset. Otherwise it is labeled indirect.

3. "Lockheed Kept Waiting—Polish Entrepreneurs Carve up Offset Benefits," *Countertrade and Offset*, 21 (5), March 10, 2003: 6.

4. "Turkey to Use $2 Billion U.S. Grant for Copters," *Defence News*, January 3, 2003, p. 8.

5. "Who's Who in NATO," *Defence News*, Summer 2002, pp. 22-25.

6. "Czech: Offset Costs—Perish the Thought!" *Countertrade and Offset*, 20 (21), November 11, 2002: 3.

7. "Hungary in Regional Allocation of Gripen Benefits," *Countertrade and Offset*, 21 (5), March 10, 2003: 5.

8. "Poland/Lockheed—The Official Version," *Countertrade and Offset*, 21 (2), January 27, 2003: 4.

9. "Offsets in Defence Trade," Fifth Annual Report to Congress, *The DISAM Journal*, Fall 2001, p. 84.

10. "Gripen Raises Offsets to Match Hungary's New Contract Value," *Countertrade and Offset*, 21 (4), February 24, 2003: 5.

11. "Poland: Lockheed Win—But Offset Values Are Downgraded," *Countertrade and Offset*, 21 (1), January 13, 2003: 5.

12. "Offsets in Defence Trade," Sixth Annual Report to Congress, U.S. Department of Commerce, February 2003.

13. According to OECD, technology intensity grows with investment in R&D.

14. "Dutch Audit—A Remarkable Document That Shows Offset Provides Substantial Measurable Benefits," *Countertrade and Offset*, 21 (3), February 10, 2003.

15. "Interview: Hatem Fawzy, Director of the United Arab Emirates Offset Ventures," *Defence News*, March 10, 2003, p. 46.

16. Through the Small Business Act, published by the U.S. government, providing the guidelines for the involvement of SMEs in supply chains coordinated by the prime contractors (non-SMEs).

17. A paper published by the European Commission on January 21, 2003, addresses the main issues and challenges for the European SMEs. Bodies representing the industry like the European Association of Aerospace Industries (AECMA) and the European Defense Industry Group (EDIG) have specific working groups providing recommendations to the European Commission concerning the role of SMEs in the supply chains.

REFERENCES

Al-Ghrair, A. and Hooper, N. (1996). "Saudi Arabia and Offsets." In S. Martin (ed.), *The Economics of Offsets—Defence Procurement and Countertrade*. London: Harwood Academic Publishers, pp. 219-244.

Cheng, N. and Chinworth, M. (1996). "The Teeth of the Little Tigers: Offsets, Defence Production and Economic Development in South Korea and Taiwan." In S. Martin (ed.), *The Economics of Offsets—Defence Procurement and Countertrade*. London: Harwood Academic Publishers, pp. 245-297.

Chinworth, M. and Matthews, R. (1996). "Defense Industrialisation Through Offsets: The Case of Japan." In S. Martin (ed.), *The Economics of Offsets: Defence Procurement and Countertrade*. Amsterdam: Harwood Academic Publishers.

Egan, C. and Shipley D. (1996). "Strategic Orientations Towards Countertrade Opportunities in Emerging Markets," *European Journal of Marketing*, 13 (4): 102-120.

Fletcher, R., Barrett, N., and Wilkinson, I. F. (1996). "Countertrade and Internationalisation—An Australian Perspective," *International Commercial Law*, February.

Gujarati, D. (1995). *Basic Econometrics*. New York: McGraw-Hill.

Hammond, G. T. (1990). *Countertrade, Offsets and Barter—An International Political Economy*. London: Pinter, pp. 6-40.

Hennart, J.-F. (1990). "Some Empirical Dimensions of Countertrade," *Journal of International Business Studies*, Second Quarter 1990, pp. 243-270.

Johnson, J. (1999). "Offsets in the International Marketplace: An Aerospace Industry View." In C. Wessner (ed.), *Trends and Challenges in Aerospace Offsets*. Washington, DC: National Academy Press, pp. 158-166.

Kinder, T. and Lancaster, N. (2001). "Absorptive Capacity—Building Absorptive Capacity in a Learning Region: A Socio-technical Model," *Science and Public Policy*, 28 (1): 23-40.

Liesch, P. W. and Palia, A. P. (1999). "Australian Perceptions and Experiences of International Countertrade with Some International Comparisons," *European Journal of Marketing*, 33 (5/6): 488-511.

Lumpe, L. (1994). "Sweet Deals, Stolen Jobs," *The Bulletin of the Atomic Scientists*, September/October.

Lundvall, B.-Å. (1992). "Introduction." In B.-Å. Lundvall (ed.), *National Systems of Innovation: Towards a Theory of Innovation and Interactive Learning*. London: Pinter Publishers, pp. 1-19.

Martin, S. (1996). "Countertrade and Offsets An Overview of the Theory and Evidence." In S. Martin (ed.), *The Economics of Offsets—Defence Procurement and Countertrade*. London: Harwood Academic Publishers, pp. 337-355.

Martin, S. and Hertley, K. (1996). "The UK Experience with Offsets." In S. Martin (ed.), *The Economics of Offsets—Defence Procurement and Countertrade*. London: Harwood Academic Publishers, pp. 337-355.

Mowery, D. (1999). "Offsets in Commercial and Military Aerospace: An Overview." In C. Wessner (ed.), *Trends and Challenges in Aerospace Offsets*. Washington, DC: National Academy Press, pp. 85-113.

Nelson, R. (1997). "How New Is New Growth Theory?" *Challenge*, 40 (5): September/October: 29-58.

Neto, H. (2002). "Uma visão estratégica da economia portuguesa" (A Strategic Vision of the Portuguese Economy,) *Global Economics and Management Review*, 7 (1): 9-20.

Nonaka, I. (1997). "The Knowledge-Creating Company." In P. Drucker, D. Garvin, D. Leonard, S. Straus, and J. Brown (eds.), *Harvard Business Review on Knowledge Management*. Boston: Harvard Business School Press, pp. 21-46.

Redlich, A. and Miscavage, M. (1996). "The Business of Offset: A Practitioner's Perspective Case Study: Israel." In S. Martin (ed.), *The Economics of Offsets—Defence Procurement and Countertrade*. London: Harwood Academic Publishers, pp. 337-355.

Scott, R. E. (1999). "The Effect of Offsets, Outsourcing, and Foreign Competition on Output and Employment in the U.S. Aerospace Industry." In C. Wessner (ed.), *Trends and Challenges in Aerospace Offsets*. Washington, DC: National Academy Press, pp. 133-156.

Taborda, J. P. (2001). "Utilização de contrapartidas associadas a grandes compras na dinamização da inovação tecnológica: Uma metodologia de estruturação de casos" (On the Use of Offsets to Foster Technological Innovation: A Methodology for Case Analysis), MSc. dissertation. Lisbon: Instituto Superior Técnico, September.

Udis, B. (1994). "Offsets in Defense Trade: Costs and Benefits," report, Department of Economics, University of Colorado, Boulder, May.

Udis, B. (1996). "US-Swiss F-5 Transaction and the Evolution of Swiss Offset Policy." In S. Martin (ed.), *The Economics of Offsets—Defence Procurement and Countertrade*. London: Harwood Academic Publishers, pp. 321-335.

Udis, B. (2001). "Costs and Benefits of Offsets in Defense Trade," invited paper presented at Western Economic Association Meetings at San Francisco, July 7.

Udis, B. and Maskus, K. (1996). "U.S. Offset Policy." In S. Martin (ed.), *The Economics of Offsets: Defence Procurement and Countertrade*. London: Hardwood Academic Publishers, pp. 357-380.

United States Department of Commerce (2001). "Offsets in Defense Trade: Fifth Annual Report to Congress," *The DISAM Journal*, Fall, pp. 77-107.

Viana, C. and Hortinha, J. (1997). *Marketing Internacional*. Lisbon: Sílabo, pp. 397-404.

von Tunzelmann, N. (1999). "Big Business, Growth, and Decline" (review article), *Journal of Economic History*, 59 (3): 787-794.

Welt, L G. B. and Wilson, D. B. (1998). "Offsets in the Middle East," *Middle East Policy*, 6 (2), www.mepc.org/journal/9810_weltetal.html.

PART III:
FOSTERING CONNECTIVITY

9

Emerging Technology Trade Triangle: Japan Joins Mexico, the United States, and Canada

R. Ray Gehani and Rashmi A. Gehani

INTRODUCTION

Can a preferential Mexico-Japan Free Trade Agreement (MeJaFTA) revive the stalled Mexican economic miracle, and reform the Japanese economy out of its decade-long conundrum? In this chapter, we propose a dynamic open-system model to examine Mexico's evolving trade relationship with Japan, in the context of Mexico's other trade ties with the European Union and the North American Free Trade Agreement (NAFTA). Many relevant elements of the technology and innovation systems of Mexico and Japan, such as capital and technology factors of production in Japan, and a young workforce in Mexico, are reviewed along with the commitment of their political leaders at the time of writing in 2003, President Vicente Fox and Prime Minister Junichiro Koizumi. Some implications for the evolving trade relationships between Mexico and Japan are proposed through the lens of complexity and chaos theory.

IMPACTS OF NAFTA

The skill levels of factors of production and national technology innovation systems (Nelson, 1993, 1996) affect international trade between countries, their returns on capital, their workers' wages, and more. Trade relations between Mexico and Japan are defined by their respective factor endowments. Japan is abundant in accumulated capital and technology, whereas Mexico has one of the youngest skilled workforces in the world. In this chapter we will examine how national environments in Mexico and Japan influence the evolving trade relations between these two countries, and Mexico's trade ties with its northern trade partners.

Contrary to what many trade policymakers generally plan or hope for, international trade ties continually evolve over time due to the growing complexity of interactions between different trading partners. Since 1994, the North American Free Trade Agreement has radically transformed trade between the United States, Mexico, and Canada. Like the establishment of the European Union in 1992, NAFTA unified the three North American nations into a common market. By 2000 the NAFTA firms had access to about 407 million people (see Table 9.1).

Table 9.1
Global Top-15 National Rankings in 2000

GDP Rank		GDP US$Bn	Population Rank	Population Million	GDP/Capita Rank	GDP/Capita US$/Capita
#1	United States	9,098	#3	276	#2	33,017
#2	China	4,854	#1	1,262	#88	3,847
#3	Japan	2,809	#9	127	#15	22,196
#4	India	1,798	#2	1,014	#136	1,773
#5	Germany	1,797	#12	83	#18	21,701
#6	France	1,327	#21	59	#13	22,360
#7	United Kingdom	1,245	#20	60	#20	20,915
#8	Italy	1,168	#22	58	#22	20,263
#9	Brazil	1,032	#5	173	#68	5,966
#10	Mexico	867	#11	100	#48	8,662
#11	Canada	708	#34	31	#10	22,635
#12	Spain	660	#28	40	#31	16,506
#13	S. Korea	638	#25	48	#38	13,429
#14	Russia	624	#6	146	#81	4,276
#15	Indonesia	598	#4	225	#109	2,662

Source: Country Review; CountryWatch.com. GDP measured by Purchasing Power Parity.

NAFTA's impact was particularly profound on the transfer of technology-intensive manufactured goods and knowledge-intensive services. NAFTA offered these technology-driven firms higher economies of scale, more ability to raise capital at competitive rates, and ability to relocate production operations to the locations that offered the highest competitive advantage in factors of production. On the other hand, NAFTA exposed many Mexican firms to severe competition from their much larger and more sophisticated American and Canadian counterparts. NAFTA's preferential economic integration altered the market entry behavior of other foreign enterprises, particularly from Japan and the European Union. Therefore, NAFTA also brought about significant changes in the trade ties between Japan, Mexico, and the United States.

About a decade after signing NAFTA, the trilateral NAFTA ties continue to develop in different directions. In early 2003, the last protective tariffs against Canadian and American agricultural produce were getting lifted in Mexico. President Vicente Fox of Mexico used the hosting of the Asia-Pacific Economic Cooperation (APEC) summit in Mexico in October 2002 to initiate a new free-trade agreement (FTA) with Japan. Mexico, with free-trade agreements with 32 countries, also initiated talks with China and New Zealand for the Closer Economic Partnership Agreement (CEPA) as a precursor to an FTA (Latin American Monitor, 2002).

Japan, on the other hand, a longstanding supporter of multilateral trade liberalization, is also steadily shifting towards preferential bilateral and regional agreements with its neighboring and distant partners (Armbruster, 2003). A Mexico-Japan Free Trade Agreement offers the promise of simultaneously developing new markets in Japan for the Mexican agricultural and manufactured goods, while also allowing Japanese manufactured goods to compete in Mexico against the American and Canadian manufactured goods.

EVOLVING INTERNATIONAL TRADE SYSTEM

Whereas most researchers have reviewed the NAFTA trilateral regional agreement with respect to its potential economic benefits for the Northern partners (particularly the United States), relatively few studies have focused attention on its evolutionary long-term effects on NAFTA's Southern partner, Mexico. Little work has been reported investigating Mexico's evolving bilateral trade ties with the rest of the world. Some observers assert that NAFTA has helped Mexico industrialize rapidly and create many jobs. On the other hand, other researchers point out the deteriorating environment (water and air pollution) in Mexico caused by poor implementation of their enacted regulatory laws.

EX-ANTE STUDY OF MEXICO-JAPAN FREE-TRADE TIES

This chapter reviews the longitudinal evolution of the international trade ties between Mexico and Japan. We do so in the context of Mexico's other major trade ties with the United States and Canada, defined by NAFTA. We compare the emergent trade relations between Japan and Mexico in the context of the increasing complexity introduced by Mexico's many bilateral, regional, and global trade ties with various world economies.

The international trade relations between Mexico and Japan are examined as a dynamic, complex, and evolving system shown in Figure 9.1. We note how the evolving national environments in Mexico and Japan influence their international trading strategies, and the use of multilateral trade globalization strategy or preferential bilateral trade strategy. The emergent Mexico-Japan trade ties, in a cyclical manner, define the two countries' national environments, and influence the divergent forces of pursuing either bilateral preferential trade strategy or multilateral trade globalization strategy in their trade agreements with other economies.

Figure 9.1
Emergent International Trade System

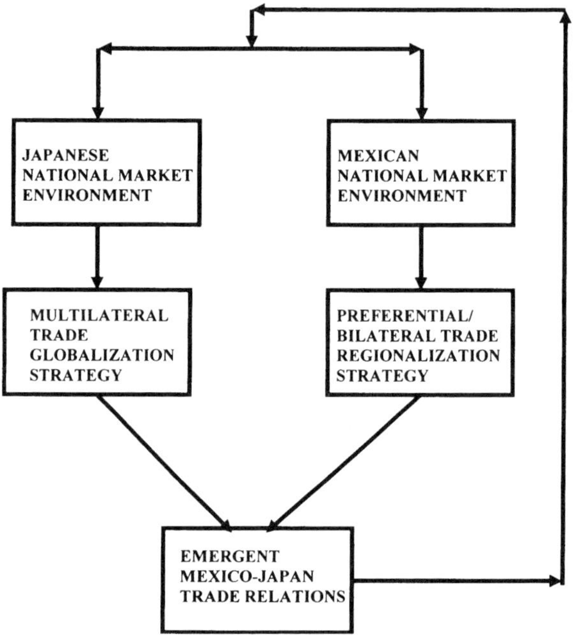

MULTILATERAL TRADE GLOBALIZATION VERSUS BILATERAL (REGIONAL) PREFERENTIAL TRADE

In any international trade relationship, a country has the strategic choice of pursuing either a multilateral trade globalization strategy or a preferential trade strategy (Bhagwati, 2000). Globalization of free trade is taking place due to the growing interconnectedness of the global trading partners through their increasing fair and free flows of factors of production, such as capital, goods, services, people, and information. This involves substantial degrees of economic, political, social, and informational openness. Many theorists differ in postulating the effects of globalization (Bhagwati, 2000). Somc postulate that globalization homogenizes different trading nations (Ohmae, 1990, 1995), whereas others support that globalization cannot overcome the cultural and indigenous divergence of the different trading nations (Berger and Dore, 1996). A third group of transformational globalization theorists suggests that countries differ in the degree to which they are integrated with the world economy

(Giddens, 1991, 2000; Held et al., 1999). In this chapter, we use the transformational globalization theory to examine the evolving trade relations of Mexico with Japan, in the context of Mexico's trade ties with United States and others.

Bilateral or regional trade relationships with preferential trade agreements (PTAs) involve two or more countries getting together with the purpose of lowering their mutual trade barriers. A free trade agreement between two or more countries is a special case of PTA, wherein most of the barriers between the agreeing parties are lowered or removed. For many years, FTAs and PTAs have proliferated in America and Europe. In East Asia these agreements have started growing only after the 1997 East Asian Financial crisis (Gehani, 1999). For example, in 2000, Singapore signed an FTA with New Zealand, and started parallel negotiations with Australia, Japan, Mexico, and United States. Japan and Korea are currently working on establishing the "Japan-Korea Economic Agenda 21" to align issues such as investments, taxes, standards, and intellectual property. Close geographic proximity is no longer a precondition for such FTAs and PTAs. Some national trade policymakers believe that PTAs can accrue higher returns compared to the potential returns from multilateral globalization of free trade.

MEXICAN AND JAPANESE NATIONAL ENVIRONMENTS

As stated before, the national environment and technological infrastructure of a country make significant impacts on the nation's selection of growth strategies for trading and technological innovation (Gehani, 1998; Nelson, 1993, 1996). National demographics, which define market size and concentration as well as the consumers' needs and quality of life, define a nation's trading requirements. Next, we examine the changing national environments in Mexico and Japan.

MEXICAN MAYHEM

With an area of 756,000 square miles directly south of the United States and a population of over 100 million, Mexico is the third largest country in Latin America and the ninth largest economy in the world (see Table 9.1). The population of Mexico is growing at the rate of 1.84% annually, making it a country with one of the youngest workforces. The share of population of Mexicans under the age of 20 is 55%, while only 4% Mexicans are 65 years of age or older. Mexico's unique *maquiladora* assembly plants near the U.S. border once employed 1.3 million skilled workers, with more than 600,000 workers in two of the six border states of Chihuahua and Baja California. Under the terms of NAFTA, only the parts and materials originating in one of the three NAFTA trading nations are allowed to enter the processing zones duty free. Parts originating from elsewhere were subject to tariffs as high as 25%.

About 60% of the foreign direct investment in Mexico, in a wide variety of industries from electronics to computers and pharmaceuticals is from the United States (Torres, 1997; Gori, 2001). Wal-Mart Stores acquired a controlling share in Mexico's largest retailer, Cifra SA. Bell Atlantic, a technology-intensive U.S. company, has full control of the Mexican cellular company Grupo Iusacell SA, and Procter & Gamble acquired Loreto y Pena Pobre, a consumer products company. In 1999 Mexico signed a preferential free-trade agreement with the European Union. Thereafter, many European technology-intensive firms—such as Siemens of Germany, Philips Electronics from the Netherlands, Thomson S.A. of France, and Telefonica of Spain—have expanded their operations in Mexico. Free trade negotiations are also progressing between Mexico and the four Nordic countries, so that in the near future the Nordic technology-intensive companies such as, Ericsson, Nokia, and Saab-Scania are also expected to invest in Mexico.

In the past, because of Mexico's proximity to the North American and South American markets, many Japanese technology-intensive MNCs also invested quite heavily in Mexican *maquiladoras*. However, due to the higher import duties and a rising cost of Mexican labor, some of these Japanese MNCs moved production of goods from Mexico to Asia, even for export to North America.

Debt Disaster After NAFTA

After signing NAFTA in 1993, Mexico was expected to reform and modernize its industrial system to join the front ranks of the world's leading trading nations. The age-old corrupt political patronage of a nationalist economic structure was expected to be dismantled. One major effect of signing NAFTA, however, was that the proportion of intrafirm export trade increased from 40% to 55%. More than three-fourths of foreign direct investments went into either *maquiladoras* or for speculating in highly discounted assets. Only one-fourth of investments went into direct investments in enhancing other domestic production plants. With rising interest rates, many of Mexico's nonexporting domestic producers could not survive the assaults from their global competitors.

Barely two years after joining NAFTA, in December 1994, the Mexican bubble popped and the economy was in ruins. As panicking foreign investors fled, the Mexican peso was devalued rapidly. The Mexican economy contracted by 7%, more than 1 million jobs disappeared, and wage incomes fell by more than one-third. Loss of jobs and falling real wages led to more than 20% of Mexicans falling behind on their loans for homes, cars, and small businesses. Banks were forced to default on their loans. Some claim that this "Tequila Effect" spread economic failures to other countries in Latin America and Asia (Rohwer, 2001).

Evolving Mexican *Maquiladora* Miracle

Since 2000, Mexico has been going through some significant changes. Real GDP fell from a 6.7% growth in 2000 to a 0.3% reduction in 2001 (see Table 9.2). Annual industrial output shrank by 3.5% in 2001 compared to 6.1% growth in 2000. Jobs have been hemorrhaging in the *maquiladora* industry, Mexico's economic epicenter. These border processing plants emerged in the 1960s, but grew rapidly in the 1980s when the U.S. companies were scrambling for cheaper ways to compete with their Japanese rivals. The Mexican laws allowed U.S. companies to export duty-free parts to Mexican "twin processing plants," for assembly by low-wage Mexican workers. The finished parts are then reshipped back to the United States.

Table 9.2
Slowdown in the Mexican Miracle (2000-2001)

	2000	2001
Total GDP, US$ billion	580.1	617.8
Population, million	99.6	101.0
GDP per Capita, US$	5,826	6,117
Real GDP Growth	+ 6.7%	- 0.3%
Annual Industrial Output	+ 6.1%	- 3.5%
Unemployment	1.9%	2.5%
Annual Average Inflation	9.5%	6.4%
Annual Foreign Exchange MXN/US$	9.46	9.34
Exports, US$ billion	166.5	158.5
Imports, US$ billion	174.5	168.3
Trade Balance, US$ billion	- 8.00	- 9.8
Total External Debt, US$ billion	148.7	144.5
Total External Debt as % GDP	25.6%	23.4%
Oil Exports, US$ billion	16.38	12.80
Average Oil Output, million barrels/day	3.45	3.56
Avg. Oil Price, US$/barrel	24.64	18.57

Source: Some numbers differ because the data were adapted from multiple sources: Latin American Monitor (2002), International Monetary Fund, Bank of Mexico, and Reuters.

As the U.S. economy slowed down, and the Chinese economy opened, the *maquiladora* plants suffered. Whereas some U.S. companies, such as Rochester, New York-based Eastman Kodak, moved many of their U.S. jobs to Mexico, over 300,000 Mexican jobs disappeared in three years, from a peak of 1.3 million jobs in the late 2000. The garment, toys, and electronics manufacturers made a gradual migration from Mexico to China (Millman, 2003). Many technology-intensive companies, such as Scientific Atlanta Inc., a Ciudad Juarez-based maker of TV-top cable boxes for North American cable service providers, cut about 4,000 Mexican jobs between 2000 and 2002 (Millman, 2003). Saturn Electronics & Engineering industries, a maker of electronic harnesses for the auto industry, slashed about 800 jobs from its roster of 2,200, and consolidated two production facilities into one.

Gradually more and more Mexican technical work started moving to China. Critics claim that the rising costs of fuel, taxes, and administrative paperwork in Mexico have pushed the share of labor costs to less than half of the overall production costs in Mexico. The Mexican peso's appreciation has also motivated many owners of assembly plants in northern Mexico to relocate to China. China has offered them five to ten years of tax breaks, resulting in a loss of billions of dollars in Mexican exports. Many blame this malaise on the lack of leadership by the popularly elected Mexican president Vicente Fox.

Mexican Presidential Leadership

In the midterm Mexican elections held on Sunday, July 6, 2003, the center-right National Action Party (PAN) of popular Mexican President Fox clearly lost much ground. Some might scoff at this by saying: "It is the economy, stupid." Relative to its potential and Mexican people's hopes, the Mexican economy is severely underperforming. With large numbers of new young persons entering the job pool every year, the Mexican economy is under enormous pressure to continue to grow at a brisk pace and create thousands of new jobs.

In the 2003 midterm elections for the 500 Lower House Chamber of Deputies and six of the 31 governorships, President Fox's PAN party held 202 seats but lost 49 seats. The former incumbent Institutional Revolutionary party (PRI) got the highest number of 227 seats, up from the 207 seats it held earlier. PRI fell short of a clear majority, but made a great recovery by improving its position with populist causes, such as fighting tax overhauls and private investments in energy (Cordoba, 2003). PRI won 34.4% of the popular vote, compared to 36.9% it got in 2000. The Democratic Revolutionary Party (PRD), positioned left of the PRI, also gained its position, by doubling its number of seats to 99 and getting 17% of the popular votes. This was still short of the 125 seats PRD won in 1997 elections. Almost 60% of the Mexican voters, however, did not take part in the election, suggesting their massive disapproval of Mexico's political process and President Fox's leadership.

There are some bright spots to the Mexican mayhem under President Fox. Ten years of tariff reductions under NAFTA have transformed Mexico into an economy unlike its much more volatile Latin American neighbors. The Mexican peso, though appreciated, is a relatively strong currency. Mexico has all-time-high foreign exchange reserves in U.S. dollars. Under President Fox, the first legislative action for a freedom-of-information act was passed in June 2003, allowing citizens to get copies of public documents (Cordoba, Luhnow, and Millman, 2003). A financial law made it easier to seize assets of defaulting debtors. Many drug dealers have been arrested or killed. These unprecedented sociocultural developments seem to balance President Fox's lackluster economic performance.

More Mexican Weaknesses

Mexico has a low tax-collection rate of about 12% of gross domestic product (GDP), which puts severe limits on government spending on much-needed roads, hospitals, and schools (Cordoba, 2003). Lack of constitutional and legal clarity on foreign and private ownership of Mexico's energy resources has translated into lack of foreign investments, low capacity, and high-energy prices (Cordoba, 2003). Some investors are moving north of the Mexican border for cheaper and abundant supplies of energy. Mexican oligopolies charge their consumers excessive prices. Mexico has uncompetitive labor laws that discourage hiring. NAFTA has dramatically changed the political economy of Mexico, and severely curtailed the government's power to intervene.

Faltering Mexico-U.S. Ties

President Fox took office in Mexico about the same time that power was transferred in the United States from the Democrat President Bill Clinton to the Republican President George W. Bush in early 2001. After returning from a visit to the United States, President Fox was slow to react to the 9/11 attacks on the World Trade Center and Pentagon. Whereas President Fox publicly read his statement of support, U.S. leaders were upset that many flags continued to fly high in Mexico while the flags flew at half-mast in other supporting capitals of the world (Cordoba, 2003). Under such criticisms of lack of support, President Fox cancelled September 16 Independence Day parties at the Mexican consulates abroad to please the Northern neighbors.

The U.S.-Mexico relationship deteriorated further once the issue of the war against Iraq emerged. Mexico, as a rotating member of the United Nations Security Council, did not support the U.S. calls for the UN forces to enforce preemptive military sanctions on Iraq (Cordoba, 2003). Mexico was also, apparently, somewhat tentative about its position for a likely second UN vote. President George Bush has reportedly expressed dissatisfaction at the lack of support received from President Fox, along with other leaders such as Spanish President Jose Maria Aznar. Both the American and Mexican administrations have maintained that despite their disagreements and frustrations over Iraq, the ties between them remain strong. There is, however, a standstill over President Fox's hopes for President Bush's support for a comprehensive immigration accord to legalize the millions of illegal Mexicans living in the United States.

JAPAN: DOMESTIC DEBT AND DEFLATION DEBACLE

Japan, despite facing some hard times, still continues to hold a prominent position in the world economy. As the second largest economy in the world, the Japanese financial numbers are still highly impressive. Within five decades of World War II defeat, Japan accumulated in 1999 about $12 trillion in personal financial assets, about $1 trillion of net overseas assets, and about $300 billion in U.S. Treasury bonds (Rohwer, 2001: 70).

The Japanese economic engine began to falter in January 1990. At the start of the 1990s, Japan's real estate assets, with a nation the size of California with about half the national population of the United States, were valued at $20 trillion, or about five times as much as the entire United States. This runaway success, however, came to an end around the same time as the Cold War ended and the Berlin Wall was brought down. During the 1990s, the Japanese GDP grew by a meager 0.5% a year. This was one-third of the GDP growth in the EU, and about one-fifth of the growth in the United States. On December 31, 1989, the Tokyo Stock Market peaked at 40,000, and lost about half of its value, or about $2.2 trillion, in the first nine months of 1990. By 1998 more than 70% value of the stock market at the start of the 1990s was gone.

Throughout the 1990s, the return on capital remained below the cost of capital in Japan. The Japanese banks, with a massive pool of capital generated by high personal savings rates in Japan, extended loans and credits to select favored industries and enterprises, with little consideration for their cash flows, profitability, or credit risk. Most banks followed the guidance of the Japanese government agencies and their implicit guarantees. Japanese banks have always relied heavily on collateral such as land and the company's shares, while asset growth and market penetration, and not earnings and profitability, drive most Japanese enterprises. Banks or affiliate companies (and not individuals) hold most of the Japanese companies' shares, and they are unlikely to trade their shares. More government intervention led to more industry cartels. Many bank loans suffered in Japan as land and stock prices plummeted. These statements provide some insights into the fall of the mighty Japanese economy.

As a result of these developments, unemployment rates in Japan rose to an unprecedented 5% in 1999, and then even higher at 5.6% in December 2001 (*Economist*, 2002). In 2001, more than 19,000 Japanese companies declared bankruptcy. Many leading Japanese companies (Kaisha) such as Toshiba, Hitachi, and Fujitsu, saw their profits collapse, and were forced to announce unprecedented layoffs. Many less-known Japanese enterprises and suppliers saw successive waves of job cuts. These cuts have hampered Japan's household spending, and have eroded national income and confidence. These deflationary trends have hindered the financial health of many Japanese banks. With a decade of fiscal deficits, Japan's overall official debt has escalated to about 130% of GDP. With persistent deflation, Japan's GDP would continue shrinking even without adding further debt.

Despite Japan's large economic size, since the early 1980s, trade policymakers in Japan have feared that a weak American economy has built the pressure for more trade restrictions against Japanese imports, with a devastating impact on the Japanese economy. One school of thought said that Japan's high competitiveness in global markets may have caused a worldwide opposition and repudiation of the principle of free trade (Christopher, 1983: 303). Many policymakers in other industrialized economies, such as in United States and Western Europe, have felt that Japan has abused the free-trade system by exporting aggressively to open markets in other countries, while protecting Japan's domestic markets. Others have resented that the Japanese have prospered too much by using well the prescribed rules of free trade (Christopher, 1983). These critics claimed that Japan's conspicuous exports have disrupted many key industries in developed and developing nations. A mood to adopt autarkic trading strategies by many national policymakers could easily bring an end to the postwar growth in global trade, and push the world back toward the "beggar-thy-neighbor" protectionist policies of the pre-World War II 1930s (Christopher, 1983: 304).

TECHNOLOGICAL INFRASTRUCTURE AND DIFFUSION OF TECHNOLOGY

Technological innovations and infrastructure play important roles in global trade and transfer of technology-intensive goods (see Table 9.3). The nations with highly dense populations tend to develop their infrastructures more extensively than the large nations with less population density (Gibbs, Kraemer, and Dedrick, 2003: 9). The underdeveloped infrastructure and distribution challenges of large nations with low population density may hamper these nations from accruing the benefits of international trade.

Table 9.3
Comparative Technological Infrastructures and Diffusion of Digital Technologies

	Japan	Mexico	United States
A. Telecommunications Infrastructure (per 1,000 persons)			
1. Telephone lines			
1995	496	94	607
2000	585	125	700
% growth	18%	33%	15%
2. Mobile subscribers			
1995	93	7	128
2000	526	142	398
% growth	464%	1,839%	210%
B. Information Infrastructure (per 1,000 persons)			
1. Personal computers			
1995	120	26	328
2000	315	51	585
% growth	162%	98&	78%
2. Internet users			
1995	16	1	76
2000	371	27	347
% growth	2,229%	2,643%	356%

Source: Adapted from Gibbs et al. (2003), and International Telecommunications Union (2001).

Mexico and Japan have significantly different technological infrastructures. In Mexico, a vast difference in the income levels of the top 20% and bottom

20% populations is less conducive to a high diffusion of advanced computer, digital, and Internet technologies. Unlike in Japan, many people in Mexico still cannot afford new innovations in digital technologies. However, the presence of many multinational enterprises, particularly in the northern states, has boosted the technological infrastructure in Mexico. This should help grow trade with Japanese enterprises across the Pacific Ocean.

EVOLVING MEXICO-JAPAN TRADE TIES

Japan began to show interest in a free trade agreement with Mexico in 2002 (Xinhua, 2003). But there are some obstacles yet to be overcome. Mexico's agricultural exports account for a significant share of about 20% of Mexico's exports bound for Japan. Japan has a very strong farmers' lobby for protection against the rice trade. Japan excluded agriculture from its free trade agreement with Singapore, as Singapore did not export large volumes of agriculture goods besides some cut flowers. The European Union was also able to exclude agricultural goods in its bilateral agreement with Mexico, because Mexico's agricultural exports to EU account for less than 10% share. The World Trade Organization (WTO) demands that the bilateral trade agreements must cover a significant majority of the bilateral trade between any two nations. Japanese Prime Minister Koizumi must, therefore, balance the anger of many Japanese rice farmers against the additional support he could get from major Japanese manufacturers and service providers. The latter group would gain if Japan is able to negotiate lower tariffs and seek more transparent investment guidelines in Mexico for them.

Preamble to a Mexico-Japan FTA

In November 1998, after a gap of six years, the previous Japanese Prime Minister Obuchi and the previous Mexican President Zedillo established the New Japan-Mexico Commission for the 21st century. The charge of this wisemen's conference of the representatives drawn from the two private sectors was to develop bilateral Japan-Mexico relations while taking into account the rapidly evolving global environment. In July 1999, they held their first Plenary Meeting in Tokyo, with subsequent consultations to be held annually, alternately in each country.

In June 2000, Japanese Prime Minister Yoshiro Mori met with the representatives of seven Latin American countries, including a special advisor to the minister of foreign affairs from Mexico and other envoys from El Salvador, Guyana, Peru, Nicaragua, Chile, and St. Lucia. Japan continued its commitment to support these Latin American countries, and sought their support in the UN Security Council over the Korean Peninsula islands.

From June 5 to 6, 2001, the newly elected Mexican President Vicente Fox made an official visit to Tokyo, and met with the Emperor of Japan and Prime Minister Koizumi. On October 27, 2002, Prime Minister Koizumi held a summit meeting with President Fox in Los Cabos, while he attended the Asia-Pacific Economic Cooperation leaders' meeting. They discussed the ways to strengthen the bilateral Mexico-Japan relations, including a free-trade agreement to be negotiated within a year, by November 2003. The second round of the Mexico-Japan consultations took place in February 2003, and the third round for strengthening the bilateral relations was held in May 2003. Both these rounds moved the bilateral agreement closer to a preferential free-trade agreement between the two Pacific Ocean nations.

Complimentary Economies

The representatives of Japan and Mexico have noted that the two nations are endowed with complimentary economic and technological conditions (Ministry of Foreign Affairs, 2002); and, that bilateral trade relation can help promote future economic development in both the countries. Mexico, with a market of more than 100 million consumers, is the ninth largest economy in the world. It is rich in natural resources and land that Japan needs yet lacks. As mentioned before, Mexico is also endowed with a large number of young and relatively skilled workers. Japan, on the other hand, is still the second largest economy in the world, with a market of 126 million people, mostly ageing fast, and high levels of accumulated technology and capital that Mexico needs.

For Japan, Mexico is one of Latin America's most promising economies that is located strategically as a gateway to North America, Latin America, and Europe. A bilateral agreement with Mexico could provide Japan access to a network of 32 countries with which Mexico has already concluded free-trade agreements. These agreements provide Mexico preferential access to the United States, Canada, the EU (15 countries), European Free-Trade Association (EFTA) (4 countries), Israel, and some other Latin American countries, representing about 60% of the world's GDP. For Mexico, Japan is a very large market for Mexican exports, and a major source of foreign direct investment and flow of technology. These would improve Mexico's employment, production, and overall competitiveness.

Mexico-Japan Trade Diversion

Despite the potential benefits of strong bilateral Mexico-Japan trade relations, their relative importance as the other's trade partner has declined in recent years. Mexico's exports to Japan, as a percent of its total exports, decreased from 1.6% in 1994 to 0.3% in 2001 (see Tables 9.4, 9.5, and 9.6). In Mexico's imports, the value of Japan's share decreased from 6.1% in 1994 to only 3.7% in 2000, then recoved to 4.8% in 2001 (Ministry of Foreign Affairs, 2002). Compared to these low figures, the share of the United States in Mexico's exports was 88.5%, and was 68.0% in Mexico's imports. The value of the EU share in Mexico's imports was 9.6%, and the EU has 3.4% share in Mexico's exports.

Table 9.4
Mexico-Japan Trade Balance (in million US$)

Year	Mexican Exports	Imports from Japan	Trade Balance	Total Trade
1993	686	3,929	-3,242	4,815
1994	997	4,780	-3,783	5,777
1995	979	3,952	-2,973	4,931
1996	1,393	4,132	-2,739	5,526
1997	1,156	4,334	-3,177	5,490
1998	851	4,537	-3,686	5,388
1999	776	5,083	-4,307	5,859
2000	932	6,480	-5,548	7,411

Source: Adapted from Mexican Ministry of Economy and data from Bank of Mexico.

Table 9.5
Mexican Imports from Japan versus Imports from United States/Canada (NAFTA) and EU (in million US$)

Year	Japan	% of Total	United States/ Canada	% of Total	EU	% of Total	Total Imports
1993	3,929	6.0	46,470	71.1	7,799	11.9	65,367
1994	4,780	6.0	56,411	71.1	9,058	11.4	79,346
1995	3,952	5.5	55,203	76.2	6,732	9.3	72,453
1996	4,132	4.6	69,280	77.4	7,741	8.7	89,469
1997	4,334	4.0	83,971	76.5	9,917	9.0	109,808
1998	4,537	3.6	95,549	76.2	11,693	9.3	125,373
1999	5,083	3.6	108,216	76.2	12,743	9.0	141,975
2000	6,480	3.7	131,582	75.2	14,745	8.5	174,473

Source: Adapted from Mexican Ministry of Economy and data from Bank of Mexico.

Table 9.6
Mexican Exports to Japan versus Exports from United States/Canada (in billion Japanese yen)

Year	Mexican Exports	% of Total	United States/ Canada	% of Total	China	% of Total	World Exports
1993	119.7	0.4	7,064.5	26.3	2,278.0	8.5	26,826.4
1994	137.3	0.5	7,336.6	26.1	2,811.4	10.0	28,104.3
1995	140.4	0.4	8,088.1	25.6	3,380.9	10.7	31,548.8
1996	205.6	0.5	9,731.5	20.3	4,399.7	11.6	37,993.4
1997	195.0	0.5	10,333.9	25.2	5,061.7	12.4	40,956.2
1998	160.5	0.4	9,781.1	26.7	4,844.1	13.2	36,653.7
1999	187.8	0.5	8,539.8	24.2	4,875.4	13.8	35,268.0
2000	257.1	0.6	8,717.4	21.3	5,941.4	14.5	40,938.4

Source: Adapted from Mexican Ministry of Economy and data from Bank of Mexico.

In terms of foreign direct investment, during 1994-2001, the share of the United States in cumulative investments in Mexico was 67.3%, compared with 18.6% investments from the EU, and only 3.3% investments from Japan.

The trade and investments from the United States and the EU have grown in Mexico because of Mexico's recent preferential free-trade agreements under NAFTA starting from 1994, and Mexico-EU FTA starting in 2000. As a result of these preferential trade agreements for the technology-intensive enterprises from the United States, Canada, and the EU, the Japanese technology-intensive enterprises (such as those exporting electric equipment and power plants) have faced disadvantageous higher tariffs in Mexico. When the Japanese enterprises located in Mexico imported parts from Japan or Association of South East Asian Nations (ASEAN) countries instead of from countries with which Mexico had preferential trade agreements, higher tariffs were imposed. The higher tariffs on imported parts diminished the competitiveness of the Japanese producers, and the higher tariffs on finished goods, such as automobiles, diminished the economic welfare of the Mexican consumers.

Whereas Mexican and Japanese representatives have assigned great importance to being consistent with the WTO rules for trade liberalization, they have felt that an early conclusion of their bilateral free-trade agreement was needed to alleviate their current obstacles. According to the APEC sources, the simple average bound tariff rate was 36.2% in Mexico and 8.7% in Japan (Ministry of Foreign Affairs, 2002). The simple average applied tariff rate was 16.2% in Mexico and 8.1% in Japan. With respect to these differences, the Mexican negotiators pointed out to their Japanese counterparts that Mexico's tariffs were about half of what Mexico was entitled to under the WTO, considering its relative development and the tariffs of other Latin countries with comparable development. Even though average tariff rates in Japan are not high, there were some tariff peaks for Japanese imports of interest to Mexico.

Trade in agriculture is a sensitive sector for the two countries. Japan imports over 60% of its food consumption, but Mexico has a negligible share in Japan's total food imports. Japanese consumers demand high standards of safety, quality, and variety throughout the year. Whereas the Mexican side hoped for further liberalization in their agricultural exports to Japan, the Japanese side admitted that this was not so easy to do in Japan. For example, mangoes from the Mexican state of Chiapas were considered free of fruit diseases by many countries, but Japan was waiting for data on the Mediterranean fruit fly in this region, and has yet to recognize this. Japanese agricultural imports and exports do not complement Mexican agricultural exports and imports. This can be considered as a factor against their preferential free-trade agreement.

Support for the Forthcoming Japan-Mexico FTA

A majority of the members of the Japan-Mexico Joint Study Group supported that a bilateral free-trade agreement would promote their mutual interests much more effectively than waiting indefinitely for a new round of WTO negotiations to conclude (Ministry of Foreign Affairs, 2002). They, however, hoped to be consistent with the WTO guidelines, such as (1) Article 24 of GATT calling for elimination of duties and restrictive regulations of commerce for "substantially all the trade," and (2) a gradual elimination of restrictive regulations of commerce within about 10 years. The two countries have proposed to include rules of origin conditions in their agreements.

EVOLVING MEXICO-JAPAN TRADE TIES AND THE CHAOS AND COMPLEXITY THEORY

To understand the evolving nature of complex trading institutions and their international relationships, a growing number of management researchers (such as Gleick, 1988; Waldorp, 1992; Lewin, 1993; and others) have recommended the use of non-Newtonian lenses of the natural science of complexity and chaos theory. According to Valle (2000), the chaos and complexity theory is the study of aperiodic dynamic behavior in nonlinear open systems (such as multilateral free trade or regional/bilateral preferential trade). Aperiodic behavior refers to the evolving and irreversible changes in open systems. Nonlinearity in open systems leads to exponential effects due to minor changes in initial conditions of the interacting entities (such as a preferential bilateral trade agreement). In simple terms, the chaos and complexity theory is the study of unpredictable, irreversible, and irregular behavior of an evolving open system. Complexity in international trade relationships is due to the large number of interacting yet independent parties that tend to organize and reorganize themselves from time to time.

The emerging new management perspective based on the chaos and complexity theory challenges the systems thinking rooted in the Newtonian linear approach and the classical Kantian theory proposed in 1790 based on control, coordination, and predictability. This linear approach was developed in simpler and relatively predictable times, with limited ambiguity, variety, and subjectivity. Instead, the emerging complex response organizations, such as many trading economies and their interorganizational dependencies, are driven by shrunken time frames. This demands an understanding of their adaptive spontaneity, evolving self-organization, and symbiotic paradoxical relationships (Lewin, 1993; Valle, 2000). As a result of these complex characteristics, even small initiatives in these open systems, such as a preferential trade agreement, can generate exponential subsequent effects, say in national productivity growth.

The chaos and complexity theory has helped many quantum physicists, astronomers, biologists, investors, and other professionals explain behaviors that were previously considered to be determined by serendipity and chance factors. In the same way, trade strategists in different countries must consider their bilateral and multilateral trade ties with other nations as an evolving open system, rather than as a planned closed system. This will be examined in detail in a separate study. In the meantime, for this chapter, we are convinced that it is futile to plan, control and coordinate highly complex and chaotic multilateral free-trade agreements—as if these were easy to manipulate simple linear relationships.

CONCLUDING REMARKS ON EFFECTIVE PREFERENTIAL TRADE ALIGNMENT

An ideal preferential trade agreement involves the economies with the most complimentary and diverse comparative advantages (Kotler et al., 1987: 250). Such an agreement generates maximum new trade, with the least likelihood of trade diversion. Japan with high levels of capital and technology accumulation, and Mexico with high levels of land and young skilled workforce, qualify for an effective preferential trade alignment. On the other hand, their unfavorable factors include geographic proximity and income differences. Japan and Mexico are not contiguous and they have vastly different income levels. These differences could lead to a mass migration of many low-end jobs from Japan to Mexico, as it did for the northern partners of NAFTA. But President Vicente Fox and Prime Minister Junichiro Koizumi seem to demonstrate the steady commitment they would need to take care of any conflicts and disputes emerging from a dynamic and evolving Mexico-Japan Free-Trade Agreement. With the breakdown in the Cancun trade talks in September 2003, the preferential bilateral trade agreement, such as the MeJaFTA, may be the last resort to achieving free trade between countries.

REFERENCES

Armbruster, William (2003). "A New Trade Strategy," *The Journal of Commerce*, April 14-20, 4 (14): 28.

Berger, S. and Dore, R. (eds.) (1996). *National Diversity and Global Capitalism*. Ithaca, NY: Cornell University Press.

Bhagwati, Jagdish (2000). *The Wind of the Hundred Days*. Cambridge, MA: The MIT Press.

Christopher, Robert C. (1983). *The Japanese Mind*. New York: Linden/Simon & Schuster.

Cordoba, Jose de (2003). "Wall Street Fears Mexican Vote Will Ensure Legislative Stalemate," *Wall Street Journal*, July 9 (Wednesday): A12.

Cordoba, Jose de, Luhnow, David, and Millman, Joel (2003). "Mexico's Fox Leaves Business Waiting for Major Reforms," *Wall Street Journal*, July 3 (Thursday): A1, A6.

Economist (2002). "The Non-performing Country," February 16: 24-26.

Gehani, R. Ray (1998). *Management of Technology and Operations*. New York: John Wiley & Sons.

Gehani, R. Ray (1999). "Architectural Concentration and the Catastrophic Financial Crises in the Newly Industrializing Economies of East Asia," *Global Focus* (formerly *Business & the Contemporary World*), 11 (2): 121-137.

Gibbs, Jennifer, Kraemer, Kenneth L., and Dedrick, Jason (2003). "Environment and Policy Factors Shaping Global E-commerce Diffusion: A Cross-country Comparison," *The Information Society*, 19: 5-18.

Giddens, A. (1991). *Modernity and Self-Identity*. Stanford, CA: Stanford University Press.

Giddens, A. (2000). *Runaway World: How Globalization Is Reshaping Our Lives*. New York: Routledge.

Gleick, James (1988). *Chaos: The Making of a New Science*. London: Heinemann.

Gori, Graham (2001). "Investors are Rushing to Mexico: Despite Slowing Growth," *New York Times*, May 25: W1.
Held, D., McGrew, A., Goldblatt, D., and Perraton, J. (1999). *Global Transformations: Politics, Economics, and Culture*. Stanford, CA: Stanford University Press.
International Telecommunications Union (2001). *Yearbook of Statistics, 1991-2000*. Geneva: International Telecommunications Union.
Kotler, Philip, Jatusripitak, Somkid, and Maesincee, Suvit (1997). *The Marketing of Nations*. New York: Free Press.
Latin American Monitor: Mexico Monitor (2002). "Japan FTA Back on Agenda," 19 (12), December: 4.
Lewin, R. (1993). *Complexity: Life at the Edge of Chaos*. London: J.M. Dent.
Millman, Joel (2003). "Mexico's Maquiladoras May Be Putting a Comeback Together," *Wall Street Journal*, July 25 (Friday): A12.
Ministry of Foreign Affairs (2002). *Japan-Mexico Joint Study Group on the Strengthening of Bilateral Economic Relations*. July, Tokyo.
Nelson, Richard R. (ed.) (1993). *National Innovation Systems: A Comparative Analysis*. New York: Oxford.
Nelson, Richard R. (1996). *The Sources of Economic Growth*. Boston: Harvard Business School Press.
Ohmae, K. (1990). *The Borderless World: Power and Strategy in the Interlinked Economy*. New York: Harper Perennial.
Ohmae, K. (1995). *The End of the Nation State*. New York: Free Press.
Rohwer, Jim (2001). *Remade in America: How Asia Will Change Because America Boomed*. New York: Crown Business.
Torres, Craig (1997). "Foreigners Snap Up Mexican Companies: Impact Is Enormous," *Wall Street Journal*, September 10: A1.
Valle, V. Jr. (2000). "Chaos, Complexity and Deterrence." Carlisle, PA: National War College.
Waldorp, M.M. (1992). *Complexity: The New Science at the Edge of Order and Chaos*. London: Viking.
Xinhua (2003). "Japan, Philippines to Continue Talks on Free Trade." July 9.

10

Participation of "The Periphery" in International Research Collaboration Based on Norwegian Experiences

Kaja Wendt

INTRODUCTION

This chapter discusses the current changes in internationalization of research based on Norwegian experiences as a periphery state in the world of research.[1] Internationalization is studied to see whether it may be a strategy to overcome scientific marginality at national, institutional, or individual levels and whether different scientific disciplines affect internationalization in different ways.

The main questions regard the characteristics of a scientific periphery and Norway as a scientific periphery. What strategies are eventually used to overcome such a status? Current changes in internationalization of research and higher education will be discussed to see how they are affecting the periphery—that is, what special challenges and possibilities do they represent? Scientific periphery is, of course, a relative term (Kyvik and Larsen, 1997). Norway is a wealthy modern industrialized nation with a well-established scientific infrastructure and a well-educated population. But as a small country on the North European periphery, with a language not well understood outside Scandinavia, and with only 4.5 million inhabitants, Norway will always be a net importer of scientific knowledge. In 2001 only 0.4% of R&D expenditures in the Organization of Economic Cooperation and Development countries were carried out in Norway. Its marginal position is not only due to size. The Norwegian R&D share is below the OECD average and is falling behind the other Nordic countries. Norwegian researchers are forced to seek collaboration to keep up with leading scientific centers. At the same time it is essential to conduct our own research, to give something back to the research community, and to be able to use the results of international research.

"Science without borders" has been an ideal for many centuries. International collaboration is seen as an inherent characteristic of the research process. Since the 1970s, the international aspects of research are both increasing and changing in terms of both more informal and formal agreements on research cooperation, establishment of large-scale facilities, more international scientific journals, an increase of internationally coauthored publications and research conferences, and there are more student and staff exchange programs. These changes are related to other international or global phenomena. The world is getting more closely connected in several ways. The pace of globalization of both world economy and science, the growth of multinational companies, the transition from manufacturing to service-based industries, and the growing importance of human capital for competitiveness are trends that influence and govern national competitiveness. New communication technologies have been developed and are widely used at the same time that traveling has become easier and less expensive. These rapid developments imply increased possibilities for cooperation.

The center-periphery concept rests upon an intuitive understanding of verticality and has been a widely used model for visualizing the scientific world (Hakala, 1998). Here verticality refers to the ranking of comparison and impact among players within a certain field of research. It implies that the center has a dominating position based on knowledge. The model deals with collaboration between countries and institutions when there is imbalance between them. In this model "ideas and publications flow from the centre to the periphery, whereas physical mobility takes place from the periphery to the centre" (Hakala, 1998: 54). There are many empirical examples of such processes among countries, regions, and fields of subjects.

The next section introduces general background and main concepts used in the study and, then the data sources are described. The center-periphery concept is used to present internationalization of research at different levels on basis of Norwegian experiences.

BACKGROUND AND MAIN CONCEPTS

In a strict sense "international" implies the involvement of more than one nation. The concept can be perceived as a quality as well as a result and a process (see Gornitzka et al., 2003). In some ways research has always been international and international collaboration has always been an inherent characteristic of the research process.

Higher education systems are currently under significant transformation processes in many Western countries (Becher and Trowler, 2001; Kim 2002).

The academic culture is under increased pressure as knowledge is considered a central tool and goal for economic prosperity of firms and countries. The higher education sector has become more dependent on external funding and tighter related to governments and business sector (Etzkowitz and Leydesdorff, 1997). As in many other countries the higher education sector in Norway is currently under revision and is encouraged to be more international and commercial (NOU, 2000: 14). The international competition also represents special challenges to the business sector in the periphery as an ever-increasing extent of technological innovation takes place internationally. Internationalization is however not a new phenomenon.

Historical Background

In Europe the forerunners of current academic research developed tightly connected with the church around the year 1000 and with the establishment of universities in the 13th and 14th centuries. Some hundred years later academic research was strongly influenced by the appearance of the modern national or territorial state (Sörlin, 1994). The universities became an issue of the territorial states and their appearance represented the first step toward nationalization of research. Sörlin (1994) describes the development of academic research in the northern parts of Europe as a process where research was forced into national boundaries along with the growth of the territorial state. Academic research became an issue of the national state. The grade structure was rather similar in Europe, and until the 17th century there were no language problems as Latin was the academic language in Europe. Mobility between the universities was seen as an ideal. As the universities often had only two or three faculties, mobility was even necessary for the ability to study a preferred subject (Wiers-Jenssen, 1999). But the decline of Latin as the common language of science was also part of the nationalization of science. It was, in other words, the nationalization of science and the universities that made it correct to talk about an internationalization of science as we use the term today.

Economic conditions as well as geographical boundaries and the political environment have always influenced research foci and individual possibilities to travel and publish. From the very start of academic activity in Europe, the church had a monopoly in terms of educating scientists (Sörlin, 1994). In the 20th century, science came to play a heavy role in deciding the outcome of World War II, and later the Cold War influenced research both in the East and the West with huge military/defense research budgets. The impact of scientific progress as a major tool for economic and political power of the nations is supported widely in the current rich literature of innovations and the knowledge-based society (e.g., Gibbons et al., 1994; Nieminen and Kaukonen, 2001). Science and knowledge as economic driving forces may give reasons for an even stronger nationalization. In spite of all national and international limitations, the scientific communication across geographical distances and political borders is a phenomenon as old as the scientific enterprise itself

(Hakala, 2002). Today, international communication and collaboration in science is probably stronger and more important than ever.

Disciplinary Differences

Several authors claim that there are disciplinary differences with respect to the "universal" dimension of research. Hence we should also expect some differences between disciplines when it comes to international orientation. Among factors internal to science important for the motivation to be international are cognitive aspects like paradigmatic status, communication language, degree of codification, degree of specialization, and the academic culture of a field (Kyvik and Larsen, 1997: 256; Hakala, 2002: 28). This implies that researchers within a relatively homogenous field with standardized forms of communication and who regard their field as international should be more inclined to collaborate internationally than others. Studies of Norwegian researchers' international contacts show differences between the humanities and the social sciences on the one hand, and the natural and medical sciences and technology on the other (Kyvik and Larsen, 1997: 171; Trondal and Smeby, 2001: 33).

Hakala (2002), who interviewed the scientific staff at three different Finnish university units, divided the disciplinary differences into "soft" and "hard" fields and pure and applied fields and looked upon variations in communication patterns, organization of research, career tracks, power, status, and funding structure. Hakala's micro-level empirical research revealed that disciplinary differences do not necessarily disappear when they experience a similar pressure to internationalize: Changes depend on the type of external pressure, on how it is experienced within the discipline, and on resources for accommodating and resisting change. Disciplinary differences also show that a scientific periphery status is not entirely determined by citizenship. First-class research is carried out within certain fields of science in the periphery; for example, geophysics receives higher citation rates than other subject fields in Norway due to the political and economic importance and hence support of this field in Norway. Disciplinary differences can probably not be seen independently from the normative question of whom the research should serve: universal ideals, national taxpayers, or commercial interests to name but a few. Hakala (2002) shows that such questions can be of importance for the researchers' way of being international. Despite disciplinary and normative differences, there seems to be one direct benefit of international contact that all researchers agree on: The mutual exchange of ideas and thoughts is the most important reward (and hence motivation) for collaboration (Melin, 2000: 37).

Even though large parts of the world are experiencing an extensive increase of collaboration, networking, diffusion, transactions, new communication technologies, and traveling possibilities, there are still important sides of the knowledge phenomena that only to a small degree are international. An academic career is still mostly national; grades, positions, and jobs are still mainly distributed nationally (Hakala, 2002). The legal systems and the social context are important factors that influence information flows (Becher and Trowler, 2001). Large business enterprises still tend to keep their R&D activities "at home"; internationalization happens reluctantly or often as an "accident" (Gulbrandsen, 2003). When national governments advocate an increase of internationalization, the motives are often to strengthen national research and not to replace it (e.g., Georghiou, 1998). It is also tempting to ask whether internationalization of research during the last decades should be understood as a "Westernization" of research. Hakala (1998) puts it this way: "[I]t would be more apt to speak about continentalisation of science, because interaction increases mainly within zones in which countries have traditionally had strong links to each others, mainly Europe and North America." This development is also visible in Norway, as the data will show.

Center-Periphery

The center-periphery concept rests as described above upon an intuitive understanding of verticality. It was the Norwegian peace researcher Johan Galtung who, since the beginning of the 1970s, foremost influenced the use of the center-periphery concept for theories in social sciences. For the development toward conceptualization Galtung (1976) differentiated between levels of social analysis: interpersonal, interdistrict, international, and interregional. In this chapter the concept is used to examine the uneven division of science, based on data concerning the Norwegian research system. The data used show that the Norwegian research system status concerning center-periphery varies on different indicators. Official government documents reveal that internationalization is seen as one important tool to deal with this status (i.e., Norges forskningsråd, 2000).

The central idea to the center-periphery concept is that by means of international research collaboration marginal countries, universities, or regions can reduce their marginality and move toward the center. The results of international collaboration in research can also be a "brain drain" from the periphery when an educated workforce (especially in developing countries) is attracted to the better working and living conditions at the centers. It has also been argued that the center-periphery model is no longer valid and that the picture has changed in direction of a network model with the creation of more centers (Kyvik and Larsen, 1997). The theories of the triple helix focus on a network approach to the current transformations in the relationship between industry, government, and universities (Etzkowitz and Leydesdorff, 1997). It is still evident that the centers of research are accumulated in certain geographical areas and this division changes slowly; the United States has taken a major position in the last century and Japan in the past decades, while Europe has had a strong position for many centuries. The slowly changing process is of course only part of the picture as new centers develop inside the borders of many countries. The new centers are often more open and have an international orientation with loose ties to a nation, disciplines, funding partners, and organizational structures than traditional research organizations. On the other hand internationalization has probably strengthened the position of regions and universities that already had a strong world position in research (Gulbrandsen, 1997). But the current flux in the research and knowledge system also provides increased possibilities to the research community, particularly in the periphery. There are new communication tools to search new research partners, more funding opportunities, and more centers and networks. The concept of a "global research market" goes one step further and describes the scientific system as a market where individual researchers or research institutions sell their know-how (consulting) and products (articles, papers, books) and if they are successful their products are bought by other researchers and paid by citations, invitation to conferences, and collaboration (Kyvik and Larsen, 1997). Still it is not hard to imagine that centers and peripheries exist in a market.

DATA SOURCES

This study relies on different sources. R&D statistics are drawn from national surveys in the Nordic countries and statistics gathered by OECD and UNESCO. Data on bilateral agreements at Norwegian universities are drawn from surveys on international research collaboration carried out at the Norwegian Institute for Studies in Research and Higher Education (NIFU) and published in Sundnes et al. (2002). Norwegian researchers' attendance at international collaborations is part of the data collected by NIFU. Surveys were carried out in 1981, 1991, and 2000 covering a wide range of problem areas, including scientific communication. The results are documented in Trondal and Smeby (2001) and Smeby and Trondal (2002). Data on foreign citizenship among researchers at Norwegian universities come from a match between NIFUs register on researcher staff (based on information from the R&D performing units in the higher education and government sector) and information from Statistics Norway on immigration. A match between the registers is possible because each individual carries a unique identification code (social security code). Data from 1991 and 2001 are used. Bibliometric data are drawn from the Institute on Scientific Information (ISI) database.

CENTER-PERIPHERY

Center-Periphery at the System Level

The location of scientific centers varies between different research fields and they also vary in number, strength, and over time. Small and poor countries have fewer opportunities to be considered as centers of research and can have difficulties attracting the best researchers. Europe, Japan, and most of all the United States must be considered being the major research centers of the world.

The last year UNESCO made a total comparison of R&D efforts was 1994. Most R&D then took place in North America with 37.9% (40.9% in 1992) of the total, in Western Europe 28.0% (29.4% in 1992), and in Japan and the newly industrialized countries 18.6% (18.4% in 1992). China was performing 4.9%

(5.2% in 1992) of total R&D, India and the Central Asian countries 2.2% (1.8% in 1992), the Commonwealth of Independent States some 2.5% (1.0% in 1992), and Latin America 1.9% (0.9% in 1992) (UNESCO, 1996, 1998). These are old data, but the relative decline of North America and Europe is found in the OECD statistics as well.

Figure 10.1
R&D Expenditure in OECD after Region, 1981-1999 (%)

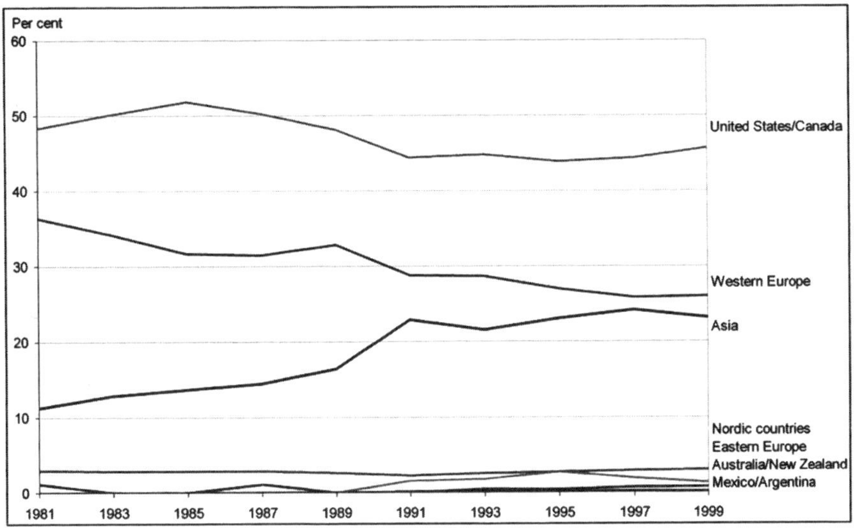

Source: OECD Main Science and Technology Indicators (MSTI) database.

In 1981 48% of the R&D expenditures in the OECD countries was used in United States and Canada. Since 1981 there has been a small decrease in the OECD share of R&D expenditures in these countries, as shown in Figure 10.1.[2] In 1991 the share was 44%, but in 1999 it increased to 46%. Canada's share was between 2 and 3%. There has also been a decline in Western Europe from 36% of R&D expenditures in 1981 to 26% in 1999. In Figure 10.1 this region includes most EU countries excluding the Nordic countries. The strongest increase has been in Asia. This region had 11% of R&D expenditures in 1981 and 23% in 1999, of which Japan constituted 18%. Latin America (only Mexico and Argentina are included in the OECD statistics) is hardly visible in this picture; less than 1% of expenditures on R&D were spent here, though there has been an increase from 0.4% to 0.7% from 1993 to 1999. In Australia/New Zealand less than 0.2% were spent the last years. The Nordic countries have had a remarkable stable share of around 3% in the period covered. Among the Nordic countries Sweden and Finland have the highest R&D share of GDP; 3.4% in Finland and 4.3% in Sweden in 2001. In OECD only Israel has a higher share with 4.4%.

The overall picture has changed slowly since the early 1980s. In spite of a decreasing share, the United States is still the lead R&D spender. Europe also has a large but declining share. Asia, especially Japan, has increased its share of total R&D expenditures in the OECD. Eastern Europe joined the OECD only after the Cold War and has had some increase in R&D expenditure since the early 1990s. The small Nordic countries have kept their position, while other countries are almost not visible in this picture. Existing data on world distribution of R&D resources entail methodological uncertainties regarding updating and coverage. The data reveal, however, a minor decrease in the share of R&D performed in the scientific centers of the world, but these are still clearly dominating the world research system.

The Norwegian share of OECD distribution on R&D has been stable at 0.4% since 1981. On a system level Norway has in other words a clear marginal position and must be classified as a scientific periphery that has neither gained nor lost position over the last decades. Norway is however part of a region with very high performances regarding R&D measured as a share of gross domestic product. This improves the periphery status of Norway, as the chances for cooperation with other R&D-intensive nations are good. The Nordic countries have geographical and cultural bonds with long traditions of regional collaboration. Today many Nordic organizations support this collaboration, especially through the Nordic Council (Wendt, 2001). Norway is however not as research intensive as the other Nordic countries and has fallen behind the development in the Nordic region, as shown in Figure 10.2.

Figure 10.2
Gross Domestic Expenditure (GERD) as a percentage of GDP; Nordic Countries and OECD Average, 1991-2001

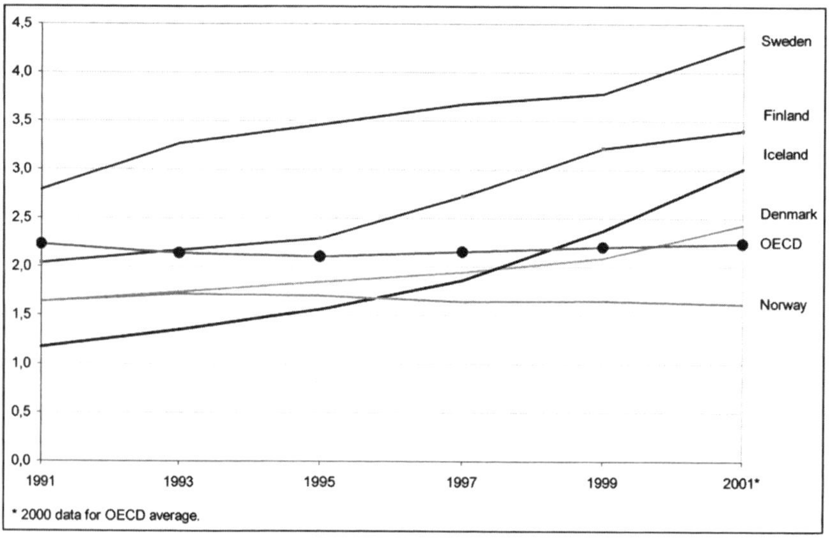

Source: OEDC MSTI and national R&D statistics for the Nordic countries.

Norway is a country provided with rich natural resources. In the last research report to the Norwegian Parliament, the government emphasized that knowledge nevertheless has been the most important factor in Norway's economical basis and development (St.meld. nr. 39, 1998-1999). An active international profile is regarded as necessary to ensure Norwegian researchers access to new developments in science and technology.

The government has formulated the goal that Norwegian research efforts shall reach OECD averages measured as a percentage of GDP within five years, that is, before 2005. Figure 10.2 is an often-used figure in the research policy debate in Norway. It shows a picture of increasing R&D efforts in the other Nordic countries during the 1990s, while Norway has not had the same positive development. The use of the gross expenditure in R&D as a percentage of GDP (GERD/GDP) goal is criticized as it is not fixed and varies with the level of GDP, which is high in Norway. But also when other indicators to measure the Norwegian research are used (e.g., growth rates and GERD per capita), Norway is falling behind the countries it is often compared with—the other Nordic countries.

One important explanation to the relatively low research effort in Norway is to be found in the business enterprise sector. In Norway 60% of all research was performed by this sector in 2001, while the OECD average was 10% higher (70% in 2000). The Norwegian business enterprise sector consists of many small and medium-sized firms in industries that traditionally finance and perform little research (Norges forskningsråd, 2002). The other Nordic countries have large international enterprises in R&D-intensive industries like pharmaceuticals, electronics, and information technology. Enterprises like Nokia and Ericsson have no counterpart in Norway. Benchmarking of Norwegian conditions for innovation (NOU, 2001: 29) reveals that on the positive side there is a high level of education in the population and there is a highly developed sector of research institutes serving the business enterprise sector. Norwegian research efforts are on the same level as other countries within some branches and per capita and in the share with higher education, while R&D as a share of GDP is low, innovation activity and the share of new products are low, and there is little cooperation between universities and industry (ibid.). Compared to other countries, Norwegian salaries for R&D personnel are very low. This can be a disadvantage when it comes to attracting a foreign workforce to Norway. The latter is a goal supported by both the industry and the government. The relatively low salaries among Norwegian researchers could however also be an advantage to international companies to make R&D investments in Norway. The share of Norwegian research conducted by technicians is on the bottom in Norway: only 28%, while the average was 46%.

In Norway, the Confederation of Norwegian Business and Industry (NHO) has recently for the first time published a strategy document on research (NHO, 2003). Also the business enterprise sector emphasizes the importance of international mobility of researchers and expresses concern about the decrease of the sectors' funding of international research cooperation. The goal to increase R&D efforts to reach OECD levels has naturally mainly economic arguments in this document: a dynamic business enterprise sector is necessary to (1) maintain the welfare system, (2) improve competitiveness, (3) develop of new industries, and (4) strengthen the national innovation system. Norway is not a typical example of a small, open economy as they locate only a minor part of their R&D activities abroad. Narula (2002) points at different possible explanations to this: The Norwegian economy is relatively introverted. In Norway there is a "technological inertia" limiting foreign investments in Norway and Norwegian investments abroad. The Norwegian innovation system is composed of industries and firms that are not very international. The Norwegian innovation system is to a relatively large degree built on exploitation of natural resources and because of "inertia" in the in R&D location, Norway has a systemic lock-in that is less internationalized than in other countries. The traditional enterprises are supported by the S&T system, while new technology-based firms seek competences in technologies that are not available domestically (Narula, 2002: 803 ff). Countries that can base their economic growth on exploitation of natural resources are not as forced as other countries to promote innovation. In the Nordic region, Finland and Sweden have had stronger incentives to develop new industry than Norway. In a small country like Norway the exploitation of the oil in the North Sea demanded a concentration of research efforts within geophysics and related disciplines at the expense of other fields of research.

Since the late 1990s the Norwegian government has introduced several major measurements to improve Norwegian research efforts:

- In 2003 a tax deduction of R&D expenses in enterprises to increase R&D efforts in Norwegian industry was introduced.
- Long-term target areas have been formulated: marine research, information and communication technology, medicine and health, and research between environment and energy (St.meld. nr. 39, 1998-1999).
- The establishment of a large research fund in 1999, with a capital of 16 billion NOK (Norwegian Kronor) and a 545 million NOK yield in 2002.
- The Research Council of Norway initiated a Centers of Excellence (CoE) scheme in 2002. The intention is to bring more researchers and research groups up to a high international standard. The centers will be devoted to long-term, basic research.
- International research cooperation is meant to increase the quality of Norwegian research and innovation in Norwegian industry and society (St.meld. nr. 39, 1998-1999: 97).
- A quality reform of the higher education sector is initiated involving changes regarding grade structure, funding, and commercial use of results (St.meld. nr. 27, 2000-2001).
- There has been increased attention toward the recruitment situation in the research system (St.meld. nr. 39, 1998-1999: 49).

There is a high awareness of the international dimension of research in Norway. The government emphasizes the quality of national research as a main motivating factor to international research cooperation. It is quite unique that the Norwegian tax reduction system includes cooperation with foreign companies.

The formulation of long-term targets can be seen as part of a small-country strategy: Norway is forced to concentrate its research efforts on areas where good results can be expected. The establishment of the centers of excellence can also be seen as part of this strategy. The Norwegian government however has been rather reluctant; Finland established the first centers in 1993, while it was not until 2002 that the first of 13 centers was opened in Norway. Within certain fields the government wishes to mobilize efforts and attract foreign researchers. The change of the grade structure is a tool to encourage international student mobility as international degrees can be approved in Norway and foreign students can achieve an international degree in Norway. In 2002 the Norwegian budget for participation in large European research organizations and programs like the EU, COST, EUREKA, CERN, ESA, EMBL and ESRF was 126 million euros (Sundnes et al., 2002). In 2002 the Nordic Council initiated the creation of a white paper for the development of an international leading Nordic region for research and innovation. The development of a Nordic knowledge market is a political ambition. These strategies can be regarded as tools to Norwegian policymakers to reduce the scientific periphery status of the country.

Center-Periphery at the Institutional Level

In most developed countries the science system and the research providers are currently under increased pressure. Among these, the universities are probably going through the most severe changes. This is an effect of the increased attention toward knowledge as a major tool to economic growth. External funding made up around 20% of total funding at Norwegian universities in 1981. In 2001 the university dependence on external funding had increased to 36%. Table 10.1 shows that total expenditures on R&D at Norwegian universities increased by 27% from 1991 to 2001, while external funding increased by 52%. Funding from abroad showed a 375% increase. It is foremost the EU funding that has caused this increase; only 3% of total R&D expenditures came from abroad in 2001. The universities are becoming more dependent on external funding sources and international sources represent new funding opportunities.

Table 10.1
Relative Changes in University Expenditures on R&D in Norway, 1991-1999 (fixed 1991 prices)

Year	R&D Expenditures Total	External Funding	Funding from Abroad
1991	1.00	1.00	1.00
1993	1.08	1.21	1.33
1995	1.07	1.10	2.37
1997	1.19	1.26	4.13
1999	1.30	1.38	4.89
2001	1.27	1.52	4.75

In Norway the government has traditionally not given the universities any specific regulations toward internationalization, but the Norwegian participation in EU research and education programs are important exceptions to this (Olsen, 1999). The internationalization was mainly based on individual contact. In the 1980s this changed and the number of formal international agreements increased. The international content of education gained more attention and institutional networks developed. To the government it has been regarded as essential that Norway, as a small country, must follow the scientific development abroad.

Figure 10.3
Agreements on International Research Cooperation at Norwegian Universities, 1996-2002 (number of bilateral agreements)

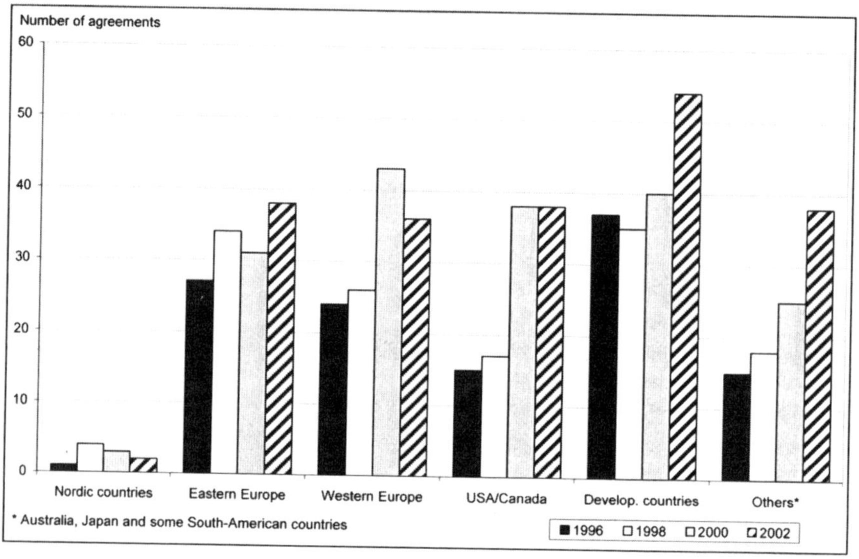

Source: NIFU.

The total number of international agreements at the Norwegian university level increased from 120 in 1996 to 205 in 2002 (see Figure 10.3). There are few Nordic agreements, showing that cooperation between the Nordic countries is still mainly informal and based upon personal contact. The number of formal agreements with Eastern Europe, Western Europe, and United States/Canada are in 2002 on almost the same level, while the number of agreements with developing countries and others has had the strongest increase over the period (Sundnes et al., 2002). The number of agreements might decrease when cooperation reaches a certain level and can be based on personal contacts, as in the Nordic countries.

Center-Periphery at the Research Performing Level

The increased governmental attention toward internationalization as a tool to secure quality of research has resulted in many strategies to increase Norwegian internationalization. Favorable student funding and attention toward academic international publishing are important strategies. The increased number of research agreements indicates that internationalization of research has led to a stronger formality of the collaboration.

Figure 10.4
Norwegian Articles With and Without International Co-Authorship, 1981-2002

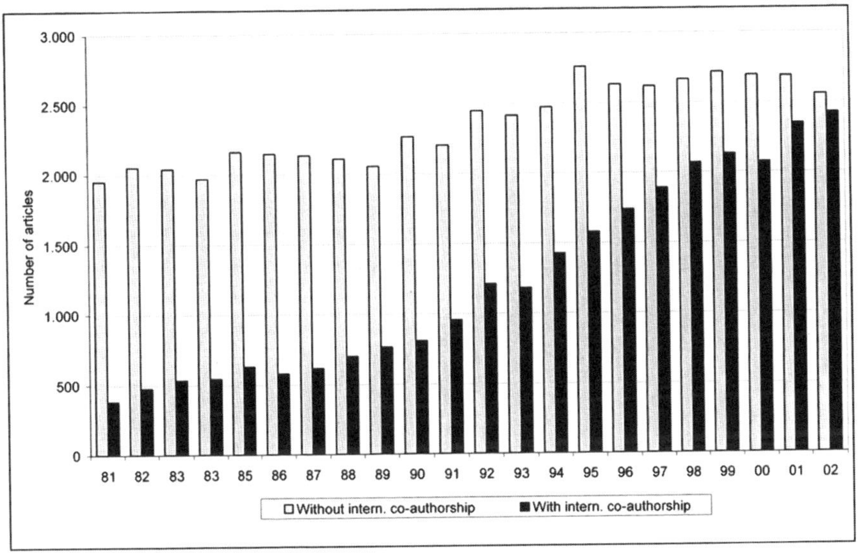

Source: Science Citation Index.

The database of the Institute for Scientific Information covered 646,000 articles in 1996 and in 2002 there were 712,000 articles. The increase in the number of articles written in the world is overwhelming. In Norway there were 11,000 co-authored articles between 1992 and 1998 and 12,700 between 1998 and 2002. The increasing internationalization of Norwegian science has developed rather fast since the early 1980s. The share of Norwegian articles with international co-authorship illustrates this quite distinctly (see Figure 10.4). While 16% of all Norwegian scientific articles comprised by the ISI database in 1981 had international co-authorship, the share had increased to 49% in 2002. Norwegian strategies to internationalize and take part in international research cooperation have hence been rather successful.

Bibliometric data provide some of the most distinct indicators of Norwegian internationalization of research. However the changes are part of an international trend where the same development can be found in other countries. Among

Norwegian researchers there has been a move toward the core of international scientific journals: Subject fields where international publishing has not been very common, like social sciences, had an increased share of ISI articles, partly because ISI covers more journals in English. Within medicine and natural sciences, which traditionally have had good ISI coverage, there has been a transfer from Nordic journals in English toward more prestigious English and American journals. The international integration in research publishing is increasing (Sivertsen, 2002: 142).

Norwegian researchers have changed their regional orientation slightly from the period 1992-1996 to 1998-2002. Most of the articles are written with European researchers and the share increased from 48% to over 50% in the second period. The share of articles written with other Nordic researchers decreased some from 23% to 22%, while the share of articles written with North American researchers fell from 20% to 18%. Research cooperation within the EU framework program has probably been very important to this development. Norway is not a member of the EU, but as an associate member through the European Economic Agreement of 1994 (EEA), it is a full member of EU research and educational programs. The Norwegian effort to join the research program has annually been around 63 million euros from 1999 to 2002 (Sundnes et al., 2002). This has in other words been a development highly supported by the policymakers. The Nordic countries are all taking part in the EU research programs and the Nordic Council had a research budget of 30 million euros in 2003 (ibid.: 27). The total share of articles written in cooperation with researchers outside North America and Europe was 8.3% between 1992 and 1998; this share had a minimal increase to 8.9% between 1998 and 2002. The overall picture provided in Figure 10.5 is a Europeanization of research, a decline of the relative importance of North America, and no significant increase of cooperation with researchers outside Europe and North America.

Figure 10.6 shows the share of faculty members having research collaboration with foreign scientists. The development regarding geographical orientation is slightly different from the patterns for international co-authorship shown in Figure 10.5. The figures do not cover exactly the same periods or regions, but still the same relative decline of cooperation with North America and the increase of European cooperation can be found. Figure 10.6 reveals that research collaboration with Nordic researchers and the rest of the world increases, even though this cooperation does not include a corresponding increase of co-authored articles.

Data on research collaboration among faculty members shows that Norwegian researchers undertook far more travels abroad in 2000 than in 1981. The total number of travels also had a more distinct increase from 1991 to 2000 than from 1981 to 1991. Research collaboration is increasingly directed toward regions outside North America, and the orientation is both European and global.

Figure 10.5
Share of Norwegian Researchers Who Co-Authored Articles in the Periods 1992-1996 and 1998-2002 by Region

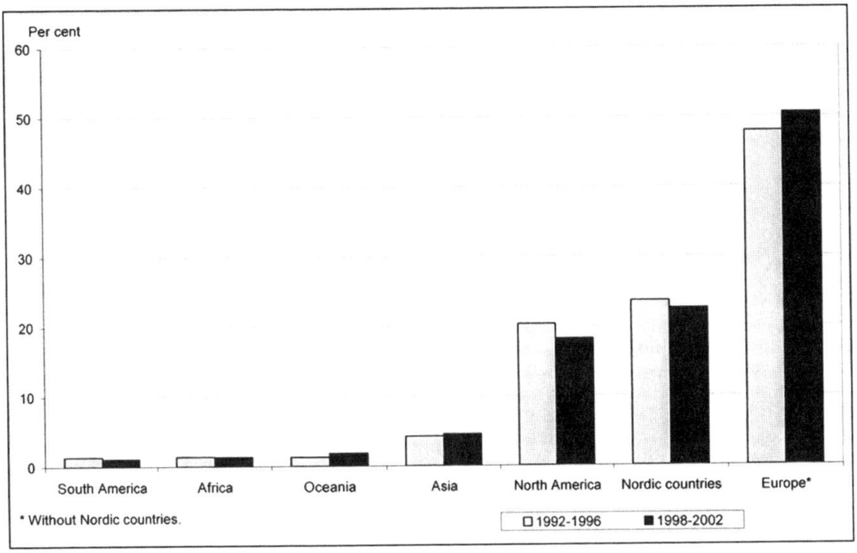

Source: National Citation Report/ISI/NIFU.

Figure 10.6
Percent of Faculty Members Having Research Collaborations with Foreign Scientists During the Periods 1989-1991 and 1998-2000, by Region

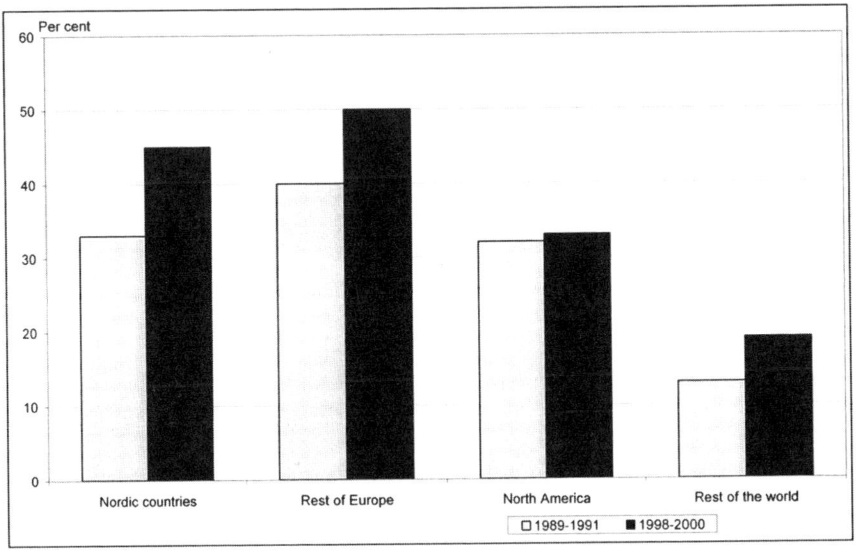

Source: NIFU.

Figure 10.7 shows that all types of professional travels increased from the period 1989-1991 to 1998-2000. Despite the rapid development of electronic publishing facilities and computer communication, personal contacts seem to have gained more importance. Maybe the different types of contact are mutually reinforcing (Smeby and Trondal, 2002: 11). International traveling among Norwegian researchers is mainly related to conferences and research collaboration. In 2000, international research collaboration equals research collaboration conducted within faculty members' own departments (ibid.).

Figure 10.7
Percent of Faculty Members Who Undertook at Least One Journey Abroad Related to Conferences, Guest Lecturing, Study and Research Visits, Evaluation Work, and Research Collaboration in 1981, 1991, and 2000

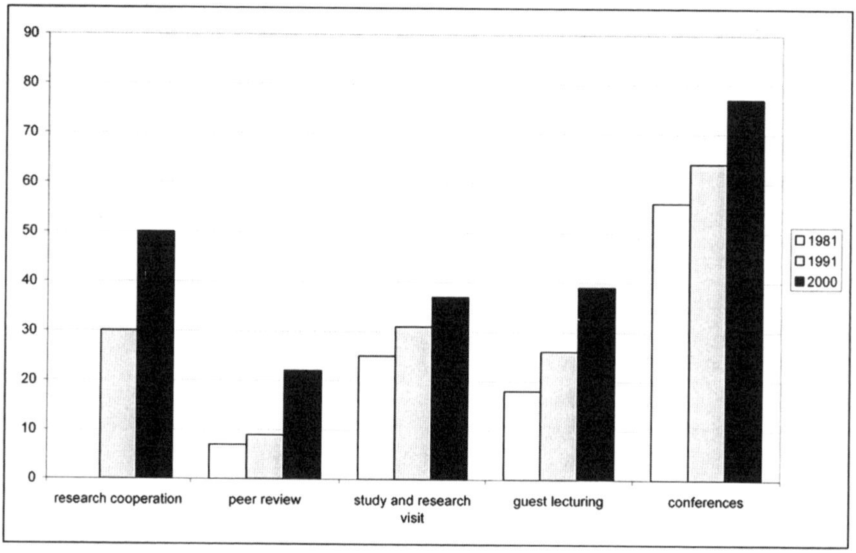

There were only small differences between fields of subject regarding the travel frequencies. The largest share of traveling abroad was found among the technologists and the lowest share among researchers within agricultural sciences. There were however no differences between male and female researchers regarding traveling abroad. Older researchers (aged 40-55) have more international contacts than younger, and professors travel more than researchers at lower positions. Professors are also traveling further distances, mostly to Europe outside the Nordic countries, but also to North America and the rest of the world, while associate professors travel to the Nordic countries and Europe. The survey reveals that Norwegian researchers have increased their journeys to all parts of the world from 1981 to 2000. The relatively largest increase was to the rest of the world, while the rest of Europe category had the smallest increase in journeys for professional reasons.

For most Norwegian scientists, international collaboration is not one-dimensional, being oriented toward one type of activity. Type of international collaboration—such as conference attendance, scholarly visits abroad, international research collaboration, and publishing in international journals—is correlated, implying that being internationally oriented in one way enhances the possibilities also for other types of international activity (Smeby and Trondal, 2002). When it comes to field of subject Hakala (2002) found that in Finland scientists within different fields often had fundamentally different interpretations of what internationalization means. The researchers in the "soft" fields often looked upon internationalization as a possibility to do comparative studies and to get new ideas. They were more skeptical to the thought of developing normative international standards of what was considered important research. Some claimed that internationalization had gone too far, as publishing in international journals gives little understanding back to the local society and culture. The researchers from the "hard" fields looked upon internationalization as a natural continuation of research done at home and regarded it as essential to rationalize the research process. These differences might decrease as all fields get more international exposure, but so far it is not clear under what conditions internationalization will lead to a more homogenous scientific world, or if collaborative work will accentuate country and/or disciplinary differences (Gornitzka et al., 2003). When it comes to geographical orientation of research collaboration, the increase in technological possibilities to travel and communicate with practically all parts of the world seem less important than traditional collaboration patterns and attractive funding opportunities.

Student Mobility

To a small country, international student mobility is essential to establish contact with scientific centers. Norwegians have long traditions of studying abroad (Sörlin, 1994). In the middle ages Norwegian students were part of the European student and research mobility that centered around Bologna, Paris, Prague, Cologne, Frankfurt, and Wittenberg. To Norwegian students, German universities were popular and later the University of Copenhagen (Wiers-Jenssen, 1999). In the 19th century many of the Norwegian "nation builders," politically, in the civil service and in the private sector got their higher education at academic institutions abroad. It was not until 1811 that the first Norwegian university was established in Oslo and students had a national alternative to going abroad to take a university degree (Wiers-Jensen, 2003). It was not until after World War II that education within most fields of subjects were offered in Norway. In the 1950s, 30% of all Norwegian students studied abroad. At the beginning of the 1970s, still half of all medical students were educated abroad. Today students abroad do not count for the same large share as the early 1970s, even though the number has increased significantly. A large number of Norwegians still earn their university degrees abroad and this is a development that today is supported by different Norwegian governments.

This has not always been so; after the World War II and even until 1980 the goal of self-sufficiency was often repeated in government proposals and parliament debates. Official politics claimed that support should be given only if an educational track had admission control in Norway. The building of national capacity was considered as more important and less expensive than buying education abroad. Another important difference was that internationalization was not considered an advantage to the same extent as it is today. As the capacity of the Norwegian education system increased, the share of students who took their degrees abroad decreased. In 1970, 5.5% of the students were abroad. During the 1980s this changed and studies abroad were no longer considered an emergency solution, but an important supplement to national institutions. The Norwegian officials began to consider Norwegian students abroad not as a solution of capacity problems but as a tool of internationalization in higher education (NOU, 2000: 14). The *share* of students abroad has been relatively stable between 5 and 7% since the 1970s, but during the 1990s the *number* of students abroad has increased, and since the mid-1990s there has been a major shift to which countries Norwegian students visit, as shown in Figure 10.8.

Figure 10.8
Number of Norwegian Students Abroad by Region, 1958/59, 1968/69, 1978/79, 1988/89, 1995/96, and 2001/2002

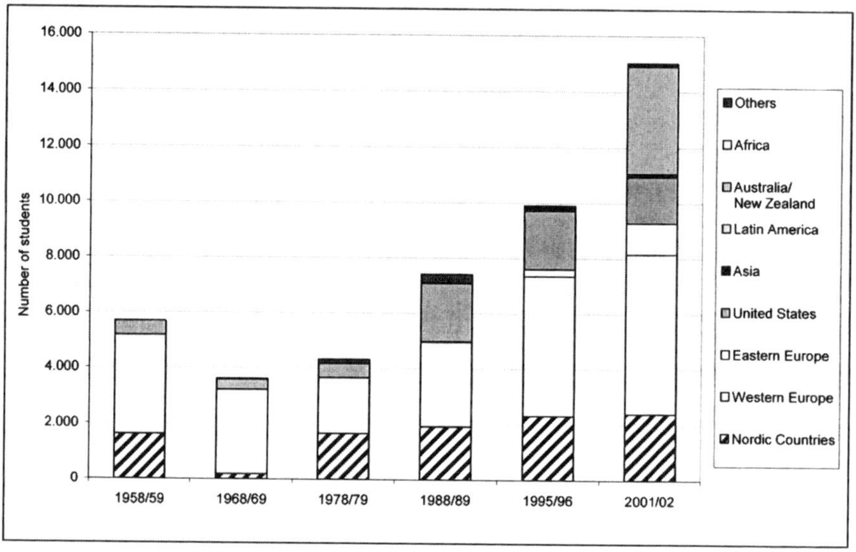

Source: The State Education Loan Fund.

Norway is a net exporter of students and has a higher share of students abroad than most European countries, including the Nordic countries. Worldwide the largest net exporters of students are not the most periphery countries, but countries that have already gone through major developments. In recent years the largest student flows go from periphery or semiperiphery states to the core or center states (Wiers-Jenssen, 1999: 14). The Norwegian student flows are characterized by a high number of students taking their whole degree abroad—so-called free movers—and public policy and limitations of capacity are the main reasons for the student flow. In this regard Norway is more like Arabian oil states than other European countries. However, when it comes to fields of learning the profile differs. The share of Norwegian students going abroad is high within technology and economics, but also the level of students within art, social sciences, and humanities is quite high. For small countries it is important to supplement specialized education tracks. The need for international experiences is recognized among students as well as employers and the government. Until 1993-1994 Norwegian students needed a professional justification to study outside Europe and North America. When the government adjusted the funding rules, the share that went to Australia increased rapidly, also due to heavy marketing of Australian universities among Norwegian students. Also Hungary, Poland, and the Czech Republic have had clear increases in the number of Norwegian students in recent years. This is mainly due to special medical or veterinary education tracks for foreigners where the education language is English or German (Wiers-Jenssen, 1999: 21). The State Education Loan Fund provides Norwegian students with favorable grants and encourages students to take part in education abroad. To internationalize Norwegian universities, foreign students are wanted at Norwegian higher education institutions. Some Norwegian institutions have received students from developing countries for several decennials through different programs.

Foreign Researchers in Norway

In 1991 10% or 670 persons, who had their first registered citizenship in a country outside Norway, held an academic position including R&D activities at one of the four Norwegian universities. In 2001 the share of persons who had their first registered citizenship in a country outside Norway had increased to 16% or 1,400 persons. In 1991 almost 4% of the persons were missing, in 2001 less than 2% of the persons were missing. The majority of these persons are probably also immigrants, so that the real number of persons with a first registered foreign citizenship is even higher.

The following categories on region were made:

- the Nordic countries (Sweden, Denmark, Finland, and Iceland): 4% of total academic staff in 2001 (3% in 1991)
- OECD countries: 8% of total academic staff in 2001 (5% in 1991)
- the rest of Europe (East European countries): 1% in 2001 of total academic staff (0.2% in 1999)
- the rest of the world rest (the majority are developing countries): 3% of total academic staff in 2001 (2% in 1991)

Figure 10.9
Total Non-Norwegian Academic Staff at Norwegian Universities by First Registered Citizenship in 1991 and 2001 (number of researchers)

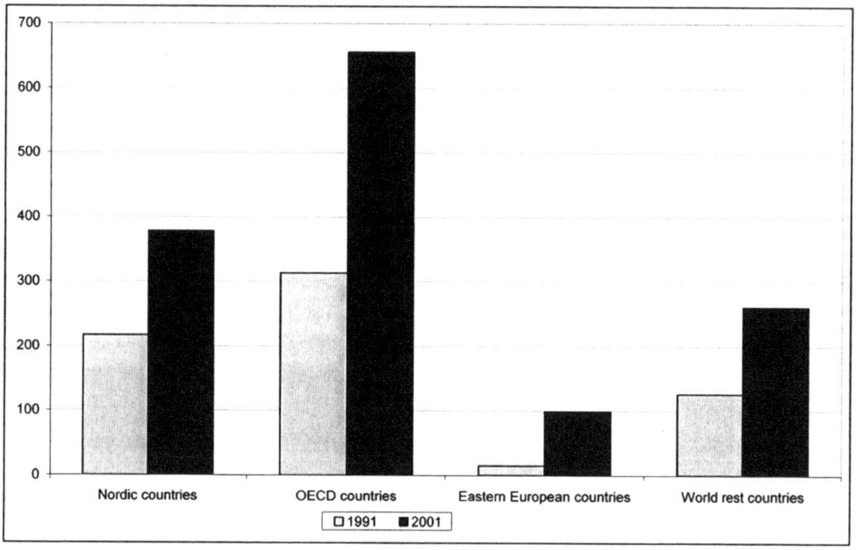

Source: NIFU/Statistics Norway.

In absolute numbers the highest increase from 1991 to 2001 was among the OECD countries. Norway has four universities situated in the eastern (University of Oslo), western (University of Bergen), middle (Norwegian University of Science and Technology), and northern (University of Tromsø, including Norwegian College of Fishery Sciences) parts of the country. The share of researchers with citizenship from other countries does not vary significantly between the institutions and lies between 14 and 19%. It is interesting that the world's northernmost university, which also is the smallest and last established Norwegian university, had the largest share of foreigners in 2001: 19%. According to this indicator the most periphery university is in other words the most international one. The University of Tromsø had the highest share of researchers from other Nordic countries (35%) and a relatively high share of R&D personnel from East European countries. Being close geographically to Russia there has been a strong tradition for exchange of knowledge with the northern part of Russia. The University of Tromsø has a national responsibility for cooperation in research and education within the Barents Sea region. The relative success in recruiting foreign researchers may also be related to the special scientific profile of the university: Research on space, fisheries, and Polar regions is gathered here. The Norwegian University for Science and Technology had the largest share of researchers from developing countries and Eastern Europe. This can be explained partly by the fact that this university is the only to offer education and research in engineering and technology. In 1991 the share of researchers with foreign citizenship was between 9 and 12% at the four universities.

Broken down by field of science and technology categorized in the OECD *Frascati Manual* (2002) the material shows some variations. In 2001, engineering and technology had the highest share of researchers with a first foreign citizenship (22%) while social sciences still was the smallest field of subject with 12%. Engineering and technology was the field with the highest share from the rest of the world category in 2001 (35%). Medical science was also attractive to countries outside Europe and the OECD; about one-fourth of the foreign researcher population in medicine at the universities and university hospitals had their origin in countries in the rest of the world category. In absolute numbers medicine had the largest increase from 1991 to 2001, from 150 to 466 persons. There are few international data on citizenship among researchers. In a global perspective engineering and natural sciences are the most common fields to study abroad, as they are more likely to fuel economic growth (Wiers-Jenssen, 1999: 13). Whether the researchers with a foreign first citizenship in Norway are former students or immigrants has not been examined.

Norway and Developing Countries

It has been shown that Norway is a periphery state in the scientific system in several ways. There is however also another dimension to Norwegian involvement with internationalization and the center-periphery concept. Being a country with long traditions regarding developing aid and a good relationship with many developing countries, it is not surprising that the international commitment is also included in papers and action plans concerning internationalization of Norwegian research (e.g., Norges forskningsråd, 2000). Research cooperation with developing countries and Eastern Europe has special importance as components in the development and building of the societies (St.meld. nr. 39, 1998-1999: 96). Norwegian development research is often defined to comprise research based on the needs of the political authorities in Norway or other user needs, and it is often understood as a rather exotic research divided from central theoretical or methodological discussions within the different fields of subjects. Development research in Norway has often been seen as part of development politics more than as a contribution to empirical diversity and abundance of research (Helland, 2001).

In 2002 a Norwegian Action Plan for Combating Poverty in the South toward 2015, was introduced (Norwegian Ministry of Foreign Affairs, 2002). The plan raises the question whether education is the most important precondition for development. The UN Millennium Development Goals (2000) do not explicitly address the need for research and higher education. The goals for gender equality, child mortality, HIV/AIDS treatment and environmental conditions cannot be reached without scientific knowledge. Focusing on higher education is important for capacity building in the public sector, to make government administration more effective and to improve the delivery of social services to the public, and to enhance economic development.

CONCLUDING REMARKS

International research collaboration and scientific publishing have increased since the 1970s. In Norway, research has become more international and collaboration more institutionalized, especially since the 1980s. This can be seen from the number of internationally co-authored articles and the increase in international funding and research agreements at the universities. The increase is seen within all fields of research although the motives differ. Within hard fields cooperation is often a natural continuation of the research process, while comparison is more often the driving force for collaboration within soft fields. There has been an increase of collaboration with all geographical regions, but European research collaboration has increased especially strongly. The EU research programs contribute to this development. The share of cooperation with North America has decreased, while the share of global cooperation with developing countries has had a small increase. In other words, the Europeanization has been stronger than the internationalization.

Despite these internationalization efforts, Norway has not been able to improve its rather weak R&D situation over the recent years. Norway belongs to the Nordic region, which is among the most research intensive in the world. However, Norway has been lagging behind the development in the other Nordic countries. Among other indicators, research efforts are below the average of the OECD countries measured as a percentage of GDP.

The poor research performance among Norwegian companies has gained much attention. Norway has based much of its economic growth on exploration of natural resources. This has caused large human resources and know-how to be tied up in oil- and gas-related industries. This is reflected in the Norwegian research system, which has up until now favored the traditional industries and to a lesser extent the emerging, more innovative, and knowledge-based industries.

The policymakers in Norway are now further encouraging internationalization in order to increase the overall importance of Norwegian research activities. This is in turn believed to establish a basis for economic growth. Internationalization is encouraged at all levels: participation in student exchange programs and EU research programs, establishment of centers of excellence, and R&D expense tax deductions. Another strategy often chosen by small countries is the concentration of efforts toward certain fields of research. In Norway, four long-term target areas have been identified.

Norwegian policymakers have traditionally emphasized the inclusion of developing countries in research collaboration. However, this can be seen more as a policy for development aid than as a research policy as such.

A major political challenge today is about establishing an overall research agenda. To reach the UN Millennium Development Goals (2000), among others to reduce world poverty by half, the participation of all nations in international cooperation in research and education seems essential.

Economic power will often determine the international position of a nations' R&D system. Poor countries have less opportunities to establish a competitive R&D system without collaboration efforts. Nevertheless, a periphery country like Norway has shown poor performance in many research disciplines despite its high internationalization efforts. This may lead to the conclusion that other political decision making may be more important to research development than internationalization efforts per se.

NOTES

1. This study is part of the larger research program on "Internationalization of Research and Higher Education" at NIFU, the Norwegian Institute for Studies in Research and Higher Education, funded by the Norwegian Research Council (2002-2004).

2. During the 1990s the OECD data also included selected nonmembers. These are included in the data used in this chapter. In 2001 Argentina, Israel, Romania, Russia, and Taiwan were included in addition to OECD members.

REFERENCES

Becher, T. and Trowler, P. R. (2001). *Academic Tribes and Territories. Intellectual Enquiry and the Culture of Disciplines*, 2nd ed. Buckingham, UK: The Society for Research into Higher Education & Open University Press.

Etzkowitz, H. and Leydesdorff, L. (eds.) (1997). *Universities and the Global Knowledge Economy: A Triple Helix of University-Industry-Government Relations*. London: Cassel.

Galtung, J. (1976). *Social Position and Social Behavior: Centre-Periphery Concepts and Theories*. Oslo: University of Oslo.

Georghiou, L. (1998). "Global Cooperation in Research," *Research Policy*, 27 (6): 611-626.

Gibbons, M. et al. (1994). *The New Production of Knowledge: The Dynamics of Science and Research in Contemporary Societies*. London: Sage.

Gornitzka, Å., Gulbrandsen, M., and Trondal, J. (eds.) (2003). *Internationalisation of Research and Higher Education: Emerging Patterns of Transformation*. Oslo: NIFU.

Gulbrandsen, M. (1997). "Universities and Industrial Competitive Advantage." In H. Etzkowitz and L. Leydesdorff (eds.), *Universities and the Global Knowledge Economy: A Triple Helix of University-Industry-Government Relations*. London: Cassell.

Gulbrandsen, M. (2003). "What Do We Know About the Internationalisation of Industrial R&D?" In Å. Gornitzka, M. Gulbrandsen, and J. Trondal (eds.), *Internationalisation of Research and Higher Education: Emerging Patterns of Transformation*. Oslo: NIFU.

Hakala, J. (1998). "Internationalisation of Science: Views of the Scientific Elite in Finland," *Science Studies*, 11 (1): 52-74.

Hakala, J. (2002). "Internationalisation of Research: Necessity, Duty or Waste of Time?", *VEST*, 15 (1): 7-32.

Helland, J. (2001). *Norsk utviklingsforskning—utviklingstrekk og utfordringer. Rapport til Norges forskningsråd Området for miljø og utvikling*. Oslo: Norges forskningsråd.

Kim, L. (2002). "Lika olika. En jamförande studie av högre utbildning och forskning i de nordiska länderna." Stockholm, Högskoleverkets rapportserie, 40 R.

Kyvik, S. and Larsen, I. M. (1997). "The Exchange of Knowledge: A Small Country in the Research Community," *Science Communication*, 18 (3): 238-264.

Melin, G. (2000). "Pragmatism and Self-Organization: Research Collaboration on the Individual Level," *Research Policy*, 29 (1): 31-40.

Norwegian Ministry of Foreign Affairs (2002). "Fighting Poverty: Norway's Action Plan 2015 for Combating Poverty in the South." http://odin.dep.no/archive/udvedlegg/01/03/pover050.pdf

Narula, R. (2002). "Innovation Systems and 'Inertia' in R&D Location: Norwegian Firms and the Role of Systemic Lock-in," *Research Policy*, 31: 795-816.

NHO (2003). *Et kunnskapsbasert og kunnskapsutviklende næringsliv: Forskningspolitisk dokument*. Oslo: NHO.

Nieminen, M. and Kaukonen, E. (2001). *Universities and R&D Networking in a Knowledge-Based Economy. A Glance at Finnish Developments*. Helsinki: Sitra, 139.

Norges forskningsråd (2000). *Internasjonalisering av norsk forskning: Utfordringer, anbefalinger og tiltak*. Oslo: Norges forskningsråd.

Norges forskningsråd (2002). *Det norske forsknings- og innovasjonssystemet—statistikk og indikatorer*. Oslo: Norges forskningsråd.

NOU 2000. 14, *Frihet med ansvar. Om høgre utdanning og forskning i Norge*. Oslo: Kirke-, utdannings- og forskningsdepartementet.

NOU 2001. 29, *Best i test? Referansetesting av rammevilkår for verdiskaping i næringslivet*. Oslo: Nærings- og handelsdepartementet.

OECD (2002). *Frascati Manual: Proposed Standard Practice for Surveys on Research and Experimental Development*. Paris: OECD.

Olsen, H. (1999). "Internasjonalisering ved de norske høyere utdanningsinstitusjonene. Omfang og organisering av formaliserte institusjonelle aktiviteter." Oslo, NIFU rapport 1/99.

Sivertsen, G. (2002). "Bibliometri i studier og dokumentasjon av forskning", i Norges forskningsråd: *Det norske forsknings- og innovasjonssystemet—statistikk og indikatorer 2001*. Oslo: Norges forskningsråd.

Smeby, J.-C. and Trondal, J. (2002). "Globalisation or Europeanisation? International Contact among University Staff," not published.

Sörlin, S. (1994). *De lärdas republik: om vetenskapens internationella tendenser*. Malmø: Liber-Hermods.

Sundnes, S., Slipersæter, S., and Wendt, K. (2002). "Norges internasjonale forskingssamarbeid—en oversikt for 2002". Oslo: NIFU skriftserie 27/2002.

Trondal, J. and Smeby, J.-C. (2001). "Norsk forskning i verden. Norske forskeres internasjonale kontaktflater." Oslo: NIFU skriftserie 17/2001.

UN Millennium Development goal (2002). http://www.un.org/millenniumgoals/

UNESCO (1996, 1998). *Status of World Science*. Paris: UNESCO.

Wendt, K. (2001). "Nordic Participation in the EU Fourth Framework Programme and EUREKA." In K. Wille Maus (ed.), *Science and Technology Indicators for the Nordic Countries 2000: A collection of Articles*, TemaNord 2001: 539. Copenhagen: Nordic Council of Ministers.

Wiers-Jenssen, J. (1999). "Utlendighet eller utflukt? Norske studenters vurdering av å studere i utlandet." Oslo: NIFU rapport 9/99.

Wiers-Jenssen, J. (2003). "Over bekken etter vann?" Oslo: NIFU, work in progress.

11

Innovation Clusters and Cooperation Networks to Foster Technology-Based Firms

Carlos Quandt and Luiz Márcio Spinosa

INTRODUCTION

Entrepreneurship and innovation have become central to the ability to promote growth and increase productivity, particularly in association with the development of knowledge-intensive industries. Knowledge and advanced skills are now perceived as the fundamental strategic resources of our age, and the essential drivers of productivity and economic performance (OECD, 1996). The divide between more developed and less developed areas is likely to be increasingly defined in terms of their relative ability to create and apply knowledge rather than their stock of capital or other factor endowments. Competitiveness in the world economy is linked not only to the development of local innovation capacity, but also to the ability to participate in widening networks of information and production resources. The challenge of developing world-class industries and technologies is even greater for developing countries and regions, which have struggled to design and implement strategies, policies, and institutions that effectively promote local innovation and facilitate knowledge diffusion.

In developing countries, the efforts to achieve regional competitiveness are often ineffective under unfavorable macroeconomic conditions, and their poorly developed institutional endowments generally make them vulnerable to strong corporate globalization pressures. The long-term viability of technology-intensive firms in such regions is compromised by their limited access to new technology and markets within and outside the region. The main problems are related to the relative isolation of individual firms with respect to external sources of knowledge and information, as well as the scarcity of strong institutions, skills, and R&D. Local efforts to promote a knowledge-based, high-technology model of development have typically focused on two related objectives: (1) to foster the creation of technology-intensive clusters in the region and to strengthen existing ones, recognizing their role as catalysts of local innovation, development, and commercialization of technology; (2) to develop wider networks to expand and complement the capabilities of local firms and clusters with knowledge and resources from other regions.

The promotion of industry clusters and networks follows one of the dominant discourses in economic development and economic geography literatures in recent times. The examples of successful regions suggest that innovative clusters are a key to regional development, provided that they are associated with wider networks and institutions that support regional innovation and economic growth. This is connected with the rise of the network form of organization as a strategy for global competitiveness, the role of information and communication technologies (ICT) in the expansion of interfirm linkages worldwide, the nature of interactive learning processes and externalities among knowledge-intensive firms, and the need to expand and revitalize the technology base of existing clusters. In the specific case of developing-country clusters, the insertion in wider networks provides access to previously inaccessible assets and a broader range of opportunities for knowledge sharing.

The empirical focus of this chapter is a set of networking initiatives related to the *Paraná World Class Program for Information and Communication Technology* industry, which has been implemented in the State of Paraná, in southern Brazil. The program, known as "W-Class," is a joint effort of local universities, government, and business associations to plan and implement actions to promote the development of software and electronic commerce in Paraná. The specific initiatives include: (1) The NTS Network (Brazil-Japan Network of Software Business and Technology), an NGO that fosters networking and business opportunities in Japan for local small and medium-sized enterprises (SMEs); (2) GameNet-PR (Parana's Network of Computer Game Enterprises), which comprises Brazil's leading producers of electronic entertainment software; (3) REDESOL (Open Source Network), a major agent of knowledge sharing concerning open source technology among firms located in ten high-tech incubators and technology parks. The cases illustrate how the competitiveness of local software producers is nurtured by local support of entrepreneurial activity and technology development, and through information exchange and capacity building among local agents. Network links with international mentoring institutions are also incorporated into the network and integrated to the processes of knowledge creation as well as sharing and building institutional and business relationships.

CONCEPTUAL FRAMEWORK: CLUSTERS AND NETWORKS

The framework is centered on the relevance of two key concepts for SME development—*innovation clusters* and *cooperation networks*—recognizing that both are emerging as significant tools to promote regional development through the activation, diffusion, and expansion of locally generated knowledge (see Figure 11.1). In the cluster/network-based approach, both concepts are joined by a focus on interactive learning and the diffusion of different types of knowledge: tacit or codified, scientific or practical, and such in different spatial and organizational settings. That also implies a focus on the emerging field of knowledge management—that is, the explicit and systematic management of knowledge and its associated processes of creation, organization, diffusion, and applications to create wealth and promote development. The analysis is also guided by some basic assumptions: (1) *SMEs* can play a key role in triggering and sustaining economic growth and equitable development in developing regions; (2) the creation of *technology-intensive firms* is essential to build local capabilities to compete in the global economy; they are also essential to strengthen academic-industry-government linkages and encourage technological innovation; (3) the region's development potential can be greatly enhanced by adopting a *cluster/network-based approach* to address its development needs and spatial imbalances, searching for cooperation and partnerships among different government levels, the private sector, and nongovernment organizations.

Figure 11.1
The Conceptual Framework

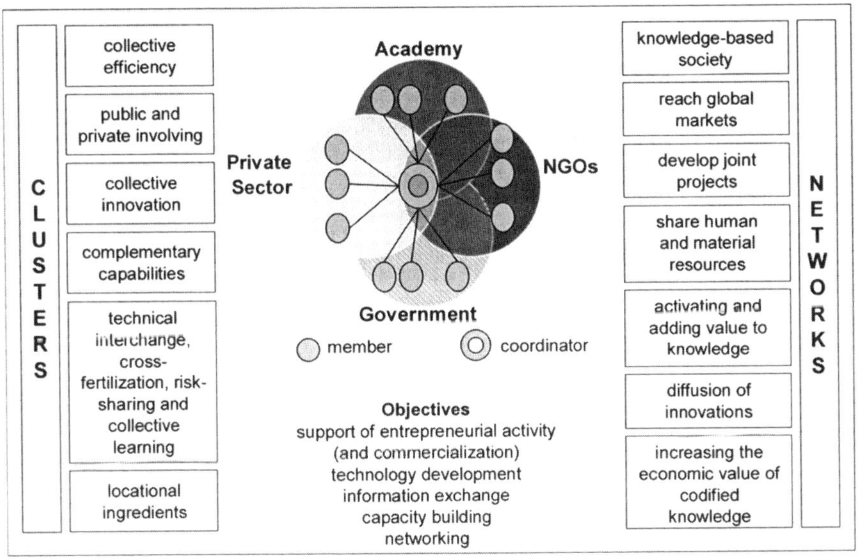

SMEs, particularly technology-based ones, have a tremendous potential to accelerate economic growth, expand their share of exports, and promote a more deconcentrated and equitable pattern of development in developing countries. However, this potential role is often not fulfilled because of their small scale. As Ceglie and Dini (1999) point out, SMEs are often unable to capture market opportunities that require a large scale of production. They are also unable to achieve economies of scale in the purchase of inputs (such as equipment, raw materials, finance, consulting services, etc.), and the creation of an internal division of labor that could foster cumulative improvements in productive capabilities and innovation.

Small size also constrains the internalization of dynamic functions such as training, market intelligence, logistics, and technology innovation. Even innovative technology-based firms tend to lack key skills and resources, such as marketing or business capabilities. Berry (1997) notes that small firms' limitations typically fall in the areas of access to technological information, and guidance on quality control; access to finance; assistance in purchase of materials or equipment, in workplace organization, in financial management, or in other determinants of effective performance; and market stability (security of demand over a period of time). More important, small-scale entrepreneurs in developing countries are often ill-prepared to look beyond the boundaries of their firms and capture new market opportunities.

It is widely acknowledged that interfirm cooperation and linkages involving SMEs in a developing economy may have a strong impact on growth and distribution performance, as demonstrated in the successful development of East Asian countries, beginning with Japan, but also including Korea, Taiwan, and others. The development of networks can improve the competitive position of SMEs and reduce the problems related to their size through mutual help. For example, firms may establish a localized network to become more specialized and complement each other's capabilities by sharing resources, pooling together their production capacities and purchasing power, thus achieving scale economies to conquer markets beyond their individual reach (Pyke, 1992). Some aspects of SME support (especially credit provision) have evolved considerably, but "linkage-inducing" policy remains largely a new and experimental area. As Berry (1997) notes, "the challenge for policy in this area is to understand the source of potential payoff to increased interfirm cooperation, the contexts which facilitate it, and the potential instruments to induce it."

THE ROLE OF INNOVATION CLUSTERS

Horizontal cooperation and the creation of external economies among SMEs in clusters contribute to generate competitive advantages through "collective efficiency." Schmitz (1995) emphasizes that "external economies are essential to growth but not sufficient to ride out major changes in product or factor markets; that requires joint action." The advantages of cooperation among SMEs are usually connected with collective economies of scale, the benefits of dissemination of information, and interfirm division of labor. These benefits tend to increase when transaction costs are low, and these in turn tend to decrease with geographic proximity and the establishment of shared infrastructure, common norms, and tacit rules for cooperation.

Therefore, an innovation cluster is characterized not only by a sectoral and geographical concentration of firms and other economic agents, which gives rise to external economies and favors the creation of specialized technical and financial services. It also comprises public and private local institutions to foster knowledge transfers and to support local economic development. This type of arrangement facilitates collective learning and innovation through implicit and explicit coordination (Humprey and Schmitz, 1995). Successful clusters depend on both the private and the public sectors (usually universities and research institutions), which join efforts to create innovative environments and to build synergies among agents with complementary capabilities. Their development is gradual and cumulative: over time, the region builds knowledge, skills, institutional support structures, specialized services, financing arrangements, infrastructure, and collective norms of cooperation and mutual trust.

Clusters are built on linkages and relationships that integrate the isolated technological capabilities of institutions, firms, and individuals into a collective, territorial asset. The establishment of mechanisms to coordinate efficiently these relationships is essential to create a supportive environment for many forms of technical interchange, cross-fertilization, risk-sharing, and collective learning. This is essentially a territorially based process, as people who share the same space discover the advantages of "learning by interacting."

As Bianchi (1993) points out, the crucial characteristic of the "Marshallian" type of cluster or *milieu* is the set of competitive and collaborative linkages among agents in a socially and historically defined agglomeration, complemented by a set of collective intangible assets that belong to the production system as a whole. The cluster benefits from its complex web of interactions because innovation rarely happens in isolation. It is an experimental, trial-and-error activity, and each agent may draw innovation inputs from a wider matrix of institutions to take advantage of a division of labor in the generation of knowledge and skills. In that sense, the cluster improves "dynamic" efficiency (or innovative capability) by reducing uncertainty through information sharing and screening, and by establishing a durable relational basis for the construction of competences (Camagni, 1995).

The literature on high-tech clusters has described diverse "locational ingredients" that are usually seen as necessary for the development of successful clusters. These ingredients can be categorized as tangible and intangible elements, as proposed by Bortagaray and Tiffin (2000). The tangible elements are: knowledge-based firms, knowledge inputs, specialized consulting services, specialized inputs, markets, cluster support, and financing. The intangible elements are: a supportive social climate, links and interactions among individuals and organizations, and quality of life for people working in the community where the cluster operates.

Among the tangible elements, the knowledge-based firm is the core element. As Kozul-Wright (1995) points out, "... the firm is in a position to fulfill a number of critical conditions for innovation: (i) by acting as an organization for storing knowledge (including tacit knowledge); (ii) as an enduring institution which can reproduce that knowledge and inculcate it in new entrants or share it with other firms, and (iii) as a social agent which can establish trust and cooperation."

In more general terms, two regional components may be considered as necessary but not sufficient conditions for a successful cluster. The first is a "critical mass" of human resources, including entrepreneurs, scientists, engineers, technicians, and skilled labor. The second is a capable scientific and technological infrastructure, or the "knowledge assets" of a region. These may include universities, public and private research labs, libraries, technological incubators, innovation centers, and science parks. The main roles of these anchor institutions are to promote technology transfers and to support networking.

In developed countries, the usually strong networks of technology transfer agents may also play an important role in promoting the functioning of local innovation clusters. For example, the Industrial Research Assistance Program (IRAP) network of Canada maintains hundreds of scientists and engineers to serve innovating firms throughout the country. In addition to the technology transfer services they offer linking to federal R&D labs, these agents also have at their disposal industrial research grants (essentially allowing them to act as early-stage angel capitalists), and they provide advice on product design, marketing, and suppliers. A major part of their work is to liaise with other cluster support agents, including those who run local business clubs and innovation support networks in the private sector.

THE ROLE OF REGIONAL AND SUPRATERRITORIAL NETWORKS

There is an increasing awareness that networks are emerging as significant tools of social change, and that access to global information resources has become an essential condition to maintain international competitiveness and develop a knowledge-based society. It is also recognized that knowledge diffusion through formal and informal networks is just as essential to economic development as the creation of knowledge itself. The network model is becoming increasingly dominant in modern productive sectors, not only for companies but also for institutions in the area of governance and development. The rise of concepts such as the networked organization and the "virtual enterprise"—which may comprise, for example, transitory teams of freelance or temporary workers organized in flexible ways to develop a specific project—has challenged directly the traditional place-bound, centralized notion of organization of production.

Cooperation networks enable firms to position themselves in the trade-off between market-related transaction costs and the high costs related to internal development of know-how. They create opportunities to reach global markets, absorb new technologies, develop joint projects, and share human and material resources. Even though high costs and risks are integral aspects of the network form of organization, it is particularly suitable for coping with dynamic processes such as systemic innovation and control over future technological trajectories.

In a network, information is transmitted horizontally, reciprocally, and iteratively, rather than following a rigid hierarchy. Hence innovation and competitiveness depend on the ability to integrate different kinds of information and to coordinate them among the different agents, types of activities, and firms. As networks evolve and become more sophisticated, a *learning process* emerges through cooperation, together with increased reliability and trust. These elements constitute a shared intangible asset that helps to reduce both "certain" and hidden costs of the interaction among the partners as well as the probability of opportunistic behavior (Bianchi, 1993).

Lundvall and Borrras (1997) point out that the fast development of information and communication technologies gives a strong impetus to the process of codification by increasing the economic value of codified knowledge. A significant portion of useful knowledge, which can be codified and reduced to information, can now be transmitted over long distances at very limited cost. According to Foray and Lundvall (1996), codification is an important process for economic activity and development for four main reasons: codification reduces the costs of knowledge acquisition and technology dissemination; codification adds commodity-like properties to knowledge and facilitates market transactions by reducing the uncertainties and information asymmetries of knowledge exchanges; codification facilitates knowledge externalization and increases the amount of knowledge that can be acquired at a given cost; and codification helps to speed up knowledge creation, innovation, and economic change.

The most important barrier to this codification trend is change. Complexity may increase the cost of codification, but this might be overcome if the knowledge remains stable. This means that tacit knowledge is still a key element in the appropriation and effective use of knowledge, especially when the whole innovation process is accelerating. When the content of knowledge is changing rapidly, it is only those who take part in its creation who can get access to it. This explains the cluster phenomena as well as the formation of industrial networks and interfirm alliances aimed at technology development. While knowledge is increasingly being codified and transmitted through communications networks, tacit knowledge is also required, including the skills to use and adapt codified knowledge, in a process of continuous learning by individuals and firms. Innovation in a knowledge-based economy is driven by the interaction of producers and users in the exchange of both codified and tacit knowledge. As access to information becomes easier and less expensive, the ability to select and use it efficiently is essential. In that sense, local agents and structures that support the use and expansion of knowledge in the economy and the linkages between them are crucial to the local ability to diffuse innovations, to absorb and maximize the application of technology to products and processes, and for developing a common cultural basis for the exchange of information.

In sum, both the territorial network and a global network linking sets of local systems have the function of activating and adding value to knowledge produced locally. The process of clustering is similar to the network model in the sense that both are technological learning systems that help to socialize innovation-related knowledge and reduce uncertainty in the environment in which innovative agents operate. A firm's ability to create knowledge is strongly related to its interaction with related firms in a process of collective learning, involving exchanges of partly codified and partly tacit knowledge. Rather than being mutually exclusive, the two concepts (the cluster-based personal contact and the wide network linkage) are deeply linked and complementary.

THE REGIONAL CONTEXT: OBSTACLES TO KNOWLEDGE FLOWS

Isolation is a great problem for most developing-country firms, particularly small ones. Technologically advanced regional systems, and industrial clusters more generally, depend on the development of territorially based networks. However, regional innovation and growth are not restricted to local sources of knowledge, capital, or other factors. In the context of this chapter, it has become apparent to regional authorities that a focus on the mobilization of local assets to build synergy and achieve competitive advantage by fostering innovation clusters will not suffice. It must be matched by a broader focus, on the ability to join increasingly wider spatial networks and to develop alliances, partnerships, and opportunities with outside firms and investors as well as science parks and incubators, universities, and research institutes.

In a previous study (Quandt, 1999), a survey of technology-based firms (TBFs) in the region was undertaken to clarify how cooperative behavior affects the process of innovation and technical change. It surveyed the linkages among the firms and between them and other agents, considering the diverse dimensions of their knowledge base, such as the degree of tacitness, diversity, and complexity. The objective was to identify gaps and bottlenecks related both to sources and channels of knowledge among regional agents, on the basis of how institutions and TBFs themselves assessed their knowledge flows. Four dimensions were taken into account. The first was whether such flows were endogenous or exogenous, and in the latter case, what was the extent of their spatial reach (within the cluster, outside the cluster but within the region or country, or outside the country). The second related to the channels and relevant interfaces through which these flows occur. The third dimension related to the magnitude of the flows, or their perceived importance to the innovation process. The fourth was the impact of the knowledge flows on the firm's different types of capabilities, such as its strategic, internal, and external capabilities.

According to the survey, the majority of the firms indicated domestic firms as their main competitors, customers, and suppliers. Surprisingly, global competition was not an important issue for most firms. Although foreign competition is important to many of them, these firms are not seeking foreign markets. In this case, the territorial concentration of specialized knowledge has propelled many small-scale efforts to exploit market niches, which rarely reach a global scale. This is also reflected in the local scope of the dominant flows of knowledge.

As far as the sources of knowledge are concerned, the most important factor perceived to affect the company's performance was access to a skilled labor pool, followed by ties to the local research base and access to qualified suppliers. The least important factor was government incentives, which probably reflects their limited availability. A significant number of TBFs reported that informal conversations and social interactions are important means to absorb competitive information. Direct observation and analysis also appear as important ways in which companies gather new knowledge, benefiting from their closeness to other firms in the area. Cooperative alliances are also very important means to acquire relevant knowledge: 65% of the firms mentioned alliances with other firms, and 20% with universities. During the course of the survey, several entrepreneurs pointed out that one of the major obstacles for university-industry collaborations is the competitive pressure brought by the pace of technological change and rapidly changing markets. They argue that the university is generally unable to respond with the flexibility and speed that would be necessary to develop a joint project and bring a product to the market before it becomes obsolete.

Localized flows of knowledge predominate among the region's firms, in connection with the predominance of local transactions and cooperative linkages. Although in some cases a large share of the components is sourced from remote locations, specific knowledge for product development is rarely obtained abroad. Cooperative arrangements are also important for access to relevant knowledge. Alliances with other firms—and to a lesser extent with universities—both locally and outside the region are highly ranked as sources of knowledge. Again, local partners play a dominant role. Finally, clients are also seen as crucial sources of knowledge for product and process development by a significant number of firms.

For the diverse agents in a cluster, opening up to wider cooperation networks implies differentiated benefits. Although the costs and risks can be high, small firms and start-ups tend to gain more, because they generally have limited access to technology networks and to international events, and they tend to have a limited ability to interface with the infrastructure due to their small size. They may also lack other key skills and resources—such as marketing or business capabilities, which is often the case for new technology-based firms.

It becomes clear, then, that wider cooperation networks and ICT may play a major role to improve the "collective efficiency" of existing clusters, by expanding the scope of knowledge search, and deepening the capability to generate and manage technical change. It is also clear that the wider level of knowledge flows requires different channels and different types of learning processes. This appears to be a significant shortcoming of many of the firms surveyed. In the specific case of technology-intensive industries, the complex environment in which firms and clusters operate highlights the importance of access to a wide range of complementary assets and competencies. Given their limited ability to take advantage of supraterritorial networks, an important function of the institution—the cluster's incubator or managing institution—would be to help screening, decodifying, and channeling relevant information into the cluster. Some of the technology incubators in the region have attempted to perform that function, with limited results.

PARANÁ'S PROGRAM OF INNOVATION CLUSTERS AND NETWORKS

The conceptual framework of this study has been applied to on an existing initiative, called the Paraná World Class Program for Information and Communication Technology industry (W-Class), a program to promote global competitiveness and exports of ICT solutions produced in the Brazilian state of Paraná (see Table 11.1). It seeks to foster three emerging ICT clusters in the state. It is focused on production and commercialization as well as the improvement of local skills, particularly among SMEs.

The W-Class strategy draws its main elements from the World Class concept from Kanter (1995). It stated goals are to develop so-called World Class assets, by investing in actions related to the promotion of seven program areas: Innovation, Entrepreneurship, Quality, Learning, Collaboration, Networking, and Funding. Within each of these areas, the program's funds and actions are directed to the development of the firms themselves, of human resources, market, government, entities, and the community in general. The W-Class Programs comprises more than 30 different subprograms or lines of action that range from education, training, and seminars to support for access to partnerships, capital, and markets. These initiatives are linked by the state's Integrated Network of Technological Information, which includes the associated Paraná state universities. Several other state and local agencies are involved in the program, in addition to private enterprises and business associations.

A large part of the program effort is directed toward the promotion of university-industry-government cooperation, because actors from all these areas have been involved since its inception. Indeed, the emphasis on networking permeates all subprograms. The actions are partly focused on each cluster, bringing together local firms, universities, and potential investors, but the emphasis appears to be increasingly shifting toward wider forms of networking, expanding the links to new sources of information and new markets.

For example, GameNet-PR (Parana's Network of Computer Game Enterprises) has identified emerging producers of electronic entertainment software in the region, with the potential to become internationally competitive. The initial undertaking was the creation of a network linking game producers, software and hardware solution providers, and marketing channels. The main objective was to facilitate exchanges of knowledge among the members. More specific actions include: product development and commercialization (linking producers with potential investors and markets, supporting start-ups), human resource development (training and courses), technological development (supporting research labs and new projects), and infrastructure development (libraries and network infrastructure).

Table 11.1
Application of the Conceptual Model

Application of the Conceptual Framework in W-CLASS		Objectives				
Network Participants	**Main Actions**	support of entrepreneurial activity (and commercialization)	technology development	information exchange	capacity building	networking
	• Integration of members and the Open Source Group		X	X	X	X
REDESOL	• Organization of workshops, showrooms	X		X		X
	• Search for funding	X	X		X	
Private sector:	• Creation and maintenance of shared intelligence	X		X		X
1 enterprise, SUCESU	• Electronic sharing of information among partners	X		X	X	X
	• Identification of potential partners	X				
Government:	• Mobilization of Open Source Community	X		X		X
TECPAR, SETI, INTEC, CITS	• Support alliances with national and international providers of Open Source technology	X	X	X	X	X
Academy:	• Support strategic alliances among Open Source community		X	X	X	X
UEL, UEM, UFPR, PUCPR	• Development of (social) public-oriented projects	X	X	X	X	X
Coordination: TECPAR		X				X

Figure 11.1
(continued)

Application of the Conceptual Framework in W-CLASS		Objectives				
Network Participants	**Main Actions**	support of entrepreneurial activity (and commercialization)	technology development	information exchange	capacity building	networking
NTS **Private sector:** 21 enterprises **Government:** TECPAR, SETI **Academy:** UEB **Coordination:** NTS NGO	• Mobilization of Japanese community in Paraná and Japan	X				X
	• Organization of missions to Japan	X		X	X	X
	• Support to implement contracts (legal issues)	X			X	
	• Creation of Japan-oriented commercialization processes	X	X		X	
	• Localization of Brazilian software to Japanese market	X	X	X	X	X
	• Development of Brazilian software demonstrations	X	X	X	X	X
	• Creation and maintenance of shared information	X	X	X		X
	• Use of channel marketing technology	X	X		X	X
	• Japanese market research	X		X		
	• Electronic sharing of information among partners	X	X	X	X	X
	• Creation of the Brazilian Enterprises Digital Portfolio (Portuguese and Japanese)	X	X	X		
	• Japanese-Brazilian translations	X		X		X
	• Market channels identification	X			X	X
	• Potential partners identification	X			X	X
	• Bilateral meetings mediation	X			X	X
	• Showrooms organization	X	X	X	X	X
	• Funding search	X	X		X	

Figure 11.1
(continued)

Application of the Conceptual Framework in W-CLASS		Objectives				
Network Participants	**Main Actions**	support of entrepreneurial activity (and commercialization)	technology development	information exchange	capacity building	networking
GAMENET **Private sector:** 14 enterprises (half incubated) **Government:** TECPAR, SETI, INTEC, INTUEL, INFOMAR **Academy:** UEL, UEM, PUCPR **Coordination:** TECPAR	• Creation of a capacity-building program (undergraduate)				X	
	• Organization of workshops, showrooms	X		X		X
	• Demonstrations of game development	X	X		X	
	• Search for funding	X	X		X	
	• Creation and maintenance of shared information	X		X		X
	• Use of channel marketing technology	X		X	X	X
	• Brazilian market research	X		X		
	• Electronic sharing of information among partners	X		X	X	X
	• Market channels identification	X				
	• Potential partners identification	X		X		X
	• Organization of meetings with publishers and distributors	X		X		X
	• Support alliances with national and international providers of game technology	X		X	X	X
	• Support alliances with Incubators		X	X	X	X
		X	X	X	X	X

The W-Class program has also supported the development of the NTS Network (Brazil-Japan Network of Software Business and Technology), an NGO that fosters networking and business opportunities in Japan for local SMEs. Similarly to other networks under the program, this one has focused on expanding market access through a scheme called channel marketing, which makes it possible for many small firms to cooperatively reach overseas markets.

Another network supported by the program has focused on Open Source Community: REDESOL (Paraná Open Source Network) has become a major agent of knowledge sharing among firms located in Paraná and the International Open Source community.

As an increasing number of SMEs benefit from these networks, the W-Class Program is being showcased as an example of how the competitiveness of software producers in developing areas may be nurtured by local support of entrepreneurial activity and technology development, and through information exchange and capacity building among local agents. As the program's networks are expanded, links with international mentoring institutions are also incorporated into the network and integrated to the processes of knowledge creation, sharing, and building institutional and business relationships.

The establishment of networks enables firms in each cluster to expand their access to knowledge and resources from a broader range of sources, in addition to an expansion of the available resources to the incubators and clusters themselves (Gibson et al., 1999). It becomes clear that a broad-based cluster/network-based system may have a strong impact in specific areas such as the following:

- Access is available to new markets and marketing strategies.
- Access to capital: There is integrated access to services such as financial planning; support for obtaining grants; opportunities for access to venture, development, and seed capital.
- Expansion of interfirm linkages: A networked approach is ideal for maximizing the impact of programs and projects, such as partnerships, alliances, and linkages to outside suppliers.
- Technological support: There is access to services such as technology assessment and forecasting, assistance on technological choices, marketing assessment of innovative projects, and access to outside technical information.
- Technology transfer opportunities: Networks may be used to stimulate investment in S&T, R&D, technology transfer, and spin-offs.
- Access to talent and know-how: Networks may help in the process of identifying and hiring skilled people across regional boundaries. It should also be noted that labor markets are essentially place-based, yet virtual technologies may boost the development of human resources in more remote locations through training centers, distance education, career planning, virtual job markets, and also support business development through the establishment of virtual entrepreneur schools.
- Strengthening local cluster governance structures: The establishment of linkages with other clusters enables a better understanding of stakeholder needs and markets and improves organization methods. As the Brazilian Reparte network demonstrates, such arrangement may also be used to disseminate best practices in business incubation to improve the performance of firms in each cluster.

- Optimizing and sharing facilities: The operational support infrastructure may be optimized and many facilities could be shared over the network, including incubators, prototype centers, pilot plants, online library, test laboratories, and online conferencing facilities.

Some of these benefits have been achieved in the case of Paraná. The establishment of localized support mechanisms—such as adequate and accessible education, training, and technological services that build on regional strengths to stimulate innovation activities—has been matched by a broader focus. Local agents are now striving to expand cooperative agreements, alliances, and consortia involving public institutions, local firms, and foreign organizations. They also recognize the need to develop and strengthen technology information networks, technology transfer mechanisms, and liaison agents to promote exchanges of knowledge among firms, research institutions, and regions.

In terms of the specific goals of the program, it is possible to observe that, in practice, the cluster/network-based initiatives have produced results in some of the areas discussed above:

- Support of entrepreneurial activity and commercialization: The different networks have facilitated access to new markets and marketing strategies, through the organization of commercial missions, the development of specific marketing processes, the development of marketing surveys that benefit the network as a whole, and entrepreneurial training.
- Technology development and technological support: The networks have provided training and access to services such as technology assessment and forecasting, and more important, access to outside technical information.
- Information exchange: This is probably the most important impact of the network-building efforts. The initiatives include the creation and maintenance of shared intelligence; the creation of several ways for members to share knowledge and lessons learned; the facilitation of access to technology transfer opportunities; shared information on funding opportunities, potential partners, possibilities for incubation and spin-offs, and so on.
- Capacity building: The program has focused mainly on training and facilitation of alliances and partnerships, as well as the access to information and the establishment of linkages with incubators and universities. Many of the supporting institutions focus on bringing sound business practices to the entrepreneur by providing training in various aspects related to business (e.g., business plans, marketing, financial planning) that the entrepreneur often neglects.
- Networking: The efforts were focused on obtaining a better understanding of stakeholder needs and markets, while encouraging all parties to participate and share knowledge, so that the network as a whole would be strengthened.

It should also be noted that the local incubators have played an important role in the promotion of these actions. They have facilitated the incorporation of professors and senior students into entrepreneurial business, and also provided the physical facilities and specialized services to help start-up firms during the first critical years of their creation. As Bortagaray and Tiffin (2000) point out, incubators "may provide a unique source of low-cost, low-risk, high potential value investment prospects for venture capitalists, thus acting as the focal point for a small cluster" in the Latin American context.

CONCLUDING REMARKS AND POLICY IMPLICATIONS

For technology-based industries, more than in any other case, knowledge and competence flows increasingly matter more than flows of ordinary goods and services. At the same time, flexibility and rapid responses in the linkages between firms and other institutions, as well as close user-producer relations, are crucial to support innovation in new and unstable technologies that involve a great deal of uncertainty. For developing-country firms, the range of knowledge flows still tends to be very limited, not only in terms of their spatial reach, but also with respect to the channels and interfaces through which these flows occur. This in turn constrains the relevance of such flows to the innovation process, because the firms face simultaneously a difficult access to a small number of sources.

The policy implications are connected particularly to the triple role of the scientific and technological infrastructure of any given region in the knowledge-based economy: It comprises the production of knowledge (research and development), its transmission (education and training), and its diffusion (or transfer) so that such knowledge may be applied by the productive sector to generate wealth and development. The relative success of the Paraná networks strongly suggests that the focus of public support to innovation should include not only strategic science and technology projects, but also specific programs to enhance knowledge diffusion.

This includes efforts to stimulate university-industry-government cooperation and more efficient mechanisms of technology transfer. It also means that technology incubators should expand their outreach role, which is usually limited to the local community including universities, local firms, and clients. Successful incubators need to be integrated into the local infrastructure but also to national and global sources of technologies and markets. Furthermore, the evidence strongly suggests that all forms of networking should be fostered at diverse spatial levels. This may range from local venture forums to bring together potential investors and local firms, to links with other regions as a way to broaden their sources of information and to develop new markets.

Even though the impacts of the program are just beginning to appear, this case illustrates how governments can contribute to the development of adaptable programs, decentralized structures, and processes, by acting as both enablers and facilitators, and brokering new kinds of networking and association. For that purpose, clusters and networks may become not only significant tools to foster SME competitiveness, but also to help establish a new model of regional development, based on the positive impacts of sharing knowledge, advanced skills and innovation.

REFERENCES

Berry, A. (1997). *SME Competitiveness: The Power of Networking and Subcontracting*. Washington, DC: Inter-American Development Bank, No. IFM-105.

Bianchi, P. (1993). "Industrial Districts and Industrial Policy: The New European Perspective," *Journal of Industry Studies*, 1 (1), October.

Bortagaray, I. and Tiffin, S. (2000). "Innovation Clusters in Latin America." Presented at the 4th International Conference on Technology Policy and Innovation, Curitiba, Brazil, August 28-31.

Camagni, R. (1995). "Global Network and Local Milieu: Towards a Theory of Economic Space." In S. Conti, E. Malecki, E. and P. Oinas (eds.), *The Industrial Enterprise and Its Environment: Spatial Perspectives*. Aldershot: Avebury, pp. 195-214.

Ceglie, G. and Dini, M. (1999). "SME Cluster and Network Development in Developing Countries: The Experience of UNIDO," International Conference on Building a Modern and Effective Development Service Industry for Small Enterprises, Rio de Janeiro, March 2-5.

Foray, D. and Lundvall, B.-Å. (1996). "The Knowledge-Based Economy: From the Economics of Knowledge to the Learning Economy." In D. Foray and B.-Å. Lundvall (eds.), *Employment and Growth in the Knowledge-based Economy*, OECD Documents. Paris: OECD.

Gibson, D., Burtner, J., Conceição, P., Nordskog, J., Tankha, S., and Quandt, C. (1999). "Incubating and Sustaining Learning and Innovation Poles in Latin America and the Caribbean." Presented at the 3rd International Conference on Technology Policy and Innovation, Austin, Texas, August 30 - September 2.

Humphrey, J. and Schmitz, H. (1995). "Principles for Promoting Clusters and Networks of SMEs." Paper commissioned by the Small and Medium Enterprises Branch, UNIDO, Number 1.

Kanter, R. (1995). *World Class: Thriving Locally in the Global Economy*. New York: Simon & Schuster.

Kozul-Wright, Z. (1995). "The Role of the Firm in the Innovation Process," Geneva: UNCTAD.

Lundvall, B.-Å. and Borras, S. (1997). "The Globalising Learning Economy: Implications for Innovation Policy." Report based on contributions from seven projects under the TSER Programme DG XII, Commission of the European Union.

OECD (1996). "The Knowledge-based Economy," OECD/GD(96)102, excerpted from the 1996 *Science, Technology and Industry Outlook*. Paris: OECD.

Pyke, F. (1992). "Industrial Development Through Small-firm Co-operation." Geneva: International Institute for Labor Studies.

Quandt, C. (1999). "Mapping Knowledge Flows in Clusters and Networks: Case Studies of Technology-based Firms in Brazil." Presented at the 3rd International Conference on Technology Policy and Innovation, Austin, Texas, August 30 - September 2.

Schmitz, H. (1995) "Collective Efficiency: Growth Plan for Small-Scale Industry," *Journal of Development Studies*, 31 (4), April.

12

Effective Model for Higher Education and Industry Interaction

Kari Laine and Matti Lähdeniemi

INTRODUCTION

In a knowledge-driven economy there is a growing need for deeper and more productive interaction between higher education and industry. A lot has been written about what should happen between units of higher education and industry to ensure collaboration. But less has been brought up about what should happen *inside* the units of higher education to make the collaboration possible. It is not enough to have a transfer office to commercialize the knowledge created in higher education. The full exploitation of the interaction requires more than that. It requires incentives and a strong interaction between the main processes in higher education. In this chapter, a literature review is done and Satakunta Polytechnic (in Finland) is used as a case study to describe the interactions between the main processes in a unit of higher education. The main processes in Satakunta Polytechnic and in this chapter are the educational, R&D, and entrepreneurial processes. In a knowledge-based economy, knowledge and technology are no longer transferred; they are more likely created in an application collaboration with industry and higher education. In this kind of knowledge-creation process, the knowledge creation, dissemination, and utilization are carried out close to each other or even simultaneously. Also, basic research and applied research can no longer be separated. Knowledge creation is, in many cases, based on a long-term partnership where trust, commitment, and mutual benefit can be achieved. Developing processes like this is a demanding task. The best practices are based on national, regional, and many other conditions. However, some positive basic guidelines can be pointed out.

METHODS

A literature review is used to describe promising practices in higher education and industry interaction and a case study of Satakunta Polytechnic describes a practical model for internal interaction aiming for effective interaction with industry.

RESULTS

According to European research (European Commission, 2001), intensive industry-science relations (ISR) can occur under the following conditions: When industrial demand is high, there are well-developed incentive schemes, there are special programs that facilitate SMEs, legislation does not constitute a barrier for interaction, there are public initiatives to foster ISR, science and technology policy follows a stringent and long-term approach of strengthening ISR. According to the same report there are many good practices for ISR, but they are specific for market and institutional environments, and fields of technology as well.

Good practices in ISR are a comprehensive, stringent, and long-term-oriented policy, a balance of technology transfer with education and fundamental research activities, a bottom-up scheme of defining joint research themes, a competition-based approach, rising absorption capacities in SMEs, direct commercialization of research results, reforms of institutional settings in public science, creation of institutes specializing in transfer, regular auditing of research strategy, direct transfer between researchers and industry, and individual remuneration of successful transfer activities and personnel mobility.

A lot of attention has been paid to IPRs, academic start-ups, joint research, and personnel exchange, but less to cooperation in curricula planning, vocational training, institutional reform, and individual incentive systems.

There can be various incentives, like individual or institutional missions and objectives, administrative and managerial support, balancing with other major objectives of science, such as education and fundamental research.

A regional innovation system is a projection of the national innovation system in the region. In the Satakunta regional innovation system the most potential players are units of higher education, research organizations, firms and their customers, suppliers and partners, regional development companies, funding organizations, and municipalities with their economic development. The polytechnic acts as a knowledge creator and transferer in the region. It differs from universities because the knowledge is in most cases created together with the customer in the application projects.

According to Porter (2001), regional wellness is an indirect result of regional competitiveness. Regional innovativeness and productiveness must be developed first. Innovativeness leads to productiveness that leads to prosperity and wellness. In this an organization of higher education should take an active role. This means being involved in the development of regional (technology) strategies and in their implementation. Knowledge must also be cumulated to "long living" research and education organizations. These organizations must have strategic partners in the region, which are usually anchor firms. In the cluster approach there are three main elements: development of the existing clusters, creating a cluster of innovative knowledge-intensive enterprises, and overlapping of clusters. Overlapping of clusters offers potential synergies in skill, technology, and partnership in time. Units of higher education and research centers institutionalize entrepreneurship and innovation, causing a steady flow of new enterprises and innovations.

Porter made an agenda for units of higher education. It consisted of recognizing the important role, creation, and support of technology transfer offices, actively participating in cluster development efforts, aligning curricula and research (R&D) to meet the needs of local clusters, and supporting company start-up efforts. His agenda for a cluster-specific institution for collaboration would promote cluster awareness, engage in the ongoing diagnosis of a cluster's competitive position, develop training and management programs, provide programs through institutes, actively participate in recruitment efforts with the government, and widen the institutional membership to include all cluster constituents. Regional innovativeness leads to productivity that creates prosperity and finally wellness. Higher education and specialized research centers are usually the driving force behind innovation in nearly every region.

After recognition of the regional role of higher education there is still plenty of need for actions and their tangible results. Commercializing research results and supporting entrepreneurship requires that the unit of higher education is entrepreneurial itself. Clark (1998) describes the entrepreneurial university as characterized by a strengthened steering core that enables it to steer itself, an expanded development periphery like professional outreach offices, a diversified funding base, a stimulated academic heartland and an integrated entrepreneurial culture.

Barnes, Pashby, and Gibbons (2002) studied critical success factors in university industry interaction. They made six major findings for critical success factors, which were:

1. method of partner evaluation is needed to ensure that partners are interested, able to support, and can actively contribute to the work
2. high-quality project management
3. trust, commitment, and continuity
4. flexible management
5. benefits achieved, fast tangible outcomes
6. mutual benefit, balance between academic objectives and industrial priorities

In a knowledge-driven economy, new knowledge is often created in long-term partnerships (Fisher et al., 2002; Wilson, 1994). In Satakunta Polytechnic, partner evaluation and the partnering process have been found to be extremely important. The more challenging the project, the more important it is to select qualified partners. The partnering process begins with finding potential partners according to one's own strategies as seen in Figure 12.1. All the criteria mentioned above are used in the selection.

Figure 12.1
The Partnering Process

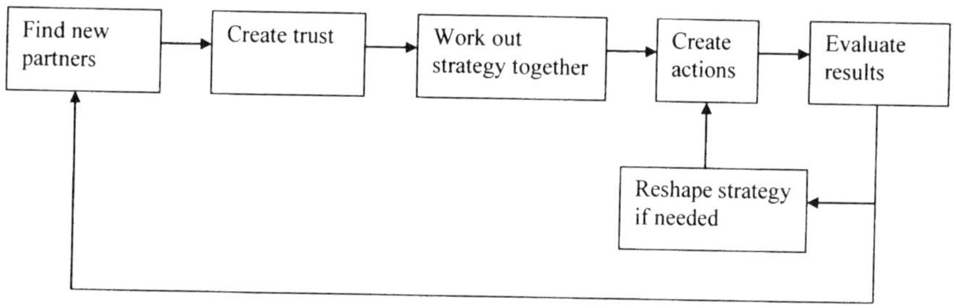

Trust can be created only with time. Wilson (1994) points out immediate problem solving, frequent contacts, honest communication, and high and wide relationships in the creation of trust. Then there is time to work out a strategy together and start actions with tangible results. It is important to evaluate the results of the partnering process and the concrete actions. Some partnerships end because of lack of support and contribution.

Figure 12.2
Increasing the Level of Interaction in Time

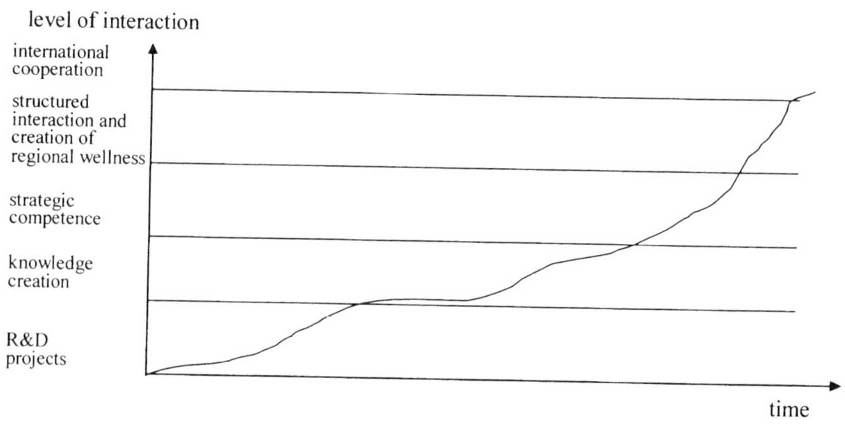

Several phases can be recognized in the interaction as described in Figure 12.2 (Saurio, 2003). In Satakunta Polytechnic, the process has had five stages. First the polytechnic started to seek funding and partners, and to start R&D projects. In the second phase creation of knowledge is truly happening because there is a market pull for creation of the knowledge. In the third phase the polytechnic has more strategic competence because it has been involved in many projects and knowledge-creation processes. This in turn leads to more structured interaction with industry and true creation of regional wellness. In the last phase the international cooperation is a quite natural consequence.

The list of facilitating factors in Satakunta Polytechnic includes strong strategy work with true focal areas of R&D, flexible repositioning and carrier rotation, critical mass in focal areas, resourced contacting and project preparation work in faculties, exact and rational investments, incentives like vision, mission, personal and team incentives, and the infrastructure. One of the major findings has been to identify the students as a main resource in R&D. Active students always connect the personnel to the projects. Positive facilitating factors for students can be calling students to start projects, calling a team of students to run a project, and opening new problem-solving contests. All these methods ensure that students will become insiders and that they will realize the possibilities in the projects and in themselves. Also, the students' enterprises are potential collaboration partners for R&D (Laine et al., 2001).

In Figure 12.3 the internal interaction is modeled. The model is based on the interaction between the educational, R&D, and entrepreneurial processes. The educational process produces high-quality specialists for the industry. The R&D process creates new strategies, products, processes, and services for local firms. In the entrepreneurial process, a new cluster of knowledge-intensive enterprises is created to absorb new knowledge and technology to the region and to act as an R&D partner for the higher education and industrial clusters. The whole system is based on regional strategic decisions and aims for regional innovativeness, competitiveness, and wellness. The effectiveness in industry interaction is internally based on the good interaction of the three main processes: educational, R&D, and entrepreneurial. In Figure 12.3 thinner arrows describe the open channels between the three main processes and the texts describe the transaction between the processes. Personnel and students are resources for the R&D process. The needed knowledge is usually created in the educational process. In many cases students have to gain new knowledge fast. Therefore study credits are also granted in the projects. Student projects and theses are used as tools to solve the customer problems. When projects are started, there grows a need to know more about the subjects the project team deals with, so it gives a motivation to learn.

In R&D projects the new knowledge is created and applied. Therefore the projects also give new content for the studies. The created new knowledge can be transferred to the learning process. New methods have been developed for this. Customer projects usually give different requirements than normal studies and exercises. They challenge the students for action and learning. They also give ideas for totally new degree and master programs. Success stories can serve as an inspiration for new students. Students work as knowledge creators and knowledge intermediaries in R&D projects.

Figure 12.3
The Process Interactions

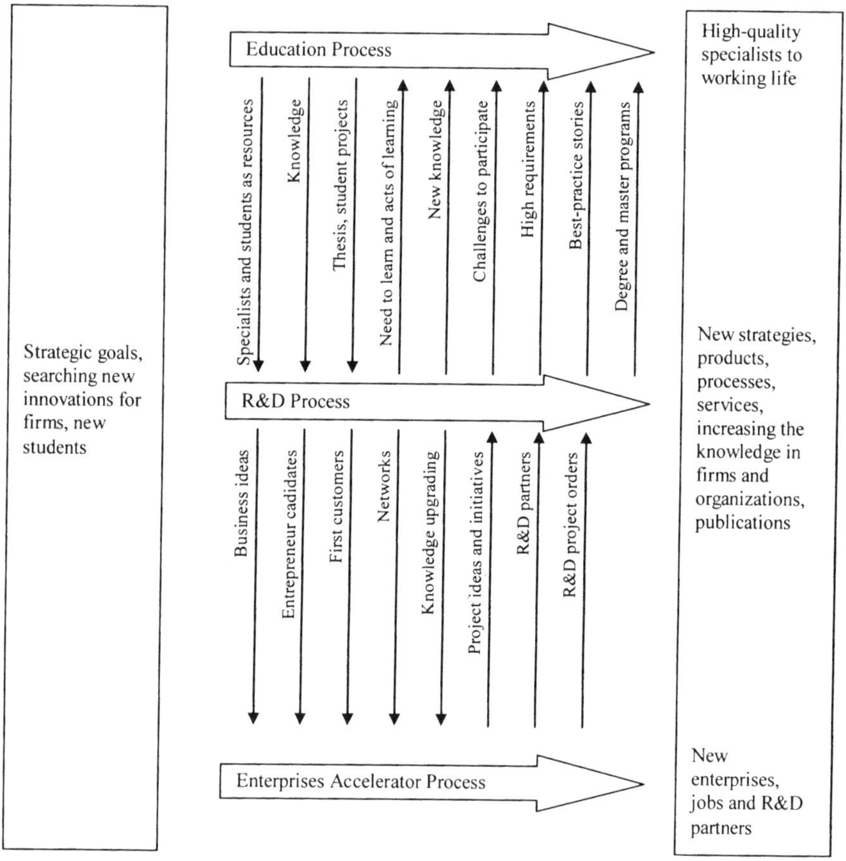

The O'Sata® R&D process creates new business ideas for the enterprise accelerator process. In student projects the most entrepreneurial students are found. They want to volunteer as project managers and accept the challenges. In the project the first potential or true customers can be found and networks with other actors and funding organizations are created. When entrepreneurs are partners of the polytechnic and participate in R&D projects they can upgrade their knowledge. Entrepreneurs also give project ideas and initiatives for the polytechnic. In some cases they even place orders for new R&D projects run by students. The O'Sata Enterprises Accelerator® process is a special process created to support the students who want to start their own enterprises during their studies. This process has been more carefully studied in Laine et al. (2001), and Lähdeniemi, Järvi, and Piiroinen (1998). All of these processes get input from the strategic goals of the polytechnic and the region, the Searching New Innovations for Firms (SNIFFer) subprocess, and new students. With the SNIFFer process new opportunities are identified and selected. This happens according to the R&D strategies and exploitation potential of regional firms.

In Figure 12.3 the arrows can be understood as open channels. There is no flow in the channels without pressure. Without the pressure the whole organization is going toward the minimum energy state. There are two possible methods to ensure flow in the channels. The first method is just to increase the pressure until something dramatic happens. A better way is to make small positive changes and see what happens. The best way is to search for the amplification. And the amplification is where the students are. The number of students is about ten times bigger than that of the personnel. Activating the students automatically activates the personnel.

There are several ways to share good practices inside the polytechnic: a centered administration of R&D projects to prevent duplicate work, cluster seminars for the personnel and industry, personnel development days and meetings, reports on the Intranet, personnel work plans that can consist of teaching and R&D projects, publications where outcomes are presented, student theses, and work placement. It is also important to disseminate the results and experiences to partners.

The role of a polytechnic in a regional innovation system is to be a knowledge creator, an intermediary, and a network builder. It is also a "neutral" force that can gather other forces for common efforts. The created internal model is based on the interaction between educational, R&D, and entrepreneurial processes.

IMPLICATIONS TO PRACTICE

The organization of higher education should to take a proactive role in the region. This means being active in the regional strategic developing processes and act as a "seeker" that finds profitable knowledge and innovations for strengthening and creating new markets for the regional firms. Qualified partners in national and international networks are very useful in this process. Effectiveness in industry interaction requires being entrepreneurial, creating an innovative learning partnership with main customers, creating strategic competence, and having internally a strong interaction between main processes. Partnerships and strategies serve as tools for creation of commitment and focus. They are also tools in looking for external funding for R&D projects.

It has been noted that entrepreneurship and good interaction with industry have much in common. There are changes that happen in technology and society. Megatrends and smaller changes have to be seen as opportunities. Then there is the time for radical selection. Unconventional methods must be experimented. Risks must be analyzed, handled, and accepted. Failures happen and they must be tolerated. Promising practices are changed to concepts that are duplicated. In Satakunta Polytechnic, true partnerships have been created and at the same time 50 new spin-off enterprises have been started. The most valuable knowledge has been created in long-term partnerships in focused research centers.

Development of regional competitiveness can take decades, but better interaction between higher education and industry can be built in years. When transferring promising practices, both the source and target innovation environments have to be understood and taken into consideration. The best practices in one environment are perhaps only promising practices in another environment.

CONCLUSIONS

There are shared concepts and promising practices for higher education and industry interaction as well as for regional competitiveness. Being interactive with industry means being entrepreneurial, finding focal areas of R&D, building partnerships, and having a good interaction between the main processes inside the unit of higher education. The model presented in this chapter is a result of a long development process. It is based on active cooperation and partnerships with industry. The model works in the Satakunta region. A remarkable ability to grow strategic knowledge, create spin-offs, and identify focuses has been detected. In this chapter the development path, results, and experiences are described. The model is applicable as a guiding principle in other higher education institutes in similar innovation environments and culture.

ACKNOWLEDGMENT

The authors highly appreciate funding from the European Social Fund.

REFERENCES

Barnes, T., Pashby, I., and Gibbons, A. (2002). "Effective University-Industry Interaction: A Multi-case Evaluation of Collaborative R&D Projects," *European Management Journal*, 20 (3): 272-285.

Clark, B. R. (1998). *Creating Entrepreneurial Universities: Organizational Pathways of Transformation*. Oxford: Pergamon/IAU Press.

European Commission (2001). "Benchmarking Industry-Science Relations: The Role of Framework Conditions." Final report, research project commissioned by the European Commission and the Federal Ministry of Economy and Labor, Austria, Vienna/Mannheim, June.

Fisher, J., Belcher, R., Cairney, T., English, B., and Harding, S. (2002). "Greater Involvement and Interaction between Industry and Higher Education," Business/ Higher Education Round Table, B-HERT Position Paper No. 7, Melbourne, January.

Lähdeniemi, M., Järvi, A.-R., and Piiroinen, H. (1998). "Activated Enterprise as a Part of Engineering Studies," The European Society for Engineering Education, 25 years Annual Conference 1998, Helsinki, Finland.

Laine, K., Lähdeniemi, M., Järvi, A.-R., and Piiroinen, H. (2001). "Innovative Companies and Networking—Technology Transfer in the Enterprise Accelerator," presented at the 5th International Conference on Technology, Policy and Innovation, Delft, the Netherlands, June 26-29.

Porter M. E. (2001). *Clusters of Innovation: Regional Foundations of U.S. Competitiveness*. Washington, DC: The Council on Competitiveness.

Saurio, S. (2003). *Yrittäjyyden edistäminen ja yrityshautomotoiminta ammattikorkeakouluympäristössä*, Satakunta Polytechnic, Report Series A, Tutkimukset 1/2003.

Wilson, L. (1994). *Stop Selling, Start Partnering*. Essex Junction, VT: Oliver Wight Publications.

13

The Interface Role on Virtual Team Project Success: A Study in the IT Sector

Lauro Noboru Hassegawa and Roberto Sbragia

INTRODUCTION

It has been some time since industrial, cooperative, or technological projects were restricted to offices and employees all in one location. Owing to the facilities of communication and the enormous growth of the global market, we have the present emphasis on international projects, for either local or multinational needs, involving people and resources spread worldwide. An international project has wider range than a normal one because of the boundaries that it crosses. Yet more new tools have been needed to deal with the new difficulties of long distances and different cultures involved. It has become very important to understand the mechanisms that permit the management of those projects, which brings a variety of new problems such as cultural differences, multiple controls, cross boundaries, transportation time, and customs regulations.

Few sectors have had so significant impact as information technology. In the early 1980s the personal computer (PC) was revolutionizing offices as a low-cost alternative. Since 1994, the development of the Internet has triggered a huge growth in the number of computer connections and Web sites. It's not much to say that globalization became possible due to the Internet. Despite having multinational companies before, they had a local character. New areas of activities appeared within a few months such as technical, commercial, providers, telecommunication, consulting, and the like, forming what is called the "new economy" or the "dot-com economy."

The IT companies developed the technology to access the Internet, and that includes computers (hardware and software). Because those companies work worldwide, they have the same needs to speed up their projects. It is natural for them to use virtual teams to implement projects. In other words, they produce this technology first hand, continually innovating and developing new products as quickly as possible.

That is how the virtual teams come to the scene. Lipnack and Stamps (1997) define "virtual team" as a group of people led by a common objective who interact through interdependent tasks. Though different from the colocated team whose members work physically near each other, the virtual team works over geographical distance, time zones, and organization boundaries, which are narrowed down and enriched as a result of the technology of communication. Gould (1997) said that virtual teams communicate with one another through computers, and might never see each other face-to-face.

Basically, what makes the difference between a virtual team and a real team is the physical distance, organization, and transposition of boundaries. Allen (1977) established that more than 15 meters of distance between co-workers would lead to little teamwork. In addition, Lipnack and Stamps (1997) say the use of technology doesn't characterize the term "virtual team" itself. The technology just represents a way or tool by which virtual teams can develop their activities. Considering effective results, it is fundamental to the success of virtual teams that the nature of those groups is exploited in a way to promote understanding in the relationship formed among the members.

Information technology (IT) includes all areas of computer science: hardware, software, telecommunication, infrastructure, and the Internet (providers and facilities). The study of virtual teams is interesting because of (1) the use of communication tools, that is, the technical aspect; (2) multinational integration, the human aspect; and (3) its dynamics that require a project management directed to cost restriction, deadlines, and quality in terms of organizational aspect.

As mentioned before, the IT sector is very unique in the way it plays both active and passive roles in the virtual interaction. Thus, the virtual team study provides an important opportunity to analyze their project management operation. In light of Lau, Sarker, and Sahay (1999), who figured out a model to assess virtual teams, effectiveness of communication is the key to their success. They pinpointed two interrelated and present dimensions of communication: the social, and the task.

While the social dimension provides the base to the members to communicate among themselves, the task dimension focuses on how well the work will be accomplished and that the deadline will be met. On the other hand, Duarte and Snyder (2001) point out the following essential factors for a successful interaction among virtual teams: human resources, training and development for the team leaders and members, standardized organizational processes, communication technology and electronic collaboration, trusting organizational culture, leadership, and leader and members competence.

Having all those aspects, which involve international projects, in mind, we have come up with the scenario focused on in this chapter through the following question (Hassegawa, 2002): "How are the technical, human, and organizational interfaces aspects associated to success or failure in international projects of Information Technology involving virtual teams?"

Thus, the technical, human, or organizational aspects frame the center of interfaces analyzed in this chapter. To better understand this term, PMBOK (2000) says the project interfaces are normally in one of the following categories:

- Organizational interfaces: formal or informal relationship among the different organizational units. The organizational interfaces can be very simple, or highly complex. For example, the development of a complex system of telecommunication may demand a coordination of those innumerous contracts during many years. However, the correction of an error in a program installed in one place may demand a little more than a simple note to the user and the production members.
- Technical interfaces: formal or informal relationship among different technical subjects. The technical interfaces happen either within the phases of a project (for instance, the drawing designed by the civil engineers must be compatible to the structure designed by the structural engineers) or among the phases of a project (for example, when a team of automobile designers pass on the results of their work to the engineers who will manufacture the vehicle).
- Interpersonal interfaces: formal or informal relationship among the members who work on the project.

Naturally, these are not the only aspects responsible for the success of a project, as we might as well consider its size or length and complexity. Regarding the length of a project, Cleland and Ireland (2000: 3.17-3.18) state: "characteristics for a small project include a single objective, one decision maker, easily defined scope and definition, an available funding, and a small team that performs the work on the project."

Sbragia (2001), however, in his studies of the complexity of a project for developing new products in the context of a Matrix structure quoted: "The term complexity here refers to the complexity experimented by the project manager on the task of managing the project, and to make it functional to the explanations on this paper. It is defined by the following indicators:

- Number of functional areas involved on the project;
- Integration intensity among the members from different functional areas;
- Cooperation difficulties among the areas engaged on the project.

Therefore, as displayed in Figure 13.1, the organizational, technical, and human interfaces are factors potentially linked to the success or failure of the international projects of IT where virtual teams take part. Different levels of length and complexity involved must be considered.

The interfaces correspond to the aspects of communication in association with different levels of relationship among the members. In that sense, the organizational interface is the link between two organizations—local and remote. The technical interface is linked to communication among the local and remote technical team on the project; and the human interface associates the personal communication in connection with the remote or local members working on the project.

Figure 13.1
Conceptual Model of the Study

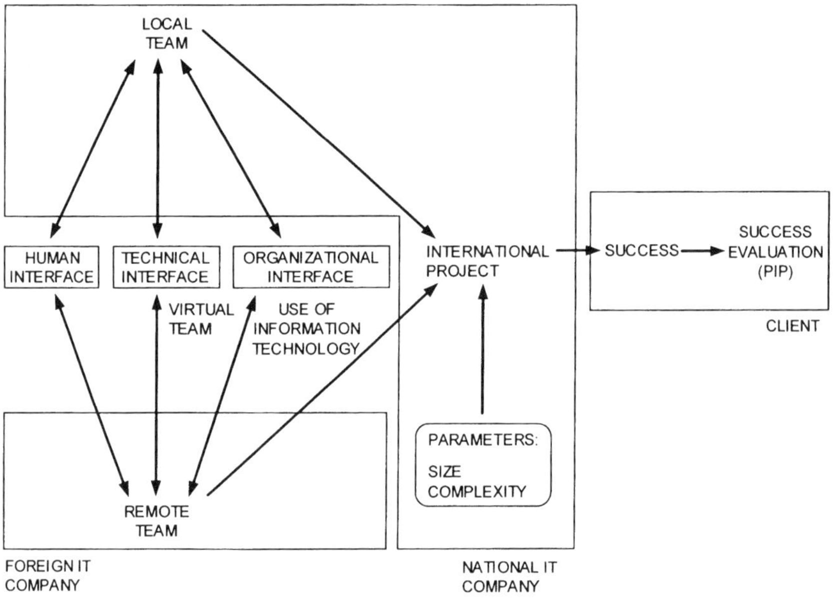

In Figure 13.1 IT is used either in national companies or in foreign ones as we assume both companies have IT as a main business. As mentioned before, a virtual team in IT is important because they are both at the same time pioneers who generate and use such high technology. In this way, it becomes a reference to other companies that work in different areas also using IT.

It is believed that the IT nature of innovation represents a tendency to be followed by the Brazilian economy. The results obtained in this chapter may be helpful and a reference to other fields that are gradually incorporating virtual teams in their methodology and work.

For this purpose, this chapter may contribute recommendations that can increase the chances for success in international projects once the key factors involved on the real and virtual teams are comprehensible. Any project as a whole is engaged with a big quantity of resources—people and time—that cannot be wasted. This is even more so when the project goes over international boundaries and gathers different companies. New problems are triggered off, consequently new methods are needed to solve those difficulties.

METHODOLOGY

Methodology Approach

Finished projects were analyzed in order to make a clearer study of their success or failure. The strategy used is to confront data referring to success and failure projects according to the relevant members of the teams and the users' (judges') feedback. The method used was Project Implementation Profile (PIP) (Pinto and Slevin, 1988). PIP is a systematic method that evaluates success. It has been used in over 400 projects; therefore, it was possible to come up with a standardized scale. Successful and failed projects were selected and put into two groups to be analyzed in terms of organizational, technical, and human interfaces, under different conditions of size and complexity. Successful cases were followed; failed cases were not followed.

Defining the Variables

Based on the literature read beforehand, there are 18 points grouped in human, technical and organizational interfaces:

Human Interface

- Human Resource Policy: Known as actions of the department that supports the local team; it also should stimulate and reward members who work with virtual teams.
- Project Leadership: The role of the leader is to stimulate integration and be supportive. Establish clear procedures and goals, allocating resources and modeling desired behavior. Notice that each team evaluates its leadership.
- Skills of Leaders and Members: Technical training is required to use the equipment required, use the technology properly to communicate, and be aware of the cultural differences.
- Team Atmosphere: There must be a feeling of solidarity and comprehensiveness between the local team to the virtual team.
- Mutual Trust: There should be a good level of trust between the local members to the virtual ones, in terms of empathy, honesty, and ethics.
- Organizational Culture: Cultural aspects from the virtual team should be considered as capacity for organization, acceptance for other cultures/values, respect for the rules, facility to empathize.

Technical Interface

- Team Training: Training can be via courses or pilot projects.
- Client Involvement: Must attend and support the client technically.
- Technology of Communication Available: There should be electronic communication and such technology for the virtual teams' projects.

- Technical Detail Planning: Level of planning and follow-ups to be focused on technical aspects.
- Ways of Communicating: All members must have access to different ways of communication, such as e-mail, telephone, video conference, fax, and so on.
- Technical Process: There should be some standardized technical process and similarities between those standards, as documents and files to be exchanged.

Organizational Interface

- Standardized Organizational Process: There must be compatibility on the organizational rules between the two teams, including standard schedule and budget.
- Organizational Structure: This is the measurement on how much the organizational structure helped the local and remote interaction; it shows if organizational conflicts occurred.
- Congruence of Objectives: Similarities among the objectives and both local and remote teams should prioritize the needs of each other.
- Overall Project Planning: There must be project planning with a budget and deadline clear to everyone.
- Top Management Support: This should prioritize the virtual team in order to have a flow.
- Presence of Feedback: Done by the virtual and local team, always checking budget, deadline, and technical details planning.

Two factors were considered as *parameters* in this study: size and complexity. The main definitions are presented below:

1. *Size*: defined by the number of people involved. We set a limit up to five people to consider it a small size. Over that, it is considered large.
2. *Complexity*: defined by the number of relations among the people involved on the virtual team. We set a limit of 10 or more interface relationships in the working team (the relationship sum between local and virtual members) to define high complexity.

Sample

A survey was made using many sources such as computer science yearbooks, specialists, journals, and professional contacts. The companies had to have the following characteristics:

- They had to be IT companies.
- They had to be located in Brazil.
- They had to be used to operating international projects.
- They had to use virtual teams.

So the sample was not probabilistic, once we had a few companies willing to take part on the research. Six companies were chosen:

- *Procomp Electronic Industry*: Founded in 1985 by four engineers, it works with banking automation. Procomp makes and sells solutions to commercial and electoral automation, in addition to computers, servers, software, and technical services. It has been in the market for 15 years and it has around 3,200 employees. It has more than 150 centers of technical assistance all over Brazil, attending 2,300 cities. Its gross sales reached US$615 million (2000). Its stock holding is with Diebold (a U.S. company). This is how the international distribution channels were opened. It was ranked 35th out of 200 largest private companies (*INFO EXAME*, 2001).

- *EverSystems*: It works with implantation and support of e-banking. It supplies software to Internet banking transactions, and recently to cell phones (WAP technology). EverSystems has around 80% of the market share. It is the leader in customizing technical solutions for Internet home/office banking. It has more than 30 successful projects over 25 clients and one installed base is used by 950,000 users. Its gross sales in 2000 were US$43 million (estimate). It has around 250 employees. It was ranked 127th out of the 200 largest companies in Brazil.
- *Digitron da Amazônia*: This company produces computer boards and motherboards. It is located in Manaus and has a branch in São Paulo. It has more than 150 employees and works with computer system integrators.
- *SPARK da Amazônia*: It makes keyboards for computers. It used to supply conventional keyboards for computers and specialized projects to commercial and banking automation. It had 35 employees during the period the project was being installed. After this project the company was split into two other companies (FABOR and MDA).
- *B&T Business*: Imports and distributes products for computer science. It has five employees. It also works with special projects of banking and corporative automation.

The projects completed by each company at the time of writing were:

- *Checkbook dispenser (DIDO)* is a project to produce some equipment for banking automation in which it has to manufacture checkbooks at ATM machines. Procomp (supplier for the banking automation) contacted the Italian CTS to supply the electronic equipment and the result was a transfer of technology from Italy to Brazil.
- *Health care terminal (TAS)* is part of a big project completed by the Health Ministry to implant a new system. It had as its objective the creation of electronic equipment to operate as a terminal to consult and validate the use of national health cards. The idea was to supply magnetic cards to identify the users in clinics and hospitals and book their exams and appointments. The role of Procomp was to supply the equipment designed and executed by First International Computer (FIC) in Taiwan. This project was based on the implementation of the terminal for a motherboard.
- *Citimail Venezuela (Citibank)* originated by Citibank in Venezuela is different from the others because the local company is based abroad and the remote company is based in Brazil. EverSystems works in Venezuela through a subsidiary. Many solutions are based on similar projects finished in Brazil such as e-mail banking (you receive your bank report and other information through e-mails). This project was featured by the transference of software technology from Brazil EverSystems to Venezuela EverSystems.
- *PC popular (PC POP)*, originated by Cybermax, looked forward to lowering costs for motherboards to accompany cheaper PCs. Digitron, the company that manufactures computer boards, had the technology to start a special project to cover the demand, because the project was simplified by a former one, and most components were implemented by Shuttle from Taiwan.

- *Reduced keyboard*: Procomp started the project to produce a smaller keyboard, similar to laptops, to be used in a special project. SPARK da Amazônia, the traditional keyboard maker, realized the necessity of this project through its supplier, Acekey. This project was also executed by engineers' services by order.
- *Electronic lock (Sargent)*: Procomp started the project because of the need to change ATM mechanical lockers to electronic ones (activated by codes instead of mechanical ones). The project was ordered by B&T, specializing in electronic products. The project was implemented by Sargent and Greenleaf. It also worked by orders.

Table 13.1 summarizes the main information about the projects analyzed, including the client, local, and remote companies involved; number of judges that evaluate their own level of performances; and the number of respondents of the survey about the variable referents to the interfaces.

Data Collection Procedures

Questionnaires for the collection of data were sent to people involved in the projects. Two stages of data collection were carried out, the first one to survey the dependent variable, through questionnaires (mostly like the PIP method) sent to three judges per project (for a total of 18 judges—1 customer, 1 project manager, and 1 member of the high administration at the vice president level). The answers were tabulated and from the data we could divide the projects in groups of bigger success and minor success, called groups of success and failure. Following that, questionnaires were distributed among involved people in the projects for survey of the independent variable, including project managers and technicians in the local and remote companies. A total of 30 respondents supplied data about their relative perceptions to the 18 aspects of interface that had been analyzed.

DATA ANALYSIS

Dependent Variable: The Success of the Projects

The dependent variable (success or failure) was obtained by the standard PIP method. Table 13.2 shows responses from three judges for each project. There were 12 PIP questions; the number used was taken from the Likert scale, up to number 7. The sums are totaled separately. Questions 1 to 5 are about the flow of the project; questions 6 to 12 are about how much the results have been used by the clients.

To obtain an added value for each project, the average of the sums by each judge was calculated. In that sense we had added values for project, client, and total for each project. Next, we can verify using the PIP scale, if we compare with answers of those 400 projects considered as references that the 50% worst ones were scored for project under 29, and clients under 39. The total sum equals 69.

Table 13.1
Profile of the Projects Considered in the Study

Project		DIDO	TAS	Citibank	PC POP	Keyboard	Sargent
Client		Bradesco	Health Min.	Citibank Ven.	Cybermax	Procomp	Procomp
Beginning of the project		na	May/2000	Nov/2000	Aug/2001	Jun/2000	May/2001
End of the project		Jul/2001	Jun/2001	Jul/2001	Jan/2002	Nov/2001	Jan/2002
Project budget (US$, estimate)		500000	150000	100000	100000	55000	80000
Local companies		Procomp	Procomp	EverSystems	Digitron	SPARK	B&T
	Country of the local companies	Brazil	Brazil	Venezuela	Brazil	Brazil	Brazil
Number of employees at the local company		3200	3200	na	150	35	5
Local team size		12	7	4	4	6	5
	Local team interacting with remote	5	5	4	4	4	3
Remote company		CTS	FIC	EverSystems	Shuttle	Acekey	Sargent
	Country of the remote company	Italy	Taiwan	Brazil	Taiwan	Taiwan	USA
Total employees in the remote company		150	na	250	870	120	55
Remote team size		12	5	2	5	7	5
	Remote team interacting with locals	5	2	2	2	2	2
Sum of remote interaction		12	10	16	6	7	6
Average of hours/week/pers spent in interaction		6.7	12.6	5.6	20	13.6	12.2
Judge composition (total)		3	3	3	3	3	3
	Directors			1	1	1	2
	Managers	3	1	2	2	1	1
	Others		2			1	
Respondent composition (total)		5	4	6	6	6	3
	Directors			1	1	1	2
	Managers	2	2	3	3	1	1
	Coordinators		2				
	Engineers	2			2	3	
	Analysts	1		2			
	Others		1				

Note: The sum of remote interactions refers to the number of interactions each local team member declared to have with remote counterparts, totaled for both remote and local teams. It could be more than the sum of the people involved because each local member could have interactions with more than one remote person, so the maximum sum would be twice the product of local and remote counterparts (each interaction declared by each side).

Table 13.2
Data Related to the Degree of Success of the Projects Studied (results of PIP method application)

Project	Jobs	Judges	1	2	3	4	5	6	7	8	9	10	11	12	Project	Average	Client	Average	Total	Average
DIDO	C	A1	3	5	6	5	6	6	4	3	2	6	7	7	25		35		60	
DIDO	TM	A2	2	2	6	6	4	6	4	3	1	2	2	2	20		20		40	
DIDO	M	A3	2	3	6	6	5	6	5	4	6	6	6	6	22	22.3	39	31.3	61	53.7
TAS	C	B1	2	4	6	7	7	7	6	5	3	7	7	6	26		41		67	
TAS	TM	B2	1	5	7	6	6	6	7	4	5	5	7	6	25		40		65	
TAS	M	B3	6	3	6	5	5	3	6	3	6	6	6	6	25	25.3	36	**39**	61	64.3
Citibank	C	C1	2	2	7	6	2	7	7	3	7	5	5	6	19		40		59	
Citibank	TM	C2	3	7	7	6	7	7	7	6	5	6	7	7	30		45		75	
Citibank	M	C3	2	2	7	6	6	7	7	6	7	5	5	6	23	24	43	**42.7**	66	66.7
PC pop	C	D1	6	6	7	5	6	5	5	6	6	5	6	5	30		38		68	
PC pop	TM	D2	7	6	6	6	5	6	6	7	6	6	6	6	30		43		73	
PC pop	M	D3	6	6	6	4	6	4	4	4	6	7	4	7	28	**29.3**	36	**39**	64	68.3
Keyboard	C	E1	5	3	5	6	4	7	7	4	6	6	6	6	23		42		65	
Keyboard	TM	E2	6	5	7	6	6	6	6	5	6	6	6	6	30		41		71	
Keyboard	M	E3	3	5	7	7	6	7	6	7	6	7	7	6	28	27	46	**43**	74	**70**
Sargent	C	F1	5	5	6	6	6	6	7	5	6	6	6	5	28		41		69	
Sargent	TM	F2	6	4	6	6	6	7	6	5	4	6	6	6	28		40		68	
Sargent	M	F3	5	5	6	6	7	7	7	7	5	7	7	7	29	28.3	47	**42.7**	76	**71**
									Use of PIP Method						100%	35		49		84
															90%	34		46		79
	Jobs positions:														80%	32		44		76
	C: Customer														70%	31		42		73
	M: Manager														60%	31		41		71
	TM: Top Management														**50%**	**29**		**39**		**69**
															40%	28		38		66
															30%	26		35		63
															20%	24		33		59
															10%	22		31		53
															0%	5		13		21

From the added value calculated for each project, we can see the project DIDO can be easily classified under 50% (failure); TAS, although it scored 39, loses its score in total and project, being included in failure projects, as is the Citibank project for the same reason. The Sargent and Keyboard projects were included in successful because they had good scores in total and client, and scores close to the cut in project. The project PC POP was included in successful because it went over the minimum score for client and project and has the score close to the total.

Analysis of Relationship Between the Interfaces and Project Success

Let us recall the question asked in the beginning, "How are the technical, human, and organizational interfaces aspects associated to success or failure in international projects of Information Technology involving virtual teams?" To examine this question we used nonparametric statistics as in Kolmogorov-Smirnov's test (Siegel, 1974), which verifies if the samples come from populations with the same distribution. If we establish a division between the groups of more or less success and apply such a test to each interface, we can verify how each interface differs from the groups of success and failure. Then we can associate them to the project of most success the aspects such as client involvement, technical details planning, and overall project planning.

The Kolmogorov-Smirnov test might be a little more efficient for small samples. For larger samples the Mann-Whitney test is more efficient (Siegel, 1974). The procedure was to repeat the test applying the unilateral Mann-Whitney test. As result, we reached the following results at the 5% significance level: human resource policy, standardized organizational process, congruence of objectives, team training, technical details planning, project leadership, client involvement, and overall project planning.

When comparing the results from both tests, all the items indicated to be significant in Kolmogorov-Smirnov's and Mann-Whitney's tests, but on the last one there were three more significant items: human resources policy, project leadership, and congruence of objectives. For our analysis, therefore, we consider Mann-Whitney's test as the most wide-ranging and detailed because we can distinguish the meaning of each item to our chapter better. A summary can be seen in the third column of Table 13.3.

We can determine the essential features of the relative order of association of aspects in one given interface with the success of the project, depending on the level of significance of each variable. In this way, in human interfaces we see the role of project leadership as highly meaningful, while human resources policy has a minor significance. Among those technical interfaces the most important and significant aspects were client involvement followed by team training and technical details planning. Regarding the organizational interfaces, the most associated aspect to success was overall project planning followed by standardized organizational process, with congruence of objectives being the least significant.

Table 13.3
Significance Table for the Relationship Between Interface Elements and Project Success

		All Projects	Project Size		Project Complexity	
	HUMAN INTERFACE	**Overall**	**Bigger**	**Smaller**	**Higher**	**Lower**
1	Human Resource Policy	S	NS	NS	NS	NS
2	Project Leadership	S	NS	NS	NS	NS
3	Skills of Leaders and Members	NS	NS	NS	NS	NS
4	Team Atmosphere	NS	S	NS	S	NS
5	Mutual Trust	NS	S	NS	S	NS
6	Organizational Culture	NS	S	NS	S	NS
	TECHNICAL INTERFACE					
7	Team Training	S	NS	S	NS	NS
8	Client Involvement	S	NS	NS	NS	NS
9	Technology of Communic. Available	NS	NS	NS	NS	NS
10	Technical Detail Planning	S	S	S	S	NS
11	Ways of Communicating	NS	NS	NS	NS	NS
12	Technical Processes	NS	S	NS	S	NS
	ORGANIZATIONAL INTERFACE					
13	Standard Organizational Process	S	S	NS	NS	NS
14	Organizational Structure	NS	S	NS	S	NS
15	Common Objectives	S	NS	NS	NS	NS
16	Overall Project Planning	S	S	S	NS	NS
17	Top Management Support	NS	S	NS	S	NS
18	Presence of Feedback	NS	S	NS	S	NS

NS = Not Significant
S = Significant ($p<.05$)

According to what was analyzed before, considering the large number of significant items, we should interpret the technical interface as the most important followed by the organizational interface, and finally the human interface being the least important. What is striking is the fact that a majority of recent authors about virtual teams considers the aspects of human interface such as mutual trust, team members competence, and fundamental to the good flow of the projects. Instead, the traditional aspects—such as technical details planning, client involvement, training and overall project planning—were considered the most important in this research, being those items not so important to most authors who studied virtual teams.

One way to attempt to grasp such facts is the emphasis given to each work considered. This chapter focused on an international project that used virtual teams as a way to accomplishment its goals. In other words we are interested in the success of the project as a whole. Thus the critical factors most associated with the projects tend to be more classical. The studies about virtual teams done by many authors focus on remote integration on the virtual team and they set a base in which the discipline of the project management was a consolidated fact, though not paying attention to the classical aspects.

Some other considerations are important. In addition to what we have already mentioned, that all the companies chosen are suppliers for IT, what made it easier for the interaction was that the companies already had former projects, so this might have minimized the aspects of culture and trustworthiness (once they had it established beforehand). In this way the study measured more the realizations of a new project in an already known atmosphere. Another relevant fact is to have studied finished projects that do not really show their failures at the time the study took place.

What was investigated were the projects of most and least successfulness. However, in reality, none of them had a real failure. Therefore, one must be cautious when looking over the results presented in this chapter.

Complementary Analysis: The Role of Size and Complexity

In this section we will consider the action of possible variables such as size and complexity of the project, in relation to the interfaces and success. The method used to evaluate what happens with the size over the relationships among interfaces and the success of the project is based on the very same process used in the analysis of the questionnaire applied for this research.

Six projects were divided into two groups. Each of these were again divided into two other groups of most successful (2 projects) and least successful (1 project). Then the Mann-Whitney was applied to verify the association between each interface aspect. In Table 13.3 we have a summary of the Mann-Whitney test to relate it to the interface project success based on project size: big size (second column) and small size (third column).

We can first notice that many interface aspects have their significance differentiated because of size. This allows us to state that size affects the association among interfaces and success. About the relative importance of interfaces we can see that the technical interface was the most important one followed by organizational and then human (as previously discussed from a visual inspection of the most important facts in each interface group). In the case of large size, the organizational is the most important, followed by human and then technical interface.

It is essential to mention that when the number of members in the teams increases, the activities for organization (as overall project planning and top management support) as the ones related to humans (mutual trust and team spirit) turn out to be the most important ones. It shows that depending on the size of the team, different priorities must be taken by the project manager in order to be successful. Again, about small projects, we can notice that the order of importance of the factors changes once again. Technical interface is the most important, followed by the organizational interface. Notice that none of the human interfaces are associated with success. One possible interpretation is the fact that only one person in small project teams is the contact between the local and the remote teams, making factors linked to human and organizational interface not as relevant.

In order to analyze the complexity effect on the relation between the interfaces and success of the project, the very same process previously described was used. However, this time complexity was used as a discriminating factor. Table 13.3 shows a summary for the results of Mann-Whitney test for the interfaces of the project success when talking about complexity: high complexity (fourth column) and low complexity (fifth column). Again, we notice that the effect of complexity alters the association with the interfaces and success. But it is totally different if compared to size. The relative order of interfaces when the project involves high complexity is human followed by organizational and then technical. In this case, the analysis is similar to the big projects. It is associated with the increased number of human interfaces, and because of that it is associated with aspects of organizational interface.

On the other hand, projects of small complexity present a striking fact: No interface could be directly associated with the success. This might be because other variables did not take part in those interfaces, such as monetary correction or logistics. Therefore, are essentially decisive to the projects' success.

CONCLUSIONS AND RECOMMENDATIONS

This chapter has analyzed the role of human, technical, and organizational interfaces on the success of IT projects using virtual teams. A group of six finished international projects, carried out by different companies, was analyzed, thru data gathered by questionnaires answered by team members. In order to be analyzed, the projects were divided into two groups of success and failure, by their ranks assigned by judges, following the PIP method, and later regrouped by size and complexity.

Besides the small number of cases—and the natural limitations of this type of study, such as low representativeness of the sample, due to intentional sampling and small control over answers from people—affecting the data reliability, some conclusions can be formulated.

The first conclusion is related to the variables most associated with the projects. From the results of the analysis we can say that the most associated aspects to the success are (significant at 5%):

Human Interfaces	*Technical Interfaces*	*Organizational Interfaces*
• human resource policy • project leadership	• team training • client involvement • technical detail planning	• standardized organizational process • congruence of objectives • overall project planning

We cannot attribute to those variables some influence from any other similar projects because of the restrictions given to the samples. Therefore, we can state that such aspects were revealed to be critical factors of success in this chapter. It was noticed that such aspects of interface most significantly associated with success are common to classical projects (i.e., nonvirtual teams). In this way, aspects considered important when using virtual team references, such as communication technology and ways of interaction, did not seem to be relevant in the analysis. That may be because all the companies chosen were IT companies.

Lau, Sarker, and Sahay (1999) tell us: "To be effective, virtual teams need to pay attention to the social and task dimensions of communication and such mediating factors as technology use, time and space, and communication pattern. We believe ideal teams are the most successful as they are effective in dealing with both dimensions of communication."

The second conclusion gives reason to the first one. The results suggest that we should consider the technical interface as most associated with success, followed by organizational interface and, finally, human interface, contradicting what most authors writing on virtual teams say: that the human interface, mutual trust, and competence among the team is fundamental for the good flow among the projects.

Duarte and Snyder (2001), state: "There are seven critical success factors for virtual teams, of which technology is only one. Others are human resource policies, training and development for team leaders and team members, standard organizational and team processes, organizational culture, leadership, and leader and member competencies."

Pinto and Kharbanda (1995) point to the same fact:

> The first study, conducted by Peter Morris (1983), was based on a large sample of British firms and their efforts at implementing new technologies. ... a point to note is that behavioral and organizational factors far outweigh technical issues in terms of their importance for implementing success. ... A separate study by Baker, Murphy and Fisher (1983) addressed a comprehensive list of those factors thought to contribute to perceived project implementation success. Their list also included a number of behavioral dimensions ..., as well as organizational issues ..., and technical factors. ... The behavioral and organizational factors are clearly most important."

Moreover, traditional aspects like technical detail planning, client involvement, and overall project planning were considered mostly present in the research.

One way to try to understand such facts is the emphasis given to each project considered. Our studies focused on results of an international project where they used the virtual team as a way to get the work done. In other words, the success of the project is seen as a whole. Thus, the factors most associated to the project tend to be the classical ones. The studies about virtual teams have their focus toward the virtual team remote integration and they presume the discipline on the project management is a consolidated fact.

The third conclusion is about the intervening variables that might occur. It was verified that size and complexity affect the association between the interface aspects and success. Even though there were some limitations in the method of the analysis—which decreased the number of samples—for the total of projects the technical interface was shown as the most important. For large projects, the organizational interface was the most important, while for the small projects the technical interface was the most important to be successful. At this point, it is important to say that the increase in the number of people on the team, as the activities for organization (overall project planning, and top management support) and the ones linked to human aspects (mutual trust, and team spirit) are highlighted. It is an interesting result because it shows that depending on the size of the team, different priorities must be taken by the project manager to reach success.

In the case of projects of high complexity, the human interface was the most important, while for projects of smaller complexity, no interface was associated with success.

Wideman (1992) defines complexity as "Project complexity: The extent to which a project, or one of its components, involves large number of parts, and/or a large number of people, to be coordinated and/or interfaced. In project management, project complexity is typically a reflection of the number of work packages involved and the number of different people required to carry them out."

One possible interpretation would be that only one person is the connection between the local and the remote team. In this sense, neither of the aspects of human or organizational interfaces were so relevant.

Additional considerations should be mentioned. Although all companies studied were IT suppliers, what made the ways of interaction easier was that most of them had already made other projects together. So the aspects of culture and trust were minimized, once they had been established in the past. Having this in mind, the research could experiment with the realization of the projects in environments where local-remote interaction had already taken place.

Applegate, McFarlan, and McKenney (1996) point to this fact:

> A company's technology base is another important factor. High-technology companies with traditions of spearheading technical change from a central research and engineering laboratory and disseminating it around the world have successfully used a similar approach with IT. Their transnational managers are used to corporately initiated technical change. Firms without this experience have had more difficulty assimilating information technology in general, as well as more problems in transplanting IT developed in one location to other settings.

More adequate research should probably be considered to study companies where there have never been any remote interactions before, as it may have a "learning effect" (where the former projects were examples to be followed to avoid problems between the remote and virtual team).

Another relevant fact is to have studied a project that was already over, because in reality it does not really show any failure. Concluding from researches, there were projects of more and less success, but none of them really failed. This must be cautiously considered when reading this chapter.

Finally, we should mention the limitations and restrictions about this research. Because of the use of one sample with reduced dimensions, we can neither make an inference to other international projects nor generalize it to other kinds of companies or even to the same type. Moreover, this chapter tries to evaluate the associations between the interfaces and the successful projects without trying to set up the casualties between them. Second, the way the data were collected is also limited in the sense we did not exercise any control about the respondents, so the reliability of the data may be seen within some restrictions. Finally, more wide-ranging and strict studies will be necessary in the future to make inferences about the results obtained here, particularly considering companies in different businesses, starting virtual teams, and so on.

REFERENCES

Allen, T. J. (1977). *Managing the Flow of Technology: Technology Transfer and the Dissemination of Technological Information within the R&D Organization.* Cambridge, MA: MIT Press.

Applegate, L. M., McFarlan, F. W., and McKenney, J. (1996). *Corporate Information Systems Management: Text and Cases*, 4th ed. Chicago: Irwin.

Cleland, D. I. and Ireland, L. R. (2000). *Project Manager's Portable Handbook.* New York: McGraw Hill.

Duarte, D. and Snyder, N. T. (2001). *Mastering Virtual Teams: Strategies, Tools, and Techniques that Succeed*, 2nd ed. San Francisco: Jossey-Bass.

Gould, D. (1997). *Leading Virtual Teams*, http://www.seanet.com/~daveg/ltv.htm

Hassegawa, L. N. (2002). "O papel das interfaces no sucesso de projetos utilizando equipes virtuais" (The Role of Interfaces for the Success of Projects using Virtual Teams,) Master Degree Dissertation, Department of Administration, FEA/USP, São Paulo.

INFO EXAME (2001). Special edition, August.

Lau, F., Sarker, S., and Sahay, S. (1999). "On Managing Virtual Teams" (March), http://www.bus.ualberta.ca/flau/Papers/cacm.htm.

Lipnack, J. and Stamps, J. (1997). *Virtual Teams: Reaching Across Space, Time, and Organizations with Technology.* New York: John Wiley.

Pinto, J. K. and Kharbanda, O. P. (1995). *Successful Project Managers: Leading Your Team to Success.* New York: Van Nostrand Reinhold.

Pinto, J. K. and Slevin, D. P. (1988). "Project Success: Definitions and Measurement Techniques," *Project Management Journal*, 19 (1).

PMBOK (2000). *A Guide to the Project Management Body of Knowledge (PMBOK® Guide) 2000 Edition.* Newtown Square, PA: Project Management Institute.

Sbragia, R. (2001). "The Interface between Project and Functional Managers in Matrix Organized NPD Projects." In T. M. Khalil, L. A. Lefebvre, and R. M. Mason (eds.), *Management of Technology: The Key to Prosperity in the Third Millennium.* Oxford: Pergamon/Elsevier, pp. 427-440.

Siegel, S. F. (1974). *Estatística Não-Paramétrica para as Ciências do Comportamento* (Nonparametric Statistics for the Behaviour Sciences.) São Paulo: McGraw Hill.

Wideman, R. Max (1992). *Risk Management Handbook.* Newtown Square, PA: Project Management Institute. http://www.pmforum.org/library/glossary/PMG_P09.htm

PART IV: MEASURING AND MODELING FOR IMPROVED UNDERSTANDING

14

Does Income Distribution Affect Innovation?

Apiwat Ratanawaraha

INTRODUCTION

Income distribution and technological innovations are two of the most important issues in the field of economic development. Many analysts have proposed economic theories and conducted empirical studies on both issues, particularly with regard to their relationships with economic growth. Income distribution plays a crucial role in many of the development models, from those of earlier economists, such as Simon Kuznets (1955) and Nicholas Kaldor (1957), to the more recent ones, such as Galor and Zeira (1993) and Alesina and Rodrik (1994). It is now widely accepted that income distribution affects economic development through various channels, such as imperfect credit markets and redistributive policy.

Technological innovations, on the other hand, have long been considered as the engine of economic development. Although the reference to technological progress is implicit in the writings of earlier economists, such as Adam Smith and Karl Marx, it was Joseph Schumpeter (1934) who first explicitly regarded innovation as the key determinant of economic growth. Since then, many economists have incorporated technological innovation into their analyses of economic growth, particularly in the recent literature of endogenous-growth models (e.g., Arrow, 1962; Romer, 1990; Grossman and Helpman, 1991).

If income distribution indeed affects the economic development process, and innovation plays a crucial role in economic development, then a causal relationship is possible between income distribution and technological innovation. Given the substantial amount of interest in both issues, it is surprising that few analysts have proposed theoretical models that explain the causal relationships between income distribution and innovation, let alone conducted an empirical study to test their relationships. Even among the more recent analyses of the issue, most tend to focus on the impact of technological innovation on income distribution (e.g., Aghion, 2002). Little attention has been paid to the other possible causal direction, that is, the effect of income distribution on innovation. In fact, to the best of our knowledge, there is no empirical study that investigates the issue of how income distribution affects technological innovations.

In light of the foregoing observation, this chapter tests empirically the causal relationship between income distribution and technological innovation. Specifically, the chapter examines whether income distribution affects the amount of investment on innovative activities, innovation outputs, and innovation productivity, and whether the effects are the same between developed and developing countries.

The chapter begins with a review of the literature and outlines the possible mechanisms through which income distribution may affect innovation. Then the descriptive statistics and the research design for multivariate regression analysis are presented. The estimated results are reported and discussed, and concluding comments are made.

LITERATURE REVIEW

This section brings together two strands of literature in conceptualizing the effects of income distribution on innovation. The first is the literature on the effects of income distribution on demand composition and market size. The second is on effects of income distribution on human capital accumulation under credit market imperfections.

Demand Composition, Market Size, and Profits

First, the literature on the effects of income distribution on demand composition suggests that income distribution affects market sizes for domestic manufacturers in the process of industrialization (e.g., Murphy et al., 1989; Baland and Ray, 1991), and that extreme concentration of wealth among the rich results in the demand for imported luxury goods rather than for domestic manufactures (Baldwin, 1956; North, 1959). Similarly, demand composition is considered in some theoretical models as the channel through which income distribution determines the rate of innovation (e.g., Falkinger, 1994; Zweimuller, 2000).

These studies base the argument on Engel's law, which states that consumer demand and preferences tend to be nonhomothetic—that is, the share of budget allocation depends on income level. When income rises, the share of basic items in total consumption tends to fall, and people demand goods of higher quality and sophistication. Unequal income distribution means that there are not many rich and middle-class people who can afford innovative manufactures, thereby affecting the level of current and potential demand and the market size for innovations. On the other hand, innovations require high front-end investment, while profits are the key stimulus to, and source of funding for, innovation activity. Unequal income distribution therefore reduces the expected profitability of innovations due to small market sizes, thereby reducing the entrepreneurs' incentives to invest in innovative activity.

Human Capital Accumulation

The other strand of literature suggests that, when credit markets are imperfect and a fixed up-front investment is required for human capital accumulation, income distribution affects human capital accumulation (e.g., Galor and Zeira, 1993; Flug et al., 1998). This statement is based on the observation that an individual's collateral, that is, his existing wealth, determines the degree to which he can access the credit markets. In an unequal society in which many people do not have access to credit markets, poor people cannot accumulate enough wealth to cover the fixed cost of human capital. On the other hand, analysts have shown that the quantity and quality of human capital affect innovation (Nickell and Nicolitsas, 1997; Roy, 1997). Linking these two arguments together, we can argue that unequal income distribution affects innovation, because it affects the accumulation process of knowledge and human capital. Under imperfect credit markets, income inequality limits the access to higher education, which requires a fixed up-front investment. When there are no credit markets for education, poor people do not have access to the education to equip them with higher-level skills.

Similarly, income inequality affects individual incentives to specialize (Fishman and Simhon, 2002). While specialization requires the investment of real resources, the incentives to invest in specialization depend on the existing wealth, because of borrowing constraints. Income inequality therefore determines the number of individuals with access to capital for investing in specialization. As human capital in innovation particularly requires specialized skills and knowledge, income inequality thus affects the stock of human capital in innovation. In sum, we can argue that when the stock of human capital affects the rate of innovation, and when credit constraints limit the access to capital, the higher the inequality in a society, the fewer people with specialized skills and knowledge, thus the lower the rate of innovation.

In addition to the arguments above, two additional channels are possible for income distribution to affect innovation: namely, the effects of income distribution on the credits for investment on innovative activity, and the concentration of investment in noninnovative sectors.

Credits for Investment on Innovative Activity

As in the case of investing on human capital, under imperfect credit markets, minimum collateral is required of potential investors in innovation. Similarly, only people with large enough wealth have access to credits. Rich entrepreneurs could easily obtain loans from banks to fund innovative projects. In contrast, potential entrepreneurs with limited asset stocks can neither fund the innovative projects by themselves nor borrow from the credit markets because they lack collateral. Income distribution therefore affects innovation by affecting the number of innovators who have access to credits. Therefore, when income inequality is great, the overall investment in innovative activity and the rate of innovation are low.

Concentration of Investment in Noninnovative Sectors

Income inequality may also affect the level of domestic investment in innovative sectors. In other words, income inequality, together with the unequal rates of returns across sectors due to capital-market imperfections, could lead to the concentration of investment in the sectors that are not innovative. This implies income distribution affects the paths of industrial development, and subsequently the rate and direction of industrial innovation.

In a standard neoclassical investment model with the premise of profit maximization, the investment decision is based on the comparison between the present values of all costs and revenues. Corporate executives invest when the net present value of the project in question exceeds zero. When there is more than one project to compare, investors choose the project with the highest rate of return.

This principle also applies to the investment decision on innovative activity. Taking into account the opportunities and risks associated with the new investment, investors are faced with two investment options: either investing in the existing products and/or processes, or investing in innovative activity. If the rates of returns of the existing products are higher than the expected rates of the new products, entrepreneurs will not invest in innovative activity. They are already content with the profits from existing production. On the other hand, if these entrepreneurs see potentially high profits from innovations, they will invest in product and process improvement.

Income distribution can affect innovation, because it affects the expected rates of returns for innovation relative to the existing products and/or processes. Under imperfect capital markets, the rates of returns to investment vary across sectors. With highly skewed income distribution, the variance is even greater. Wealth is concentrated in the hands of the few rich, who invest in the sectors that would give them the highest returns. In countries with great income inequality, businesses are not very diversified and capital is concentrated in only a few sectors. This phenomenon occurs when the rates of return in existing businesses are higher than the expected rates of return in new businesses. This is particularly true when the existing industries are natural-resource-intensive and the new industries are manufacturing. How equally resources are divided within the primary sector affects the industrialization path and development strategy that a country may take. Income distribution thus influences the transition from primary product asset formation to knowledge-based asset formation.

Empirical evidence supports this argument. Leamer et al. (1999) find that countries with low Gini coefficients tend to have more employment in manufacturing than those countries with high Gini coefficients. They show that natural-resource-intensive sectors, particularly agriculture, absorb capital that might otherwise flow to manufacturing. This depresses workers' incentives to accumulate skills and delays industrialization. With the extreme concentration of wealth and imperfect capital markets, capital owners-cum-entrepreneurs are likely to choose natural-resource sectors over manufacturing, because the expected rates of returns are higher in natural-resource sectors. Latin American countries are good examples for this phenomenon.

A comparison of land distribution between East Asia and Latin America sheds more light on the issue. Unlike in Latin America, land distribution in East Asia is relatively equal, thanks to early land reforms. This implies that income inequality in East Asia is less likely to be rooted in the land-intensive, nonmanufacturing sector, and that nonmanufacturing sectors do not generate high Ricardian quasi-rents. The attractiveness of the manufacturing sector is thus high relative to the nonmanufacturing sectors, thereby encouraging the formation of national firms and core competencies in the manufacturing sector. Given that innovations occur more in the manufacturing industries than in the natural-resource-based industries, more investment in manufacturing industries is likely to result in more innovations. In sum, income inequality affects the path of industrial development and the subsequent industrial innovations.

EMPIRICAL ANALYSIS

Hypothesis and Regression Analysis

In this section, the effects of income distribution on innovation are empirically estimated. The main hypothesis is that countries with more equal income distribution are more innovative than those with less equal income distribution. The operational hypotheses are that countries with more equal income distribution: (1) spend more on innovative activity, (2) produce more innovative outputs, and (3) are more productive in creating innovations.

Multiple linear regression models are specified, employing the Ordinary Least Squares (OLS) estimation method. Several sets of models are specified according to the hypothesis to be tested. To examine the differences between developed and developing countries, each set contains three submodels that use three different sets of data: (1) for all 72 countries in the sample; (2) for 22 developed countries; and (3) for 50 developing countries (see Appendix for list of countries).

Variables and Data

The variables for regression analyses and the data are discussed below. The summary of variables and data sources is given in Table 14.1.

Dependent Variable 1: Expenditure on Innovative Activity

To test the impact of income distribution on the investment on innovative activity, R&D expenditures are used as the measure for innovation investment. This measure is widely used in many studies at the firm level (e.g., Scherer, 1965) and those at the country level (e.g., Amsden and Mourshed, 1997). One problem in using this measure in a cross-country study is that not all countries are like the United States, which requires that firms disclose their R&D expenditures. Some firms may want to keep their R&D expenditures confidential. Nevertheless, the measure still reflects the general level of innovation investment in a country.

In this chapter, two types of aggregate data are used: (1) gross domestic expenditures on R&D as the percentage of GDP (GERD); and (2) the private sector's expenditures on R&D as the percentage of GDP (PERD). The data are from United Nations Educational, Scientific, and Cultural Organization (UNESCO).

Dependent Variable 2: Innovation Output

To test the hypothesis that income distribution affects innovative outputs, the total number of patent applications in one year (PTENT) is used as the dependent variable. Admittedly, patents are not perfect as the measure for innovative outputs. First, not all innovations are patented and certain industries tend to patent more than others. Moreover, from an economic perspective, not all patents can be considered innovations, which are supposed to have some economic value. Most of the patents issued each year are worthless and never used (Shepherd, 1979). Second, there is a problem of "blocking," in which firms file numerous patents on variants of the original patent. They do so, not because these are new innovations, but because they could block a competitor's attempt to circumvent the original patent. Third, patent laws in different countries may affect cross-country comparisons. Some types of innovation, such as biological innovations, are not patentable in many countries.

A better measure of innovation output is innovation counts, which are the actual numbers of innovations produced by firms. In the United States, this information is available for each four-digit Standard Industrial Classification Code (SIC), available from the U.S. Small Business Administration. Unfortunately, the data are not available for other countries to conduct a cross-country comparison.

Table 14.1
Summary of Variables

Variable		Measure	Year	Source
Innovation				
PTENT	Level of innovative output	Total number of patent applications by residents	1999	WIPO
PTENTCAP	Productivity of innovative output	Total number of patent applications by residents per million inhabitants	1999	WIPO
GERD	Total investment in innovative activities	Gross expenditure on R&D as percentage of GDP	Various years (1990-1998)	UNESCO
PERD	Private investment in innovative activities	Private expenditure on R&D as percentage of GNP	Various years (1990-1998)	UNESCO
Income Distribution				
GINI	Income distribution	Gini coefficient in percentage	Various years (1990-1996)	UNDP
MID40	Middle-class share	Share of aggregate income of the 2nd, 3rd, and 4th quintiles	Various years (1990-1996)	World Bank
Income Level				
INCOME	Income level	GDP per capita (1,000 US$)	1998	World Bank
Market Size				
GDP	Economy size	GDP (100 billion US$)	1998	World Bank
POP	Population size	Population size (100 million)	1998	World Bank
POPDENSE	Population density	Population (per square kilometer)	1999	World Bank
Demand Growth				
GDPGRWTH	Economic growth	GDP growth	1990-1998	World Bank
INDGRWTH	Industrial growth	Average annul growth of industry value added	1990-1998	World Bank
CSMGRWTH	Demand growth	Average annual growth of household final consumption expenditure	1990-1998	World Bank
Credit-Market Imperfections				
CREDIT	Credit-market imperfection	Credits from the banking system to nonfinancial private sector as a share of GDP	1997	IMF
Human capital				
RDPERSON	Human capital stock for innovation	Number of personnel in R&D per 100 million	Various years (1990-1998)	UNESCO

Constrained by data availability, the number of patent applications filed by a country's residents is used as the proxy of innovative outputs. The total number of patent applications is more relevant to this chapter than the number of granted patents. The criteria and process of granting patents are independent of the mechanism through which income distribution affects investment incentives and innovative outputs. Patent applications, on the other hand, reflect innovators' desire to recoup their investment in producing the new products or processes.

Dependent Variable 3: Innovation Productivity

The third hypothesis is that income distribution affects not only the absolute level of innovative output, but also the productivity of it. To test this hypothesis, another set of regressions is specified that use innovation productivity as the dependent variable. The log value of patent applications per million inhabitants (lnPTENTCAP) is used as the proxy for innovative productivity.

Independent Variable 1: Income Distribution

This chapter employs two types of income distribution measures. The first measure is the Gini coefficient, the most widely used measure of income distribution in empirical work. The data are from the World Income Inequality Database of the United Nations Development Program (UNDP). The negative sign is expected for the estimates.

Although Gini coefficients are the most widely used measure of income distribution, there are some circumstances under which a Gini coefficient is not the appropriate measure. For instance, it is inappropriate to compare two patterns of income distribution based on Gini coefficients, when their Lorenze curves intersect. In other words, two different patterns of income distribution could have the same Gini coefficients.

Therefore, in addition to the Gini coefficients, this chapter also employs the share of aggregate income of the middle-class population, namely, the aggregate income of the 2nd, 3rd, and 4th quintiles of the population (MID60). Countries with similar Gini coefficients may have different sizes of middle-class population. This measure is used to test the robustness of the empirical models. The hypothesis is that the larger the middle-class population, the greater is the country's equality. The data for the income share are available from the World Bank's *World Development Indicators*.

Independent Variable 2: Income Level

Another independent variable, which is also a control variable, is income level (INCOME). According to Engel's law, poor people spend most of their income on food, clothes, and other subsistence goods. It is only when their income rises above a certain level that they can afford similar products with better quality and more diverse kinds of products. Furthermore, according to the quality-ladder hypothesis, consumers value product quality and quality improvement according to their income levels (Glass, 1999). As quality improvement requires some sort of innovation on the producers' part, it can be hypothesized that the higher the income level, the higher the consumer demand for innovative products and the more investment on innovative activity to produce those products. More innovation investment then results in a greater number of innovation outputs than less innovation investment. A positive sign is expected for the estimate of variable. GDP per capita is used to measure income level. The data are from the World Bank's *World Development Report 1999.*

Independent Variable 3: Market Size

The literature suggests market size as another determinant of innovation. In order for increasing-returns innovation to break even, sales must be high enough for innovators to recoup fixed initial costs. Therefore, market size can be another important factor that influences entrepreneurs to invest in innovative activity.

In this chapter, three different measures are used to proxy for market size, namely gross domestic product (GDP), population size (POP), and population density (POPDENSE). A large economy and population size suggest a potentially large market for innovation. A concentrated population, on the other hand, may help to create a large market for innovation, because consumers of innovation are located within the same areas, thereby reducing transportation costs.

There are other reasons for including population and population density as the additional independent variables. The literature suggests two general views regarding the effect of population on innovation. First is the demand-driven view, which argues that high population density creates relative scarcity of resources, hence the necessity for innovation. In other words, "necessity is the mother of invention." Boserup (1981) empirically tests this argument, focusing on the relationship between population density and innovative practices in agriculture. Lee (1988) proposes a mathematical model that captures the interaction between induced population growth and induced technical progress. Although the demand-driven argument seems logical, a caveat exists. As Ray (1998) points out, population growth can be attributed mainly to income growth, and it is a combination of both population and income that is likely to drive innovation, not just population alone.

The second view is supply-driven, which holds that large population size means a large pool of potential innovators and therefore a large stock of ideas and innovations. The larger the population, the higher is the probability that the society will be able to innovate. This argument assumes that everybody has an independent chance of developing an innovative idea. This view is implicit in several models of endogenous technological change, such as that of Grossman and Helpman (1991). As Romer (1990) points out, the cost of inventing a new technology is independent of the number of people who use it. In other words, technology is nonrival. Given the nonrivalry of technology and an assumed constant share of resources devoted to research and development, an increase in population will lead to an increase in technological change. Based on that proposition, Kremer (1993) shows empirically that among the societies without technological transfers from other societies, those with larger initial populations have had faster technological change.

Independent Variable 4: Demand Growth

As Zweimuller (2000) argues, innovators may decide to invest more in innovative activities if they can expect the demand for their innovation to grow in the future. This expectation could be based on the growth trend in the past. In this chapter, three different measures are to represent demand growth: (1) GDP growth (GDPGRWTH), (2) average annual growth of industry value added (INDGRWTH), and (3) average annual growth of household final consumption expenditure (CSMGRWTH). Positive signs are expected for the estimated coefficients.

Descriptive Statistics

The patterns of income distribution and innovation vary greatly in different countries. Table 14.2 shows the summary of descriptive statistics of the data. It is clear that there is a huge gap between the country with the highest Gini coefficient, South Africa, and the one with lowest Gini, the Czech Republic. The same is true for the other measure of income distribution, that is, the aggregate income share of the 2nd, 3rd, and 4th quintile of the population (MID60). The gaps are also wide for the measures for innovation, namely, gross R&D expenditures (GERD) and private R&D expenditures (PERD), level of innovation output (PTENT), and productivity of innovation (PTENTCAP). Particularly striking is the difference in the level and productivity of innovation.

Table 14.2
Summary Statistics for Income Distribution and Innovation Measures

Variable	Mean	Standard Deviation	Minimum	Maximum	Number of Observations
Income Distribution					
GINI	37.85	10.25	22.88	60.90	72
MID60	51.62	14.86	32.30	56.30	72
Innovation Spending					
GERD	1.07	0.89	0.07	3.76	64
PERD	0.57	0.64	0.00	2.37	57
Innovative Output/ Productivity					
PTENT	12,031	46,830	1	360,364	72
PTENTCAP	250	493	0	2,853	72

Note: Number of observations varies according to data availability.
Source: *Statistical Yearbook 1999* (Paris: UNESCO).

With patent applications as the measure of innovative output, Japan is the most innovative country, with more than 360,000 patent applications per year, while Zambia and many other countries excluded from the sample have none. In terms of innovative productivity, measured by patent applications per million inhabitants, there are also huge differences among countries. Japan is the most productive in producing innovations, with more than 2,800 patent applications per million inhabitants, while the average productivity for the sample of 72 countries is only 250.

Table 14.3 presents the statistics by region. Industrialized countries, on average, have more equal income distribution and larger proportion of middle-class population than developing countries. Among developing countries in the sample, the average income distribution in Eastern Europe is the most equal, followed by Asia. In fact, East European countries are, on average, more equal than industrialized countries, possibly because of the legacy from the era of socialism. As is well known, income distribution in Latin America is greatly skewed and is the most unequal of all regions. In terms of R&D expenditures, industrialized countries clearly spend much more than developing countries. Among developing countries, both the gross and private R&D expenditures in Asia are higher than in other regions. This is also true in terms of innovative output and productivity. The total number of patent applications and the number of patent applications per 1 million inhabitants are the greatest in industrialized countries, followed by the groups of countries in Asia and Eastern Europe.

Table 14.4 reports the sources of funding for research and development. Developed countries rely more on funding from the private sector, with more than half of the gross R&D expenditures come from private enterprises. In contrast, developing countries rely on the public sector, with more than 50% of R&D expenditure from their respective governments. Among developing countries, there are clear differences among regions. Latin America relies heavily on public sources, while its R&D funding from the private sector is extremely low. Africa's reliance on overseas funding is also striking, with 35% coming from overseas. The only group of developing countries with spending patterns similar to developed countries is the newly industrializing countries (NICs) of Asia. On average, 40% of their R&D funding comes from business enterprises, a level not far below that of developed countries. A noteworthy example is South Korea, in which 84% of the R&D funding comes from private enterprises in 1994. On the other hand, R&D in Latin American NICs relies mainly on funding from the governments.

Estimation Results

Gross R&D Expenditures

In the first set of regressions, the dependent variable is the gross expenditure on R&D as percentage of GDP (GERD). The basic specification to be estimated in this section is:

$$GERD = \alpha + \beta_1\, GINI + \beta_2\, INCOME + \beta_3\, GDP + \mu.$$

Table 14.3
Summary Statistics by Region

Region	Industrialized Countries		Developing Countries							
			Asia		Latin America		Eastern Europe		Africa	
	Mean	St. Dev.	Mean	St. Dev.	Mean	St. Dev.	Mean	St. Dev.	Mean	St. Dev.
Income Distribution										
GINI	32.30	4.80	38.93	6.67	50.73	6.16	29.27	3.65	46.55	13.35
MID60	55.18	10.24	58.76	23.71	43.33	16.83	51.49	3.19	42.85	7.92
Innovation Spending										
GERD	1.88	0.85	0.81	0.86	0.42	0.29	0.75	0.36	0.46	0.34
PERD	1.02	0.63	0.53	0.80	0.09	0.09	0.29	0.24	0.38	-
Innovative Output										
PTENT	32,198	81,162	86,150	17,019	313	527	2,527	5,413	130	200
PTENTCAP	539	586	373	718	13	23	65	42	3	3
Number of Countries	22		13		15		16		6	

Source: *Statistical Yearbook 1999*.

Table 14.4
Sources of Funds for Research and Development by Region, 1999

	Number of Countries in Sample	Sources of Funds			
		Business Enterprises	Government	Education and Nonprofit Organizations	Overseas
Developed countries	22	50.4	40.2	3.0	6.5
Developing countries	55	20.1	53.4	14.9	5.3
Eastern Europe	16	28.5	59.9	5.5	6.2
Asia	16	27.7	60.6	7.6	4.1
- NICs	7	40.4	41.4	12.3	5.9
Latin America	12	7.1	77.8	5.5	9.7
- NICs	4	21.0	69.7	5.1	4.2
Africa	11	7.0	49.2	0.9	35.0
- w/o South Africa	10	3.0	54.8	0.9	31.9

Note: NICs = Newly Industrialization Countries. In this sample, Asian NICs include South Korea, Singapore, Hong Kong, Malaysia, Indonesia, Philippines, and Thailand. Latin American NICs include Mexico, Brazil, Chile, and Argentina.
Source: *Statistical Yearbook 1999*.

Table 14.5 presents the estimation results of the full sample. As hypothesized, income distribution (GINI) has a strong negative impact on gross R&D expenditures (GERD). Income level (INCOME) is also significant. For the variables that represent domestic market sizes, the estimate for GDP is positive and significant at the 0.1 level, whereas POP is insignificant. In a comparison of the standardized coefficients of all independent variables, income level (INCOME) exerts the strongest effect on GERD. None of the estimates for the three proxies for demand growth shows significant results. This means the demand growth at the national level does not affect the gross R&D expenditures.

Population density is another variable that may affect the gross innovation expenditures. However, the regression results are inconclusive. Although the estimate for POPDENSE is significant, the sign for the estimate is negative, which is contrary to the hypothesis that the more populated countries spend more on innovation.

The regressions that use shares of aggregate income as the measure of income distribution yield similar estimation results (Model 7, Table 14.5). The estimate for the independent variable MID60 shows a positive and significant result. This confirms the hypothesis that the larger propotion of the middle-class population in a country, the more the country spends on innovation.

Table 14.6 reports the estimates for developed countries and developing countries subsamples. The results clearly show the differences between developed and developing countries. The variables GINI and GDP are significant only in the models for developing countries subsamples, while INCOME is significant for both subsamples. POP and GDPGRWTH remain insignificant in all regressions.

Private R&D Expenditures

In the previous section, gross R&D expenditures (GERD) were used as the measure for investment on innovative activity. The measure includes all sources of funding from domestic public and private institutions, as well as overseas sources. Much of the R&D expenditures from the public and nonprofit sectors may not necessarily be driven by profits and business opportunities. Therefore, another set of models is specified, in which private R&D expenditures as percentage of GDP (PERD) is used as the dependent variable.

The estimate results appear to be very similar to the results when using GERD as the dependent variable (Table 14.7). All the estimates for the key variables, namely GINI, INCOME, and GDP, are significant, and the signs for coefficients are as expected. POP and GDPGRWTH are not significant.

The effects of income distribution on private R&D expenditures are more evident in developing countries than in developed countries (Models 5 and 6, Table 14.7). The estimates for GINI, INCOME, and GDP are all significant in the regressions that use the developing countries subsample. Only INCOME is significant for the developed countries subsample.

Table 14.5
Estimation Results for Regressions on GERD, Full Sample

Independent variable	Model 1	Model 2	Model 3	Model 4	Model 5	Model 6	Model 7
Constant	1.200***	1.12***	1.272***	1.271***	1.252***	1.077***	-0.395
	(3.744)	*(3.744)*	*(3.741)*	*(3.606)*	*(3.741)*	*(3.366)*	*(-0.952)*
GINI	-0.019**	-0.018**	-0.022**	-0.023**	-0.021**	-0.016**	
	(-2.552)	*(-2.552)*	*(-2.578)*	*(-2.642)*	*(-2.576)*	*(-2.139)*	
INCOME	0.050***	0.057***	0.048***	0.048***	0.049***	0.054***	0.052***
	(6.804)	*(6.804)*	*(6.317)*	*(6.228)*	*(6.589)*	*(7.047)*	*(6.570)*
GDP	0.013*		0.013*	0.013*	0.013*	0.011	0.011
	(1.733)		*(1.932)*	*(1.796)*	*(1.819)*	*(1.577)*	*(1.578)*
POP		0.029					
		(0.766)					
GDPGRWTH			0.014				
			(0.725)				
CSMGRWTH				0.025			
				(0.688)			
INDGRWTH					0.008		
					(0.581)		
POPDENSE						-0.01*	
						(-1.778)	
MID60							0.018**
							(2.073)
Adjusted R2	0.594	0.584	0.598	0.593	0.598	0.615	0.620
SEE	0.557	0.577	0.565	0.581	0.568	0.553	0.559
F	24.060	30.498	27.384	22.499	24.405	26.115	24.703
No. of observations	64	64	64	60	64	64	59

Notes: t values are shown in parentheses. SEE stands for standard error of the estimate. *** signifies $p < 0.01$, ** = $p < 0.05$, and * = $p < 0.1$.

Table 14.6
Estimation Results for Regressions on GERD, Subsamples

Independent variable	Model 1 Developed countries	Model 2 Developing countries	Model 3 Developed countries	Model 4 Developing countries	Model 5 Developed countries	Model 6 Developing countries
Constant	1.605	1.133***	1.527	1.112***	1.595	1.165***
	(1.002)	*(4.188)*	*(1.061)*	*(3.861)*	*(1.074)*	*(3.915)*
GINI	-0.027	-0.020***	-0.027	-0.017***	-0.029	-0.020***
	(-0.731)	*(-2.972)*	*(-0.717)*	*(-2.528)*	*(-0.713)*	*(-2.772)*
INCOME	0.045*	0.045***	0.047*	0.047***	0.046*	0.043***
	(1.880)	*(3.679)*	*(2.005)*	*(3.656)*	*(1.831)*	*(3.331)*
GDP	0.011	0.072*			0.012	0.067*
	(1.131)	*(1.988)*			*(1.166)*	*(1.171)*
POP			0.331	0.018		
			(1.120)	*(0.557)*		
GDPGRWTH					0.017	0.006
					(0.140)	*(0.353)*
Adjusted R2	0.197	0.350	0.196	0.288	0.158	0.325
SEE	0.769	0.453	0.769	0.460	0.784	0.460
F	2.715	8.360	0.704	6.535	1.987	5.938
No. of observations	22	42	22	42	22	42

Notes: t values are shown in parentheses. SEE stands for standard error of the estimate. *** signifies $p < 0.01$, ** = $p < 0.05$, and * = $p < 0.1$.

Table 14.7
Estimation Results for Regressions on PERD

Independent variable	Model 1 Full sample	Model 2 Full sample	Model 3 Full sample	Model 4 Full sample	Model 5 Developed Countries	Model 6 Developing countries
Constant	0.652**	0.570 **	0.761 **	0.591**	0.740	0.530**
	(2.451)	*(2.037)*	*(2.707)*	*(2.170)*	*(0.788)*	*(2.147)*
GINI	-0.04**	-0.012*	-0.018**	-0.012*	-0.024	-0.013**
	(-2.273)	*(-1.967)*	*(-2.559)*	*(-1.982)*	*(-0.989)*	*(-2.366)*
INCOME	0.032***	0.037***	0.029***	0.034***	0.041*	0.045***
	(5.273)	*(6.455)*	*(4.714)*	*(5.274)*	*(2.611)*	*(3.790)*
GDP	0.011**		0.012**	0.010*	0.011	0.078*
	(2.135)		*(2.264)*	*(1.902)*	*(1.464)*	*(1.993)*
POP		0.025				
		(0.556)				
GDPGRWTH			0.014			
			(1.163)			
POPDENSE				-0.0001		
				(-1.014)		
Adjusted R2	0.559	0.524	0.562	0.559	0.382	0.424
SEE	0.428	0.445	0.429	0.428	0.496	0.363
F	24.656	21.532	18.953	18.759	5.331	9.109
No. of observations	57	57	57	57	22	35

Notes: t values are shown in parentheses. SEE stands for standard error of the estimate. *** signifies $p < 0.01$, ** = $p < 0.05$, and * = $p < 0.1$.

In sum, the estimation results indicate that income distribution, income level, and the size of the economy are the significant determinants of innovation. However, income distribution and the size of the economy affect innovation expenditures only for developing countries. Income level affects R&D expenditures in both developed and developing countries.

Innovation Output Level

With the full sample, the estimate results in Table 14.8 show that GINI, INCOME, GDP, and POP are all highly significant as the determinants of the absolute number of patent applications (lnPTENT). The estimates for other variables, such as POPDENSE, were not significant.

However, when the full sample is split into two subsamples, GINI is significant only for the developing countries subsample, while other variables remain significant in all regressions. The result suggests that income distribution significantly affects the innovation output level of developing countries, but not developed countries.

Innovation Productivity

The results of full-sample estimations suggest that GINI and INCOME are the significant determinants of innovation productivity (Table 14.9), whereas GDP and POP are not. However, the effects of GINI on innovation productivity are significant only when using developing countries subsample, but not developed countries subsample. The estimates for other variables, including POPDENSE and GDPGRWTH, were not significant.

Inputs and Outputs of Innovation

This section examines the relationship between inputs and outputs of innovative activity. Following Scherer (1982) and Acs and Audertsch (1988), the analysis assumes that innovative outputs are related to innovation-inducing inputs in the previous period according to a log relationship. Two sets of regressions are specified, using the log values of total number of patents (lnPTENT), and the log values of total number patents per 1 million inhabitants (lnPTENTCAP), as the dependent variables. The independent variables are the log values of gross expenditures on R&D (lnGERD), private-sector expenditures on R&D (lnPERD), and number of personnel in R&D (lnRDPERSON).

The estimated coefficients reported in Tables 14.10 and 14.11 suggest that the number of R&D personnel and gross R&D expenditures significantly affect the level and productivity of innovation outputs in both developed and developing countries. Private R&D expenditures, however, affect innovation only in developed countries. This is not surprising, considering that in most developing countries, private firms do not invest in innovation.

Table 14.8
Estimation Results for Regressions on lnPTENT

Independent variable	Model 1 Full sample	Model 2 Full sample	Model 3 Developed countries	Model 4 Developing countries	Model 5 Developed countries	Model 6 Developing countries
Constant	-0.795	-1.852	-1.688	-0.856	-2.056	-1.251
	(-0.861)	*(-1.905)*	*(-0.685)*	*(-1.013)*	*(-0.871)*	*(-1.159)*
GINI	-0.082***	-0.065***	-0.050	-0.0.99***	-0.049	-0.079***
	(-3.757)	*(-2.828)*	*(-0.785)*	*(-4.880)*	*(-0.801)*	*(-3.098)*
INCOME	0.084***	0.142***	0.083*	0.083**	0.093**	0.128**
	(3.903)	*(6.601)*	*(2.040)*	*(2.138)*	*(2.410)*	*(2.583)*
GDP	0.086***		0.070***	0.701***		
	(4.116)		*(4.091)*	*(5.974)*		
POP		0.380***			2.113***	0.321**
		(3.334)			*(4.349)*	*(2.563)*
POPDENSE		-0.0003				
		(-1.588)				
Adjusted R2	0.549	0.523	0.568	0.549	0.594	0.289
SEE	1.715	1.762	1.303	1.501	1.264	1.884
F	28.942	19.924	10.208	20.073	11.223	7.380
No. of observations	72	72	22	50	22	50

Notes: t values are shown in parentheses. SEE stands for standard error of the estimate. *** signifies $p < 0.01$, ** = $p < 0.05$, and * = $p < 0.1$.

Table 14.9
Estimation Results for Regressions on lnPTENTCAP

Independent variable	Model 1 Full sample	Model 2 Full sample	Model 3 Developed countries	Model 4 Developing countries	Model 5 Developed countries	Model 6 Developing countries
Constant	5.642***	5.752***	3.226	5.805***	3.126	6.044***
	(6.845)	*(6.931)*	*(1.623)*	*(6.250)*	*(1.595)*	*(6.348)*
GINI	-0.096***	-0.096***	-0.007	-0.108***	-0.006	-0.107***
	(-4.979)	*(-5.032)*	*(-0.139)*	*(-4.959)*	*(-0.114)*	*(-4.890)*
INCOME	0.136***	0.139***	0.114***	0.194***	0.116***	0.193***
	(6.928)	*(7.714)*	*(3.453)*	*(4.428)*	*(3.586)*	*(4.364)*
GDP	0.012		0.009	0.161		
	(0.629)		*(0.652)*	*(1.240)*		
POP		-0.105			0.239	-0.107
		(-1.048)			*(0.590)*	*(-0.954)*
Adjusted R2	0.636	0.640	0.329	0.476	0.390	0.469
SEE	1.563	1.555	1.050	1.684	1.053	1.695
F	42.360	43.030	5.515	15.847	5.467	15.438
No. of observations	72	72	22	50	22	50

Notes: t values are shown in parentheses. SEE stands for standard error of the estimate. *** signifies $p < 0.01$, ** = $p < 0.05$, and * = $p < 0.1$.

Table 14.10
Estimation Results for Regressions on lnPTENT

Independent variable	Model 1 All countries	Standardized Coefficients (Beta)	Model 2 Developed countries	Standardized Coefficients (Beta)	Model 3 Developing countries	Standardized Coefficients (Beta)
Constant	-0.906		-0.468		0.500	
	(-0.654)		*(-0.317)*		*(0.243)*	
Ln GERD	0 875**	0.336	-1.928**	-0.617	0.910*	0.365
	(2.274)		*(-2.298)*		*(1.818)*	
Ln PERD	0.222	0.153	2.054***	0.958	0.178	0.140
	(0.989)		*(3.465)*		*(0.661)*	
LnRDPERSON	0.845***	0.533	0.988***	0.693	0.678***	0.465
	(6.567)		*(8.261)*		*(3.523)*	
Adjusted R2	0.767		0.893		0.593	
SEE	1.154		0.659		1.323	
F	56.965		56.701		15.543	
No. of observations	52		21		31	

Notes: t values are shown in parentheses. SEE stands for standard error of the estimate. *** signifies $p < 0.01$, ** = $p < 0.05$, and * = $p < 0.1$.

Table 14.11
Estimation Results for Regressions on lnPTENTCAP

Independent variable	Model 1 All countries	Standardized Coefficients (Beta)	Model 2 Developed countries	Standardized Coefficients (Beta)	Model 3 Developing countries	Standardized Coefficients (Beta)
Constant	-1.211		-4.040		0.163	
	(-0.675)		*(-0.867)*		*(0.071)*	
Ln GERD	0.992**	0.386	-1.627*	-0.765	1.303**	0.484
	(2.364)		*(-1.795)*		*(2.270)*	
Ln PERD	0.190	0.132	1.689**	1.157	0.102	0.074
	(0.843)		*(2.308)*		*(0.386)*	
LnRDPERSON	0.817***	0.414	1.413**	0.473	0.585*	0.323
	(3.410)		*(2.538)*		*(1.874)*	
Adjusted R2	0.757		0.776		0.638	
SEE	1.164		0.650		1.346	
F	54.081		24.160		18.612	
No. of observations	52		21		31	

Notes: t values are shown in parentheses. SEE stands for standard error of the estimate. *** signifies $p < 0.01$, ** = $p < 0.05$, and * = $p < 0.1$.

Comparisons of the standardized coefficients of all independent variables show that private R&D expenditures exert the strongest effect on innovation outputs and innovation productivity in developed countries. On the other hand, in developing countries, the number of R&D personnel has the strongest effect on innovation outputs, while gross R&D expenditures have the strongest effect on innovation productivity. The standardized coefficients of the full model indicate that, in general, human capital in R&D exerts the strongest effects on innovation output level and innovation productivity.

Credit Market Imperfection and Human Capital Stock

The results in the previous section indicate the importance of the stock of R&D personnel in producing innovation. The literature reviewed earlier suggests that income distribution may affect human capital accumulation, when credit markets are imperfect. This section analyzes the issue further by examining one of the mechanisms through which income distribution may affect innovation: namely, the effect of income distribution on human capital development under imperfect credit markets.

A set of models is specified that includes the variables that proxy for credit market imperfections, namely credits to the nonfinancial private sector (CREDIT). Following Smith (2001), the proxy for credit market imperfections is credits from the banking system to the nonfinancial private sector as a share of GDP (CREDIT). This measure reflects the borrowing constraints on the part of consumers and firms. The data are obtained from the IMF's *International Financial Statistics 1997* (line 32a).

The specification estimated in this section is:

$$\text{lnRDPERCAP} = \alpha + \beta_1 \text{ GINI} + \beta_2 \text{ INCOME} + \beta_3 \text{ GDP} + \beta_4 \text{ CREDIT} + \beta_3 \text{ CREDIT*GINI} + \mu.$$

The dependent variable is the log value of total number of R&D personnel per million inhabitants. To capture the interaction between income distribution and credit-market imperfections, the interaction term CREDIT*GINI is added to the regression. As credit markets become more perfect, the borrowing constraints become lessened. The variable CREDIT is therefore expected to yield a positive sign.

The estimation results in Table 14.12 suggest that GINI, INCOME, CREDIT, and CREDIT*GINI are significant variables. However, the sign for the coefficients for CREDIT is not a positive sign as expected by the hypothesis. The contradictory results may be because CREDIT is a poor proxy for credit-market imperfections, particularly in the credit markets for educational investment.

Table 14.12
Estimation Results for Regressions on ln(RDPERCAP)

Independent variable	Model 1 All countries	Model 2 Developed countries	Model 3 Developing countries
Constant	9.075***	10.656	9.822***
	(14.008)	*(6.721)*	*(-10.796)*
GINI	-0.073***	-0.091	-0.089***
	(-4.336)	*(-1.808)*	*(--3.942)*
CREDIT	-2.878***	-8.409**	-5.916**
	(-2.274)	*(-2.518)*	*(-2.316)*
GINI * CREDIT	0.059***	0.170*	0.110*
	(3.410)	*(-2.518)*	*(2.010)*
INCOME	0.074***	0.034	0.113***
	(3.410)	*(0.372)*	*(3.713)*
GDP	0.007	-0.052	0.074
	(0.068)	(-0.321)	*(0.924)*
Adjusted R2	0.599	0.424	0.456
SEE	766	0.838	0.883
F	18.051	3.948	7.023
No. of observations	58	21	37

Notes: t values are shown in parentheses. SEE stands for standard error of the estimate. *** signifies $p < 0.01$, ** = $p < 0.05$, and * = $p < 0.1$.

CONCLUSION

As summarized in Table 14.13, the empirical results suggest that, in general, countries with more equal income distribution spend more on innovative activity, produce more innovative outputs, and are more productive in creating innovations than those with less equal income distribution. Other significant determinants of innovation include income level and market size as measured by GDP and population size.

However, the effects of income distribution on innovation investment, innovation output level, and innovation productivity are limited to developing countries. On the other hand, income level affects all three measures of innovation in both developed and developing countries. GDP affects innovation investment only in developing countries, but affects innovation output level in both developed and developing countries. GDP does not affect innovation productivity. Population size is a significant determinant of innovation output level, but not a determinant of innovation investment or innovation productivity.

Table 14.13
Summary of Key Estimation Results

	Income Distribution	Income Level	GDP	Population Size
Innovation Investment				
Full sample	√	√	√	x
Developed Countries	x	√	x	x
Developing Countries	√	√	√	x
Innovation Outputs				
Full Sample	√	√	√	√
Developed Countries	x	√	√	√
Developing Countries	√	√	√	√
Innovation Productivity				
Full Sample	√	√	x	x
Developed Countries	x	√	x	x
Developing Countries	√	√	x	x

Note: "√" signifies significant estimation results. "x" signifies insignificant results.

The empirical results are inconclusive as to whether credit-market imperfection is the mechanism through which income distribution affects human capital stock for innovation. This is possibly due to the poor proxy for credit-market imperfections. The analyses of the relationship between inputs and outputs of innovation indicate that countries with more personnel in R&D and greater gross R&D expenditures have higher output level and are more productive in creating innovations than those with less of each of these factors. Private-sector expenditure on innovation is a significant determinant of innovative output level and productivity only in developed countries.

According to Sokoloff (1988), in the absence of property rights incentives, the size of the market is the most important determinant of the rate of growth of technological innovation. The empirical results in this chapter add income distribution and income level to the list of determinants, especially in the case of developing countries. On the demand side, income distribution affects innovation only in developing countries, because it affects the aggregate demand compositions and therefore the domestic markets for innovations. Furthermore, as investors in innovation wish to capture as much entrepreneurial rent as possible, the prices for innovation are generally set at levels too high for many consumers. Low income levels in developing countries, therefore, make the market size for innovation appear even smaller to innovators. In contrast, in the case of developed countries, the average income level is high, and the market size is large enough to accommodate innovations. The effect of income distribution on innovation is therefore insignificant.

On the other hand, on the supply side, the production of innovations requires a minimum stock of knowledge and human capital. As indicated by the analysis of the relationship between innovation inputs and outputs, human capital stock for R&D appears to be the most important determinant of innovation output level and productivity. Arguably, developing countries in general do not have the minimum human capital stock, especially in R&D. As knowledge and human capital are essentially embodied in humans, its aggregate stock therefore would be larger if its accumulation is widely spread among individuals in the society. Further research is necessary to examine whether and how income inequality affects the accumulation and diffusion of knowledge throughout the society.

Furthermore, if knowledge displays increasing marginal productivity, as demonstrated by Romer (1990), the production of innovations from increased knowledge must entail increasing returns. This means the greater the stock of knowledge a country has, the more innovations and the more productively it can produce. On the other hand, if the effect of income distribution on innovation varies proportionally to the market size and human capital stock, then there should exist a threshold point from which the effect of income distribution on market size and human-capital stock become insignificant as compared with the effects of human-capital stock on innovation. In other words, the larger the stock of human capital is in a country, the smaller the effect of income distribution on innovation output and productivity. It could also be interpreted that there exists a critical threshold for market-size effects on innovation, beyond which income distribution is irrelevant.

The empirical results in this chapter support the argument. In developing countries, market sizes and human-capital stock are generally small. The effect of income distribution on human-capital stock, and thus innovation output and productivity, looms large. On the other hand, the stock of knowledge and human capital in developed countries is much larger than in developing countries. The effect of income distribution on innovation output and productivity is therefore insignificant.

In conclusion, income distribution affects innovation expenditure, innovation output, and innovation productivity by affecting the aggregate demand composition and human-capital accumulation. Because the market size and the stock of human capital are relatively small in developing countries, income distribution has significant effects on the size of the market and the stock of human capital and hence on innovation. Developed countries, on the other hand, have large markets and large stock of human capital. Income distribution does not significantly affect demand compositions and human-capital accumulation and innovation.

As the literature suggests, credit constraints restrict human-capital accumulation. However, as income increases, credit constraints become insignificant, and the effect of inequality on human-capital accumulation subsides. This argument explains why the empirical evidence indicates that income distribution affects innovation only in developing countries. Although our empirical results do not find credit-market market imperfections as the specific channel through which income distribution affects human-capital stock for innovation, future research that incorporates a better proxy of credit constraints may shed more light on the issue.

Finally, another channel for income distribution to affect innovation should be included in future research. That is, the effect of income inequality on the selection of sectors for domestic investment. Initial income distribution may determine the development path that affects the subsequent level of industrial innovation. In other words, if a country has a highly skewed income distribution, and, as a result, invests in only noninnovative sectors, then it is likely that the country will become less innovative than it would be with more equal income distribution.

REFERENCES

Acs, Z. J. and Audertsch, D. B. (1988). "Innovation in Large and Small Firms: An Empirical Analysis," *The American Economic Review*, 78 (4): 678-690.

Aghion, P. (2002). "Schumpeterian Growth Theory and the Dynamics of Income Inequality," *Econometrica*, 70 (3): 854-882.

Alesina, A. and Rodrik, D. (1994). "Distributive Politics and Economic Growth," *The Quarterly Journal of Economics*, 109 (2): 465-490.

Amsden, A. H. and Mourshed, M. (1997). "Scientific Publications, Patents and Technological Capabilities in Late-Industrializing Countries," *Technology Analysis & Strategic Management*, 9 (3): 343-359.

Arrow, K. J. (1962). "Economic Welfare and the Allocation of Resources for Invention." In R. R .Nelson (ed.), *The Rate and Direction of Inventive Activity: Economic and Social Factors*. Princeton, NJ: Princeton University Press.

Baland, J. M. and Ray, D. (1991). "Why Does Asset Inequality Affect Unemployment—A Study of the Demand Composition Problem," *Journal of Development Economics*, 35 (1): 69-92.

Baldwin, R. E. (1956). "Patterns of Development in Newly Settled Regions," *Manchester School of Economic and Social Studies*, 24: 161-179.

Boserup, E. (1981). *Population and Technological Change: A Study of Long-Run Trends*. Chicago: University of Chicago Press.

Falkinger, J. (1994). "An Engelian Model of Growth and Innovation with Hierarchic Demand and Unequal Incomes," *Ricerche Economiche*, 48: 123-139.

Fishman, A. and Simhon, A. (2002). "The Division of Labor, Inequality, and Growth," *Journal of Economic Growth*, 7: 117-136.

Flug, K., Spilimbergo, A., and Wachtenheim, E. (1998). "Investment in Education: Do Economic Volatility and Credit Constraints Matter?" *Journal of Development Economics*, 55. 465-481.

Galor, O. and Zeira, J. (1993). "Income Distribution and Macroeconomics," *Review of Economic Studies*, 60 (1): 35-52.

Glass, A. J. (1999). "Price Discrimination and Quality Improvement," Mimeo, Ohio State University.

Grossman, G. M. and Helpman, E. (1991). *Innovation and Growth in the Global Economy*. Cambridge, MA: MIT Press.

Kaldor, N. (1957). "A Model of Economic Growth," *Economic Journal*, 67: 591-624.

Kremer, M. (1993). "Population Growth and Technological Change: One Million B.C. to 1990," *The Quarterly Journal of Economics*, 108: 681-716.

Kuznets, S. (1955). "Economic Growth and Income Inequality," *American Economic Review*, 45 (1): 1-28.

Leamer, E. E., Maul, H., Rodriguez, S., and Schott, P. K. (1999). "Does Natural Resource Abundance Increase Latin American Income Inequality?" *Journal of Development Economics*, 59 (1): 3-42.

Lee, R. (1988). "Induced Population Growth and Induced Technical Progress," *Mathematical Population Studies*, 1: 265-288.

Murphy, K. M., Shleifer, A., and Vishny, R. (1989). "Income-Distribution, Market-Size, and Industrialization," *The Quarterly Journal of Economics*, 104 (3): 537-564.

Nickell, S. and Nicolitsas, D. (1997). "Human Capital, Investment and Innovation: What Are the Connections?" Centre for Economic Performance & Institute of Economics and Statistics, Oxford University.

North, D. C. (1959). "Agriculture in Regional Economic Growth," *Journal of Farm Economics*, 51: 943-951.

Ray, D. (1998). *Development Economics*. New York: Oxford University Press.

Romer, P. M. (1990). "Endogenous Technological Change," *Journal of Political Economy*, 98: S71-S103.

Roy, U. (1997). "Economic Growth with Negative Externalities in Innovation," *Journal of Macroeconomics*, 19 (1): 155-173.

Scherer, F. M. (1965). "Size of Firm, Oligopoly and Research: A Comment," *Canadian Journal of Economics and Political Science*, 31: 256-266.

Scherer, F. M. (1982). "Inter-Industry Technology Flows and Productivity Growth," *Review of Economics and Statistics*, 64: 627-634.

Schumpeter, J. (1934). *The Theory of Economic Development: An Inquiry into Profits, Capital, Credit, Interest, and the Business Cycle*. Cambridge, MA: Harvard University Press.

Shepherd, W. G. (1979). *The Economics of Industrial Organization*. Englewood Cliffs, NJ: Prentice Hall.

Smith, D. (2001). "International Evidence on How Income Inequality and Credit Market Imperfections Affect Private Saving Rates," *Journal of Development Economics*, 64: 103-127.

Sokoloff, K. L. (1988). "Inventive Activity in Early Industrial America: Evidence from Patent Records 1790-1846," *Journal of Economic History*, 58: 813-850.

Zweimuller, J. (2000). "Schumpeterian Entrepreneurs Meet Engel's Law: The Impact of Inequality on Innovation-Driven Growth," *Journal of Economic Growth*, 5 (2): 185-206.

APPENDIX: LIST OF COUNTRIES IN THE SAMPLE

Developed Countries	Developing Countries			
	Asia	Latin America	Eastern Europe	Africa
Australia	China	Argentina	Azerbaijan	Egypt
Austria	Hong Kong	Bolivia	Belarus	Ghana
Belgium	India	Brazil	Bulgaria	Kenya
Canada	Indonesia	Chile	Croatia	Morocco
Denmark	South Korea	Colombia	Czech Republic	South Africa
Finland	Malaysia	Costa Rica	Estonia	Zimbabwe
France	Pakistan	Ecuador	Hungary	
Germany	Philippines	Guatemala	Kazakhstan	
Greece	Singapore	Honduras	Latvia	
Ireland	Sri Lanka	Mexico	Lithuania	
Israel	Taiwan	Nicaragua	Poland	
Italy	Thailand	Panama	Romania	
Japan	Turkey	Peru	Russia	
Netherlands		Uruguay	Slovenia	
New Zealand		Venezuela	Ukraine	
Norway			Uzbekistan	
Portugal				
Spain				
Sweden				
Switzerland				
United Kingdom				
United States				

15

At the Crossroads for Binational Development: Cameron County, Texas, and Matamoros, Mexico

David V. Gibson and Pablo Rhi-Perez

The *Texas/Mexico border region* stands at a crossroads. In one direction lies a future that follows the current course of action...enrollments in colleges and universities do not keep pace with booming population growth...regional workers are not able to support a growing and globally competitive economy that is necessary for a sustainable quality of life....

In the other direction lies a future which follows a new path... the region accepts the challenge... college and university enrollments and graduations increase... Educational institutions excel through programs of excellence and advancements in research... the economy is advanced by a highly trained and capable workforce and by innovations created through R&D... individuals are challenged, their minds are expanded, and they develop a growing interest in the world around them.

Adapted from *Closing the Gaps*
Texas Higher Education Plan, 2002

INTRODUCTION

This chapter provides quantitative and qualitative data on Cameron County, Texas, and Matamoros, Mexico to assess assets and challenges for accelerated cross-border development.[1] A key organizing principle is that the Texas-Mexico border region in general and the Cameron County/Matamoros border region in particular is at a crossroads in terms of regional leadership, business and industry development, the creation of jobs and opportunities, education and workforce training, and an improving and sustainable quality of life.

Cameron County, the southmost tip of Texas, is located on the Gulf of Mexico and has three border crossings to its closest neighboring city, Matamoros, and one at Los Indios (Figure 15.1):

- Gateway International Bridge
- Private B&M Bridge
- Veteran's International Bridge at Los Tomates
- Los Indios Bridge (Lucio Blanco)

Figure 15.1
Map of Cameron County & Northeast Mexico with Indication of Four International Bridges

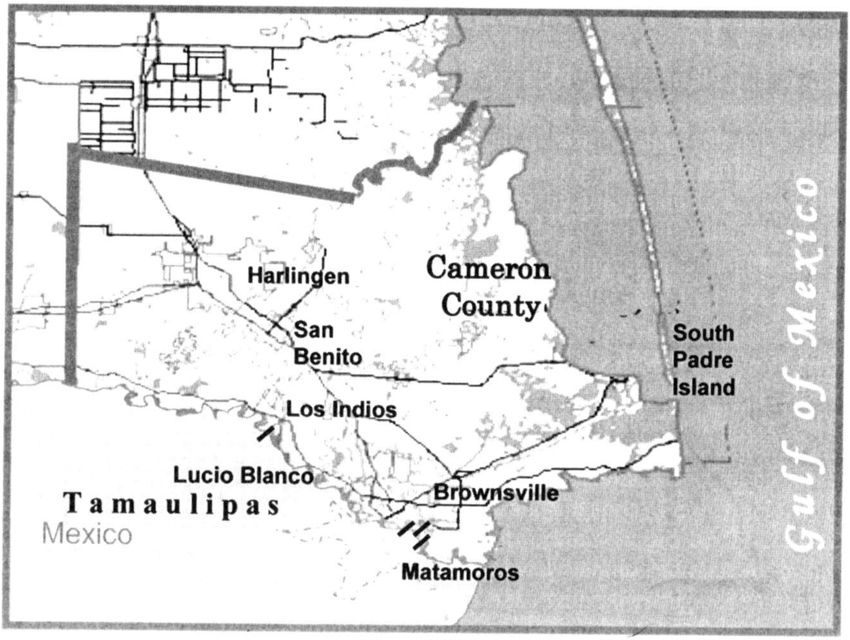

In 2004 the number of pedestrians going north from Mexico to Texas through Brownsville's border crossings surpassed 3.5 million. Some of this pedestrian traffic represents people living in Matamoros and working in Cameron County; but the bulk of this number is composed of tourists. The high volume of cross-border pedestrian and traffic flow is largely the result of the rapid growth of *maquiladoras* (Mexican manufacturing operations that are usually foreign owned with a great percentage located in Northern Mexico) in the 1980s and the passage of North American Free Trade Agreement (NAFTA) in 1994.

While benefiting from a strong binational economic and cultural heritage, established and growing educational assets, and a rapidly growing young bilingual and biliterate workforce, the border region is challenged as never before: One direction leads to regional decline; the other direction results in enhanced regional prosperity, as shown in Figure 15.2 (Gibson and Rhi-Perez, 2003; Rodriguez and De Los Reyes, 2002; Rylander, 2002; Brazier and Gibson, 2001; Patrick and Garcia, 2000; Chapa and Eaton, 1996).[2]

Figure 15.2
Lower Rio Grande Valley: Crossroads for the 21st Century

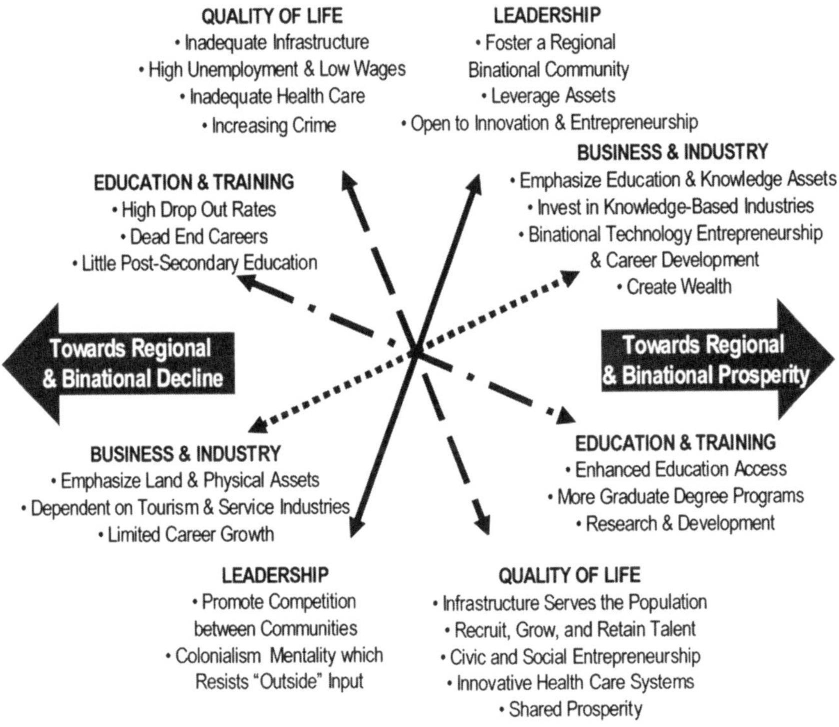

Source: IC^2 Institute, The University of Texas at Austin.

Business and Industry: These are challenged by the existence of two societies—traditional, resource-based and 21st-century knowledge-based—that maintain different values and visions for the Valley. The former emphasizes the importance of land and physical assets; the latter emphasizes education and innovation. One direction sees Cameron County's service industries and tourism as the most important avenue for regional job creation in the coming decade. The other direction emphasizes the importance of knowledge-based industries (e.g., life sciences, transportation and logistics, value-added *maquiladoras*) and binational entrepreneurship for 21st-century wealth creation.

Education and Workforce Training: One direction sees continued high rates of school dropouts, dead-end careers, and low rates of postsecondary education. The other direction sees enhanced access to postsecondary education, growing graduate programs, and the establishment of centers of research excellence.

Leadership: In one direction community leaders tend to position themselves to compete rather than cooperate with their neighbors, encourage win-lose scenarios on a city-by-city basis, and advocate a colonialism mentality where the contributions of "outsiders" are resisted. In the other direction, community leaders work together to leverage intellectual, physical, and cultural assets binationally.

Quality of Life: One direction sees continued population growth that strains regional infrastructure, continued high unemployment and low median income, increased health care challenges, and increased crime. These negative trends are exacerbated with a "brain drain" as many of the most capable and educated workers and professionals leave the Valley to pursue their careers and to earn higher salaries. The other direction sees the Lower Rio Grande Valley being "branded" as a region that (1) is able to grow, recruit, and retain "the best and the brightest," and (2) works to achieve a better life for all citizens.

It is important to emphasize that these border challenges are being confronted at a time when Mexico and Texas face considerable challenges, including:

- Increased competition in a global economy where low value-added manufacturing jobs are moving south to Latin America and offshore to China and India, chasing the world's lowest wages.
- Education and workforce needs of the 21st century require highly skilled workers and advanced degrees.
- Budget deficits limit government spending across a range of important and needed programs such as education and health care.
- Environmental and natural resource challenges are centered on water conservation.
- Enhanced binational cooperation is frustrated by perceived threats of global terrorism and U.S. homeland security.

CREATING REGIONAL OPPORTUNITIES

> Explaining South Texas has always been a challenge: Where others see lines on a map, we see limitless horizons of opportunity. Where others see two cultures, side-by-side, we recognize our home: bicultural, bilingual—a combination of the best transformed into one.
>
> Mary Rose Cardenas, Chairman
> Texas Southmost College, Board of Trustees, 2002

This chapter emphasizes that the border region needs visionary leaders that focus on cross-border cooperation:

- In one vision of the future community leaders compete for limited resources to benefit individual cities and institutions.
- In the other vision of the future leaders work together regionally and binationally to leverage limited resources to build a better future for all: One Region – Un Futuro.

While this chapter advocates regionally based business and technology entrepreneurship as key to accelerating cross-border economic development, we also emphasize the importance of civic and social entrepreneurship to facilitate regional and binational cooperation and the leveraging of critical resources to help solve common challenges, as shown in Figure 15.3 (Gibson and Rhi-Perez, 2003).

Figure 15.3
Three Types of Entrepreneurship Needed in the Mexico-Texas Border Region

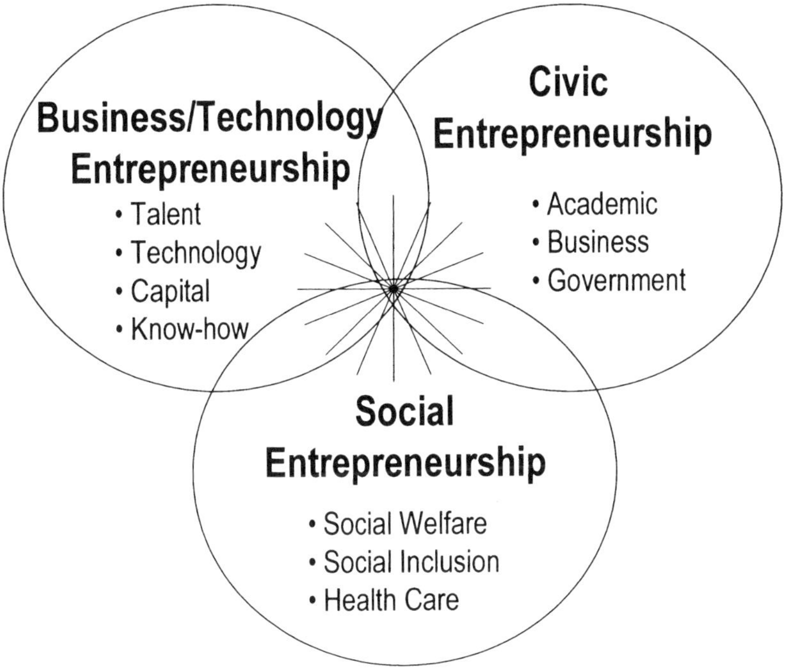

Source: IC² Institute, The University of Texas at Austin.

Business/Technology entrepreneurship centers on linking talent, technology, capital, and business know-how at the regional level for the creation of wealth from new business processes and products.[3] The objective is to build a regionally based "smart infrastructure" for new industries as well as for enhanced national and global competitiveness of established industries.

Civic entrepreneurship is concerned with networking regional academic, government, and business leaders to identify and solve community challenges.

Civic entrepreneurs utilize innovative approaches and partnerships to leverage public and private assets to solve challenges and to create new infrastructures for accelerated, sustainable public/private development.

Social entrepreneurship focuses on creative and innovative ways to improve society through shared prosperity and social inclusion. Social entrepreneurs network to link public and private sectors including nonprofit and nongovernment organizations (NGOs) and foundations to solve challenges for an improved quality of life for all.

DEMOGRAPHICS: POPULATION OVERVIEW

> ...counties located on the North Mexico-South Texas border region are consistently worse off economically, unemployment is higher, birthrates are higher, the overall poverty rate is higher, population growth is higher and the number of children in poverty is higher... at the same time the border counties have lower per-capita income, lower annual pay, and a lower growth rate in annual pay.

Carole Keeton Rylander, Texas Comptroller
Bordering on the Brink, March 27, 2001

> There is no choice in whether this region will grow or not. If we closed the border tomorrow—every bridge in South Texas and all the way to El Paso and if we stopped migration—we would still have 230% growth in 30 years. That is the nature of the current demographics. The only choice we have is whether we will be proactive or reactive to these changes....We need to do a better job of educating this population. If we don't, the Great State of Texas will be in a steady demise—economically, socially, and politically.

Dr. Juliet Garcia, President
The University of Texas Brownsville/
Texas Southmost College
Interview, August 28, 2002

The border region is confronted with a rapidly growing young and largely Hispanic and bilingual workforce that is underskilled and undereducated for meeting the demands of the 21st-century workplace. According to the U.S. Census Bureau, between 1990 and 2000 the U.S. population increased by 13% and Texas increased by 23%. Cameron County's population increased 29% (over twice the national rate). Hidalgo County, Cameron County's neighbor, had an even larger increase of 48%. Mexico's population increased 20% and the state of Tamaulipas increased by 22% (see Figure 15.4).[4]

Brownsville, Cameron County's largest city, grew from 98,962 in 1990 to 139,722 in 2000: a 41% increase. Matamoros' population grew from 303,293 in 1990 to 418,141 in 2000, a 38% increase. Cameron County's minority Anglo and black populations are projected to have zero growth, while there is expectation for dramatic growth the Hispanic population to 400,000 by 2015, and above 600,000 by 2040. Matamoros' current immigrant and native populations are projected to more than double by 2040, producing a dramatic population growth to 650,000 by 2015, and above 1,200,000 by 2040.

Figure 15.4
Population Growth Overview: 1990-2000

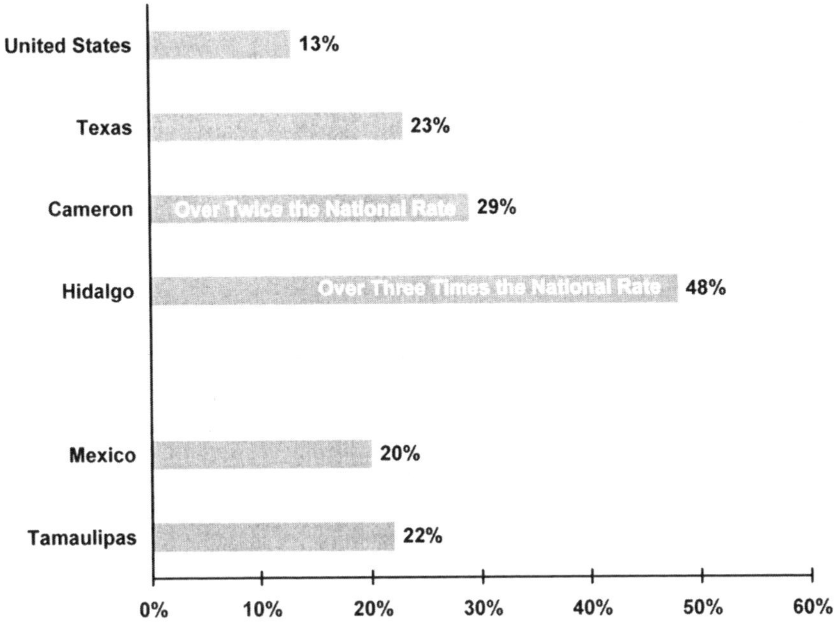

Source: U.S. Census Bureau, 2001 Supplementary Survey, and Mexico 2000 Census INEGI databases.

In 1990 the highest growing age bracket in Matamoros was ages 15 to 30, a reflection of the impact of *maquiladoras* in attracting a young child-bearing population from Southern Mexico and Latin America. This situation is illustrated in the 2000 census by growth in the 25-to-40 age bracket, and the corresponding age brackets of children under 5 years old and 5 to 9 years old. Matamoros has a younger population than Cameron County with a median age of 23; 58% of the population falls within the Mexican workforce ages of 15 to 55. Cameron County's population has a median age of 29; 69% of the population falls into the U.S. workforce ages of 15 to 64. The border population is young as the median age of the U.S. population is 35. The median age of the Texas population is 32. The median age of Cameron County is 29, and Matamoros has a median age of 23. Matamoros has more children age 14 and under than Brownsville has in its entire population (see Figure 15.5).

Figure 15.5
Population by Age: Brownsville, Texas versus Matamoros, Mexico

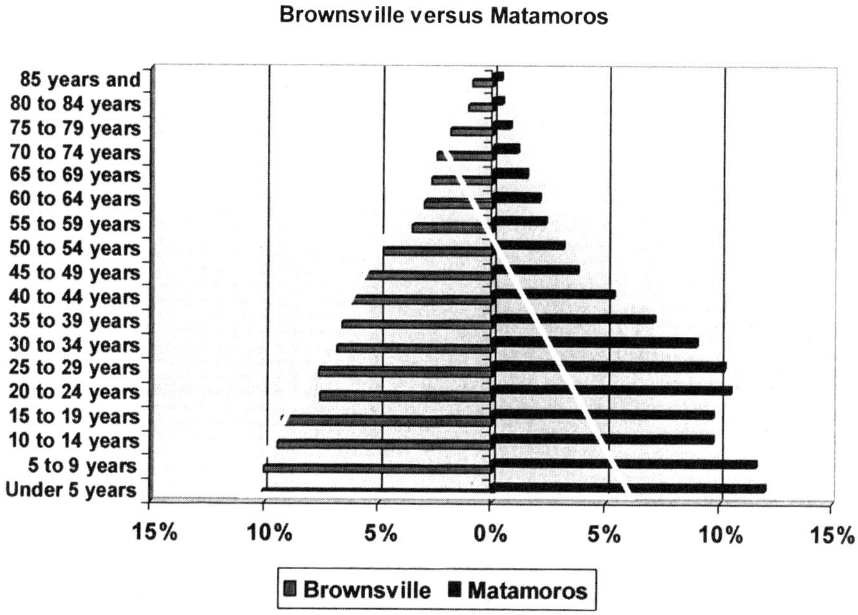

Sources: U.S. Census Bureau, Mexico 2000 Census INEGI.

ECONOMIC PROFILE

Maquiladoras in Northern Mexico, during the 1990s, offered reliable and relatively well-paid employment opportunities for many Mexicans and for immigrants from South America. These salaries were significantly lower than the U.S. minimum wage and this disparity motivates Mexican workers to seek employment across the border in Texas. And this migration often continues north as professionals and skilled workers in South Texas elect to move north for higher wages and increased career opportunities. Cameron County's annual median household income of $22,959 falls $12,490 short of Texas' $35,449, but is higher than that of nearby Hidalgo, Willacy, and Starr counties (see Figure 15.6). Matamoros' annual median household income of $10,570 is considerably less than the Texas counties to the north and $24,897 less than the state of Texas in general.[5]

Figure 15.6
Median Household Income Comparisons: South East, Texas, and Matamoros, Mexico

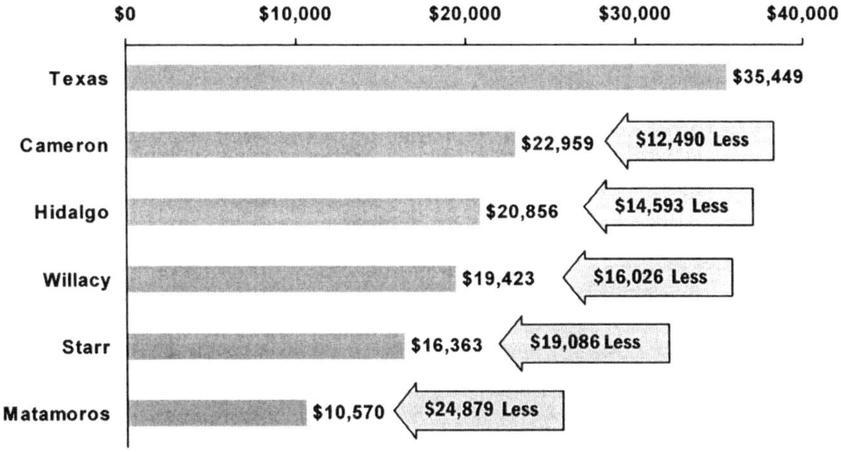

Source: Gibson and Rhi-Perez (2003).

No matter how they are measured, unemployment rates along the border of South Texas can be described in *multiples* of the state rate. These rankings show Cameron and Hidalgo counties are not merely "lagging" statewide (Table 15.1); they rank worst and second-worst in the nation for a range of key poverty-related indicators—worse than "inner city" counties such as Bronx County, New York; Orleans Parish, Louisiana; and Los Angeles County, California. Furthermore, "inner city" counties tend to be surrounded by more prosperous areas in direct proximity while Cameron and Hidalgo counties share borders with Texas and Mexican rural counties that are considerably worse off economically (Estrada and Hardebeck, 2001).

Table 15.1
Unemployment Statistics December 2002

	Labor Force	Employment	Unemployment	Percentage
Texas	10,715,009	10,103,679	611,330	5.7%
Cameron County	135,879	121,955	13,924	10.2%
Hidalgo County	223,438	193,815	29,623	13.3%
Willacy County	5,919	5,006	913	15.4%
Starr County	22,856	17,874	4,982	21.8%

Source: Texas Workforce Commission.

U.S. census data shows that the poverty level in Cameron County is 33%, over twice the state rate of 15%; and the poverty rate increases for Hidalgo, Willacy, and Starr counties. South Texas' economic profile underscores this reality, as Cameron County ranks as having (of "all" counties in the United States with over 250,000)[6]:

- the highest percentage of people living below the poverty level
- the highest percentage of children under 18 living below the poverty level

If current trends continue, what's ahead for Cameron County and the South Texas Border Region include (adapted from Murdock, 2000):

- a growing unskilled, undereducated population that cannot meet the demands of a technology-based workplace
- lost ground in the highly competitive global marketplace
- average household income will decline by $4,000 in constant dollars by 2030
- more public spending on prisons, welfare, Medicaid

BINATIONAL SURVEY

During 2001-2002, a survey on the importance of and challenges to regional economic development was mailed to 4,500 respondents in Cameron County. The same survey was administered in Spanish during face-to-face interviews with 100 Matamoros respondents.[7] While not a scientific survey, useful insights on regional and cross-border economic development were gained by comparing respondents' answers from Cameron County and Matamoros.[8] The survey focused on three main issues concerning job creation, economic development, and wealth creation in Cameron County and Matamoros in the next five to ten years, as follows:

- the importance of regional industries: established and emerging
- the importance and effectiveness of regional economic development factors
- the importance and effectiveness of regional economic development strategies

A key finding of the survey was that national, city, and institutional borders are transparent in the face of regional education, economic, and health care challenges and opportunities. Leaders on both sides of the Rio Grande voice similar concerns and priorities for the region's future economic development and sustainable quality of life.

Leading Regional Industries

A list of 40 industries was constructed to reflect the current industry structure of the border region. Respondents were asked to indicate the importance of these industries for job creation, economic development, and wealth creation in the coming 5-10 years. The highest percentage of Cameron

County respondents considered health services as an important industry for job creation, economic development, and wealth creation, followed by education services, and conventions & tourism (see Figure 15.7). These three service industries are followed by commercial, residential & heavy (highways, bridges) construction, trucking & warehousing, air transportation, retail trade, telecommunications, business services, and banking & financial services by over 50% of the respondents. In short, in terms of economic development, most Cameron County respondents stress the importance of service industries. The highest ranking technology-based industry is electronics & electrical equipment with a 48% respondent rate.

Figure 15.7
Cameron County: Important Established Industries for Job Creation, Economic Development, and Wealth Creation in the Next 5-10 Years

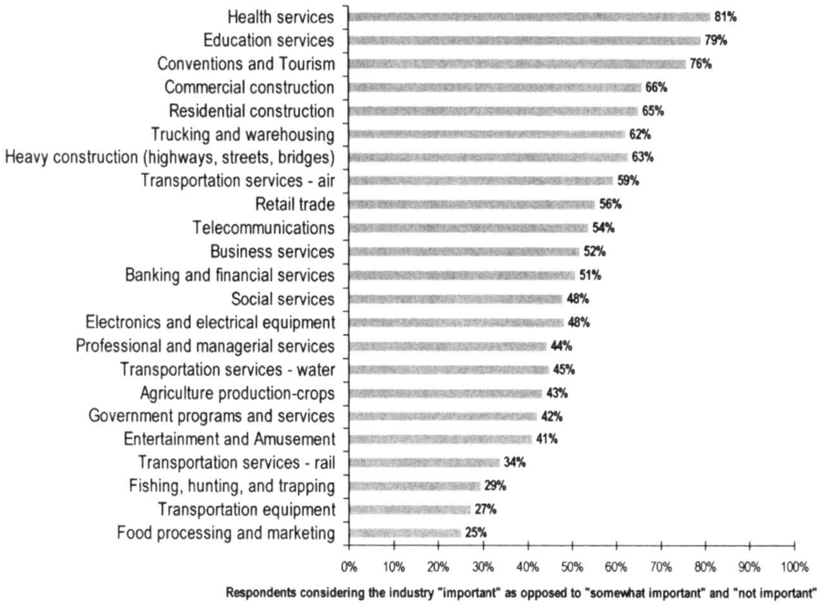

Source: Gibson and Rhi-Perez (2003).

Some 73% of Matamoros respondents considered education services as the most important industry for job creation, economic development, and wealth creation in the next 5-10 years, followed by health services, and tourism, entertainment & amusement (see Figure 15.8). These three industry sectors are followed by energy products & distribution, telecommunications, banking & financing services, transportation (trucking, rail, ship, air), seaports, airports, ground ports & international bridges, and government programs & services. The highest ranking technology-based industries are manufacturing (39%) and electronics (34%).

Figure 15.8
Matamoros: Important Established Industries for Job Creation, Economic Development, and Wealth Creation in the Next 5-10 Years

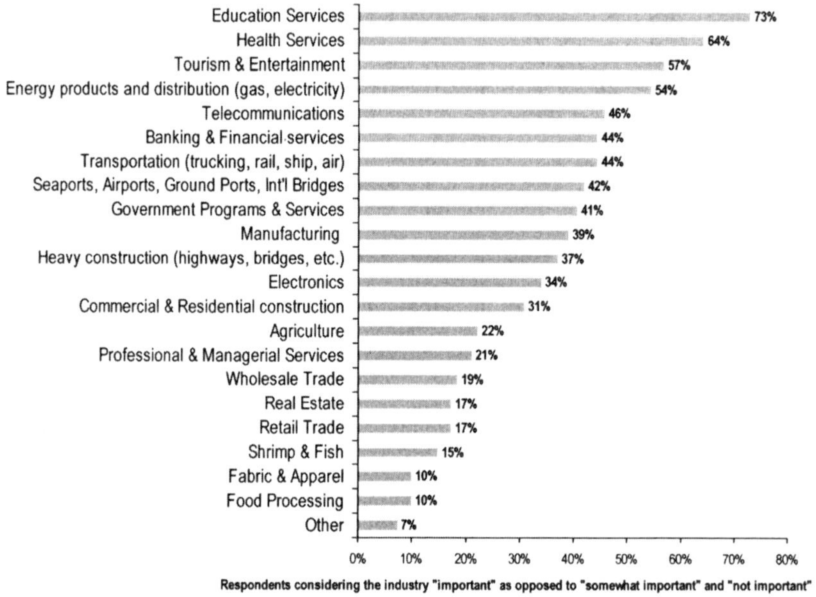

Source: Gibson and Rhi-Perez (2003).

In summary, there is general agreement between Cameron County and Matamoros respondents when asked to rank the top established industries for economic development. For both groups the top three industries are health services, education services, and tourism (conventions & entertainment). Cameron County ranks heavy construction, commercial construction, and trucking and warehousing next, while Matamoros respondents follow the top three rankings with energy utilities, electronics, and manufacturing.[9]

Top Five New and Emerging Technology Industries

For Cameron County community leaders the most important regional new and emerging industries for job creation, economic development, and wealth creation in the next 5-10 years are in order of preference: medical technologies, energy efficiency & conservation, telecommunications, advanced shipping & logistics, computer & information technologies, and agriculture technologies (see Figure 15.9).

Figure 15.9
Cameron County: Ranking of New and Emerging Industries for Job Creation, Economic Development, and Wealth Creation in the Next 5-10 Years

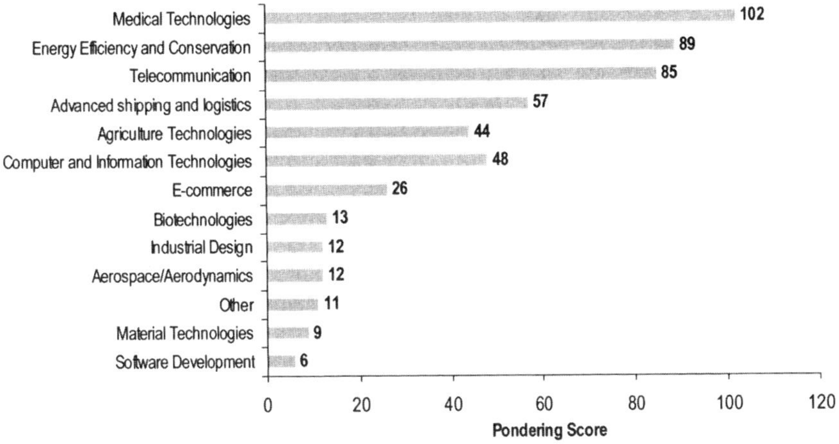

Note: The Pondering Score was attained by multiplying by 5 the number of persons that selected rank 1, by 4 the number that selected rank 2, and so on. These results gave a pondered value to each of the five ranks.
Source: Gibson and Rhi-Perez (2003).

For Matamoros community leaders, the most important new and emerging industries for job creation, economic development, and wealth creation are, in order of preference: environmental conservation, advanced food & agriculture, energy efficiency & conservation, advanced telecommunications, and computer & information technologies (see Figure 15.10).

Figure 15.10
Matamoros Ranking of New and Emerging Industries for Job Creation, Economic Development & Wealth Creation in the Next 5-10 Years

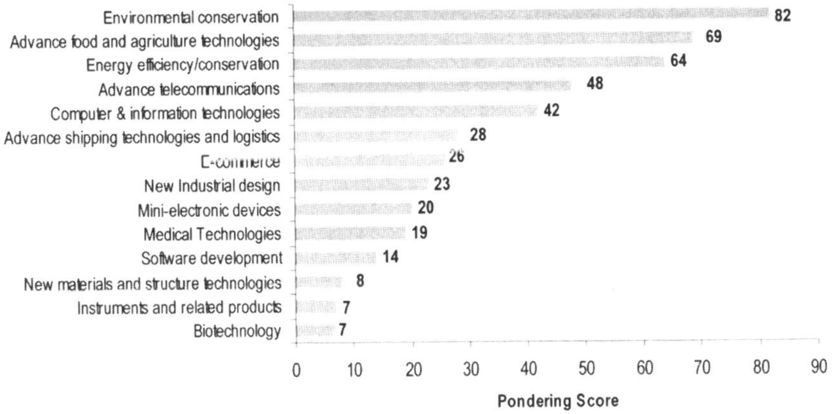

Source: Gibson and Rhi-Perez (2003).

In summary, when asked to rank the top five emerging technology industries, Cameron County and Matamoros respondents agree on the high ranking of advanced food & agricultural technologies, energy efficiency & conservation, advanced telecommunications, and computer & information technologies. The major difference is that Cameron County respondents rank medical technologies highest and advanced shipping & logistics considerably higher than Matamoros respondents, while Matamoros respondents rank environmental conservation technologies high.

Key Factors in Economic Development

Cameron County and Matamoros community leaders were asked their opinions about the degree of importance of 24 key factors in terms of job creation, economic development, and wealth creation in the next 5-10 years. The top nine factors for both Cameron County and Matamoros are listed in Table 15.2.

Table 15.2
Rankings of Key Factors in Economic Development

Cameron County Ranking	Matamoros Ranking
Quality of K-12 education	Quality of college & university education
Quality of college & university education	Affordable & available energy
Affordable & available water supplies	Quality of K-12 education
Quality of technical & vocational education	Affordable & available water supplies
Skill of entry-level workforce	Health Services
Utilities	Quality of technical & vocational education
Skill of managerial & professional workers	Tax Incentives
Affordable & Available Energy	Skill of entry-level workforce
Telecommunications	Industrial Parks

Source: Gibson and Rhi-Perez (2003).

In summary, both Cameron County and Matamoros community leaders consider education (university, college, and K-12) and affordable & available water supplies as well as the quality of technical & vocational education and skill of the entry-level workforce as *important factors* for regional economic development in the coming 5-10 years. This finding was corroborated by the responses to open-ended questions that emphasized that the growth, retention, and relocation of new industries would be severely restricted without a properly educated and trained workforce as well as affordable and available water supplies.

Effectiveness: Cameron County

Cameron County respondents were asked to rank the effectiveness of their community in providing key economic development factors. Cameron County respondents give the top two effectiveness ranks to utilities (60%), public services, affordable & available energy (54%), and affordable & available housing (54%) as being effectively provided. At the bottom of the effectiveness list, all below 31%, are cross-border infrastructure, skill of managerial & professional workforce, industrial & university research & development, and skill of entry-level workforce.

Figure 15.11 represents the combination of perceived importance and effectiveness of each of these factors. Important factors with a low rank of effectiveness (high ineffectiveness) are highlighted if they are located above the "critical index" curve.[10]

Figure 15.11
Importance versus Effectiveness of Cameron County Factors

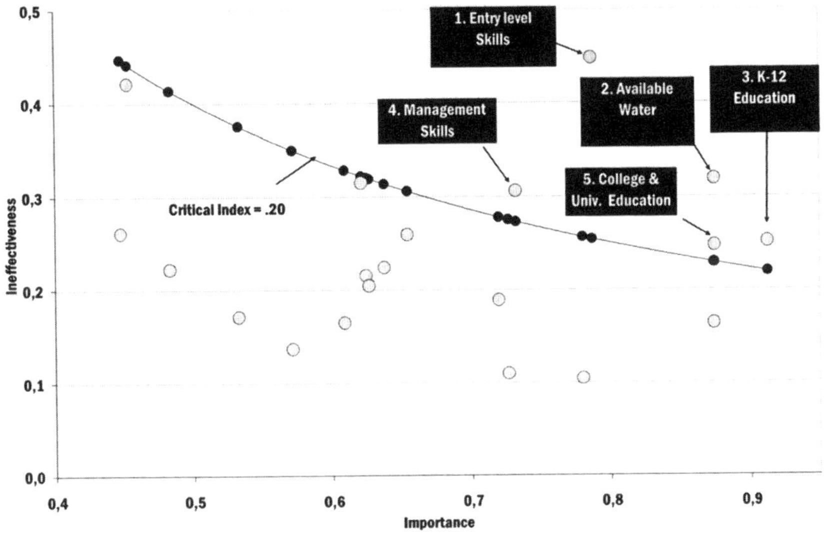

Source: Gibson and Rhi-Perez (2003).

This analysis illustrates that while Cameron County respondents consider skill of entry-level workforce as one of the most important factors for economic development, providing such a workforce is considered the least effective attribute of the region. Other factors with high importance coupled with low regional effectiveness are:

- affordable and available water supplies
- quality of K-12 education
- skill of managerial and professional workforce
- quality of college and university education

In summary, the factors ranking highest in importance for Cameron County's economic development also rank highest in terms of region's general lack of effectiveness in providing these same factors.

Effectiveness: Matamoros

Matamoros respondents were also asked to rank the *effectiveness* of their region in providing factors in economic development. The factor considered most effectively provided is industrial parks (46%), followed by telecommunications (34%), quality of life (32%), Internet availability (31%) and quality of K-12 education (30%). However for Matamoros the most striking finding is the generally low level of "all responses" regarding effectiveness in providing these key factors.

A critical index was calculated for each factor by multiplying the percentages of the "very important" responses by the percentages of the "not very effective" responses (see Figure 15.12). This analysis illustrates that survey respondents consider affordable & available energy as the most important factor for economic development, yet providing such energy is one of the least effective attributes of the region. Other factors with high importance coupled with low regional effectiveness include:

- financing and capital access
- business incentives
- public services (police, fire, etc.)

Figure 15.12
Matamoros: Critical Index of Factors (Importance versus Effectiveness)

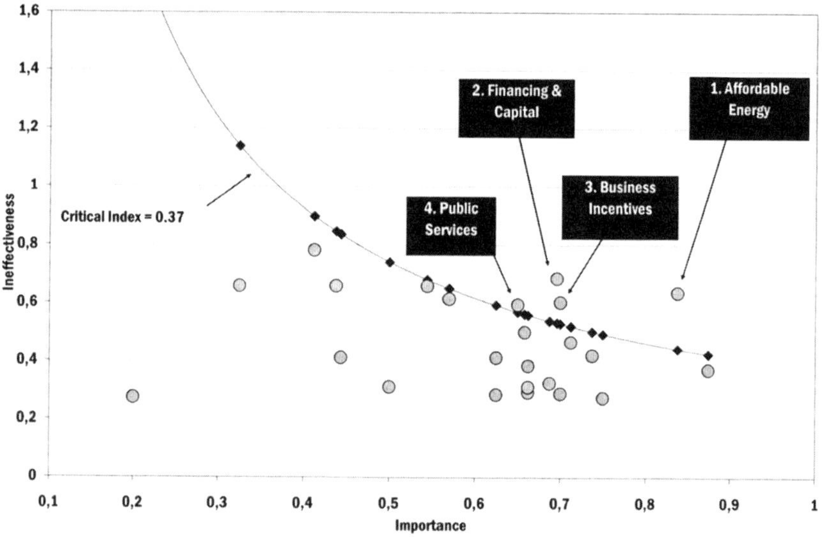

Source: Gibson and Rhi-Perez (2003).

STRATEGIES FOR ECONOMIC DEVELOPMENT

Cameron County and Matamoros community leaders were also asked to consider the importance of economic development strategies as well as the region's ability to perform these activities. In general, both regions value similar strategies as being key to job growth and economic development (see Table 15.3).

Table 15.3
Ranking of Economic Development Strategies Perceived, by Cameron County and Matamoros Respondents, as Being Key to Regional Development

Cameron County Ranking	Matamoros Ranking
Retention/expansion of existing industries/business	A "Can do" attitude for the region
Promotion/support of local start-up industries/businesses	Retention/expansion of existing industries/business
Regional economic development plans focusing on job creation	Promotion/support of local start-up industries/businesses
Promotion/support local entrepreneurs high-tech value-added industries	Promotion/support local entrepreneurs high-tech value-added industries
Relocation of industries/businesses from outside the region	Regional economic development plans focusing on job creation
Regional economic development collaborations on U.S. side of the border	Cross-border economic development collaborations

Source: Gibson and Rhi-Perez (2003).

Community leaders from both Cameron County and Matamoros stress the importance of the retention and expansion of existing firms as well as the promotion and support of local start-up industries and businesses and local technology-based entrepreneurs. The main difference is that Matamoros respondents place greater emphasis on (1) cross-border economic development collaborations and (2) the further development of *maquiladoras*.

Strategic Effectiveness: Cameron County

Some 58% of Cameron County respondents are pleased with the strategy for development of the *maquiladora* industry followed by the effectiveness of free trade zones (54%), retention & expansion of existing industry businesses (43%), and cross-border economic development collaborations (40%). Cameron County respondents list as least effective promotion/support of local start-up industries and businesses (30%), leveraging community assets (26%), promotion/support of local entrepreneurs in high-tech value-added industries (24%), access to venture capital (22%), and the promotion & support of new business incubators (16%).

Figure 15.13 represents the combination of the perceived importance and effectiveness of strategies. A single point on the graph represents each strategy's importance and its regional effectiveness according to respondents' opinions. Strategies with high importance coupled with low community effectiveness include:

- promotion/support of local entrepreneurs in high-tech high-value added industries/ businesses
- access to venture capital
- promotion/support of local start-up industries/businesses
- regional economic development plans focusing on job creation

Figure 15.13
Cameron County: Critical Index of Strategies (Importance versus Effectiveness)

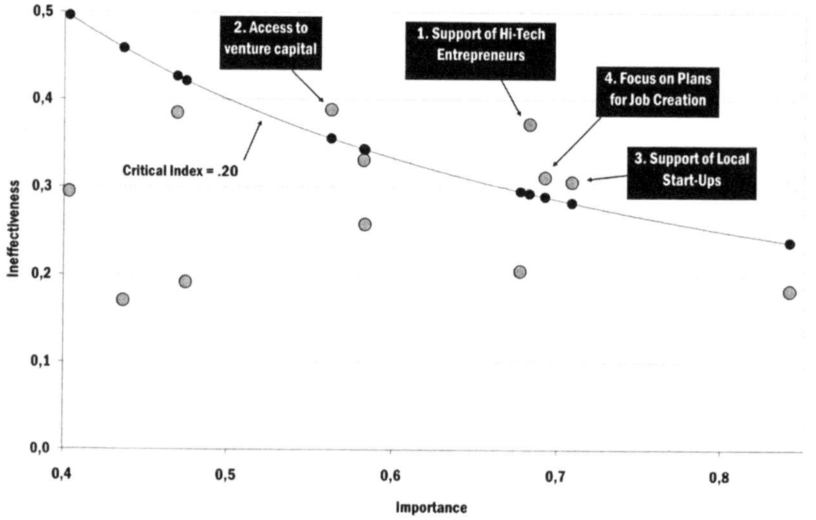

Source: Gibson and Rhi-Perez (2003).

Strategic Effectiveness: Matamoros

Matamoros respondents are generally not impressed with their region's effectiveness in providing economic development strategies. About 40% of respondents are pleased with the effectiveness of strategies for the development of the maquila industry; however, less than 33% of the respondents think the region is effective in economic diversification and 25% or less believe the region is effective in all the other listed economic development strategies. At the bottom of the "effectiveness list" are:

- a "can do" attitude for the border region (15%)
- leveraging of community assets (15%)

- cross-border collaboration (15%)
- a regional strategy targeting economic development for job & wealth creation (13%)

Figure 15.14 presents Matamoros's most critical economic development strategies that are also considered least effectively provided as follows:

- a "can do" attitude for the border region
- regional strategy—economic development plans focusing on job creation
- promotion/support of local entrepreneurs in high-tech high-value added industries/ businesses access to venture capital
- promotion/support of local start-up industries/businesses
- economic development collaboration of cities on the Mexican side of the border

Figure 15.14
Matamoros Critical Index of Strategies (Importance versus Effectiveness)

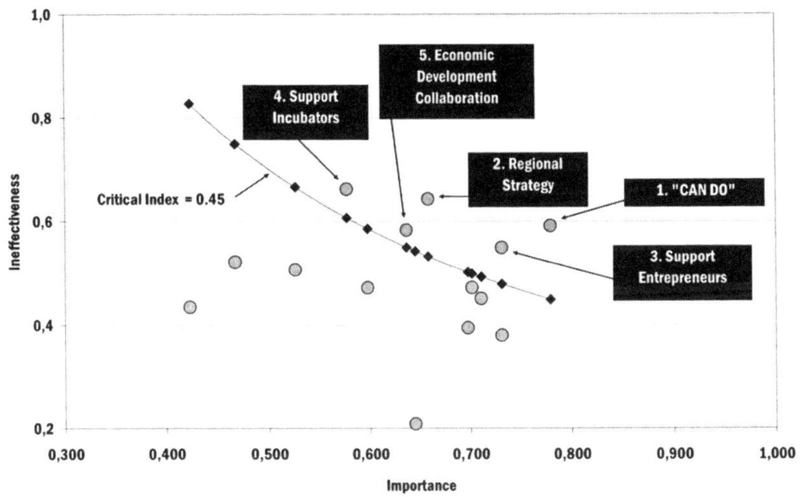

Source: Gibson and Rhi-Perez (2003).

RECOMMENDATIONS

A key organizing principle of this chapter has been that the border region in general, and Cameron County, Texas, and Matamoros, Mexico, in particular, is at a crossroads in terms of business and industry development, education and workforce training, regional leadership, and maintaining and enhancing an accessible quality of life for all the region's inhabitants.

By highlighting select demographics we emphasize regional challenges of high unemployment, birthrate and population growth, and children in poverty, coupled with low per capita income, annual pay, and growth rate in annual pay. Both Cameron County and Matamoros community leaders' opinions focused on current and future challenges to regional economic development, including the key importance of quality education and workforce training, providing needed support for entrepreneurs, and the lack of regional and cross-border cooperation. Based on these data and analyses, this chapter concludes with four recommended initiatives that are at the center of the crossroads and that are targeted to bring academic, business, and government leaders and citizens together to for enhanced regional and cross-border development.

1. regional and cross-border collaboration on targeted technology-based industry clusters for accelerated economic development
2. education, training, and research linked to regional and cross-border economic development
3. fostering regional and cross-border entrepreneurship
4. fostering cross-border and global value-added partnerships for accelerated development

INITIATIVE 1: REGIONAL AND CROSS-BORDER COLLABORATION ON TARGETED TECHNOLOGY-BASED INDUSTRY CLUSTERS FOR ACCELERATED ECONOMIC DEVELOPMENT

Because of their centrality to the border region in a manner that is unmatched by other national or global locations, this report suggests four targets for regional and cross-border collaboration for accelerated economic development:

- value-added manufacturing and *maquiladoras*
- transportation services, logistics and distribution
- border security
- health services and life sciences

Value-Added Manufacturing and *Maquiladoras*

Without adequate numbers of highly skilled labor, the region will stagnate and fall further behind in its ability to create wealth and higher value jobs and to employ and retain its young workforce. Cameron County's manufacturing sector as a whole and Matamoros *maquiladoras* in particular can no longer compete in the global economy solely on the basis of inexpensive labor. If current trends continue, the Lower Rio Grande Valley region will continue to see increasing competition for emerging industrial powers, especially those in Southeast Asia and China. As a result, the Cameron County-Matamoros region needs to devise a cluster-based strategy built on cross-border cooperation targeting higher value-added niche industry clusters and components (De Los Reyes and Rodriguez, 2002; Lazcano, Leal, and Rhi-Perez, 2001).

Maquiladora operations that focus on labor-intensive and low value-added operations are moving from Mexico to Asia, primarily China and India. The message is clear: Neither Matamoros or the surrounding border region can rely on low-wage workers to attract and retain a manufacturing industry. The best way for area manufacturing to compete in the 21st century is thought more value-added and just-in-time manufacturing. Manufacturing clusters need to be developed in such industries as electronics, plastics, and steel stamping, supported by such industries as software, design, and logistics, along with supportive infrastructure including quality technical education, industrial parks, and professional services of finance, marketing, and technology development.

As stated by Rolando Gonzalez-Barron, President, Consejo Nacional de Maquiladoras de Exportacion de Mexico (August, 2002), "Maquiladoras can be competitive in the global economy if their production and manufacturing operations are assisted with adequate financial support, new technology development, and marketing—these three are needed in addition to quality production and manufacturing. The fastest way to grow is through joint ventures with other companies for access to technology, markets, and capital" (cited in Gibson and Rhi-Perez, 2003).

Transportation Services, Logistics and Distribution

The Cameron County/Matamoros region has significant geographic and logistical assets to build a world-class research and testing laboratory for multimodal transportation for the 21st century. These assets include:

- Cameron County and Matamoros being bicultural and binational neighbors with the greatest U.S. proximity to major Mexican cities and manufacturing
- the location of the seaport of Brownsville and nearby Mexican intercoastal ports
- adjacent railroads with access to the U.S. and Mexican interiors
- nearby air cargo and passenger terminals
- four international bridges

This unique set of binational assets and the relative small size of the associated facilities provides an ideal setting for research and innovation on cutting-edge transportation and logistics technologies and processes associated with such cross-border issues as enhanced supply-chain management, trade expansion, and national security. Improved logistics and distribution efficiencies—including decreasing time and cost while increasing security—are crucial to the sustainability and growth of border industries (Hartnett, 2002; Rodarte, 2002).

A regionally based Binational Transportation and Logistics Research Center could promote global and seamless intermodal transportation and logistics systems through education programs, research, and outreach activities. It could serve as a "think and do" tank in partnership with regional, national, and global industry in such areas as:

- port-of-entry security technologies
- supply-chain management
- data systems integration and management

- cargo surveillance systems
- building cross-border partnerships and the leveraging of assets

Border Security

Targeted industry clusters for accelerated economic development—manufacturing and *maquiladoras*; transportation services, logistics and distribution; and health services and life sciences—are all impacted by border security. As the U.S. government turns new scrutiny and resources to border processes, they articulate the need for new technology development and a highly skilled workforce (U.S. Department of the Treasury, 2001). Border-based education and research centers could benefit from focusing on the concept of a secure border (see Figure 15.15).

Figure 15.15
Enhanced Texas-Mexico Cross-Border Industry and Security Needs

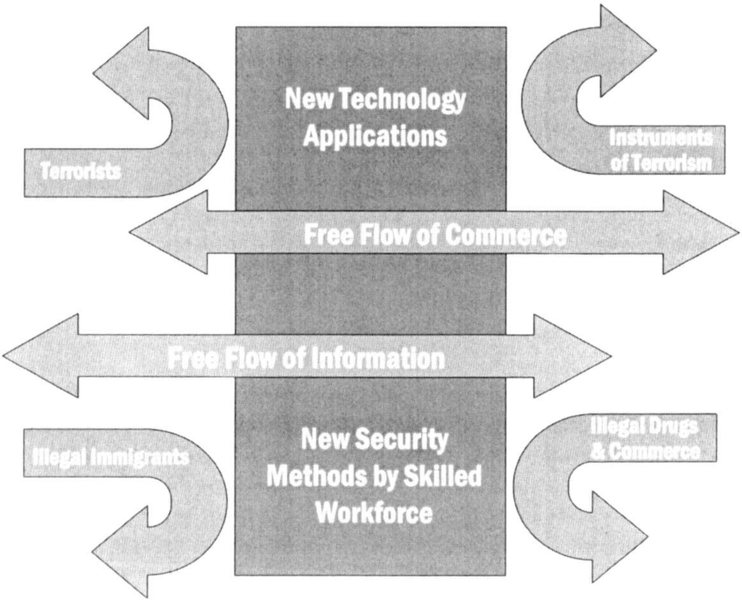

Health Services and Life Sciences

The Lower Rio Grande Valley needs to better link and leverage regional assets on both sides of the border to meet border health care challenges and to link with emerging regional health care clusters including education and training, retention and development of health care businesses, and border-specific health care challenges (Aranda, 2002; Bell, 2002; Guzman, 2001a, 2001b; Warner and Hopewell, 1999; Zavaleta, 2000).

On the one hand, the Lower Rio Grande Valley lacks an important life science infrastructure including significant numbers of large and small firms, R&D funding and innovation, and professional support including capital formation. Other regions in Texas such as Dallas-Fort Worth, San Antonio, and Houston-Galveston-The Woodlands and other regions in the United States such as California and the Northeast are more competitive in terms of such regionally based assets for the development of life science clusters (Duca and Yucel, 2002).

On the other hand, the Lower Rio Grande Valley's geography and population provide a world-unique research and education "laboratory" on a broad range of important national and global health care challenges. Border realities include diseases that reflect agrarian Third World conditions; a predominance of young and older binational residents, including their respective health care needs; poverty and malnutrition coupled with a lack of adequate quantities of fresh water, inadequate housing, and colonias; and a lack of needed quantities of health care professionals (Furino, 2001; Furino and Miller, 2002; Henton and Seline, 1998).

Quality faculties have already been attracted to Regional Academic Health Center (RAHC) programs that offer exceptional research opportunities leading to career-building contributions to health care practice. With quality faculty come exceptional students and research funds. From these activities, and with important regional support structures, can come technology spin-off activities leading to company start-ups and the development of a life science industry cluster. There is a window of opportunity for the Lower Rio Grande Valley to establish binational, world-class research centers focused on border region health care challenges. Such centers of excellence would also be important in establishing a new vision of health care business and manufacturing in the Valley.

INITIATIVE 2: EDUCATION, TRAINING, AND RESEARCH LINKED TO REGIONAL AND CROSS-BORDER ECONOMIC DEVELOPMENT

> While legislators, higher education leaders and supporters, and interest groups will have to expend energy on traditional challenges [in 2003], the framework for their debates should be that of "Closing the Gaps." Every significant proposal to add, expand or reduce appropriations, authority and programs should be measured against the contribution it would make to close the four gaps in participation, success, excellence, and research....We cannot afford not to succeed.
>
> Don W. Brown, Commissioner of Higher Education Coordinating Board, Austin American Statesman, *Insight*, September 1, 2002, p. H5

On the one hand, both Cameron County and Matamoros respondents emphasize the importance of quality education: college and university, and K-12, as well as vocational training. On the other hand, they also report that their respective regions are challenged to provide needed quality education and training for 21st-century jobs and careers. For example, the University of Texas at Brownsville/Texas Southmost College (UTB/TSC) is challenged to double its enrollment to 20,000 by 2010. UTB/TSC requires state funding to build needed facilities and to hire needed faculty to grow existing programs and to build new graduate programs and centers of research excellence. Texas State Technical College (TSTC) also relies on the State of Texas to provide needed financial resources. Both UTB/TSC and TSTC are challenged to recruit, train, and retrain faculty and to develop new technology-based programs for 21st-century Texas-based industry. While university-based research centers would be a big boost to the regional economy, full economic success can be achieved only if the local workforce is educated and trained to fill the new jobs and careers as they become available (Sharp, 1998; Tech-Prep Texas, 2001).

In light Texas and Mexican budget restrictions, this chapter advocates the development of "Partnerships for Excellence" between regional and binational businesses and academia to accelerate the growth of targeted industry clusters as well as building binational centers of education and research excellence. Such Partnerships for Excellence have proven to be viable strategies for universities to:

- increase education and research excellence
- contribute to regional economic development
- brand a region as an important national and international technology and entrepreneurial center

Partnerships for Excellence can also benefit the larger community in terms of increased tax income, real estate development, sales of consumer goods and services, and national and global alliances for accelerated technology-based growth. Such Partnerships for Excellence could target the development of R&D centers of excellence in such areas as:

- a Center for Binational Entrepreneurship (CBE) that could be linked to emerging regionally based industry clusters of manufacturing/maquiladoras, transportation & logistics, health services, and life sciences; such a center could promote binational entrepreneurial competitions for university, college, and high school students
- a Binational Transportation & Logistics Research Center (BTLRC) that could be linked to supply-chain management, advanced telecommunications, and cross-border security including health care concerns such as bioterrorism
- world-class manufacturing including technologies and processes associated with value-added *maquiladora* binational and global operations
- border health care challenges with links to emerging centers of education and research excellence at the RAHC, which is attracting excellent faculty that is recruiting excellent students and building research funding; over time these activities should lead to spin-out and start-up technology-based business ventures

INITIATIVE 3: FOSTERING REGIONAL AND CROSS-BORDER ENTREPRENEURSHIP

Cameron County does have regionally based binational entrepreneurial success stories.[11] A rather basic but important way to promote regional entrepreneurial effort is to celebrate such successes. On the one hand, Cameron County and Matamoros respondents emphasize the importance of business retention and expansion, promotion of local entrepreneurs, and the development of regional economic development plans. On the other hand, respondents on both sides of the border also give low marks to their region's effectiveness in providing support for entrepreneurs, leveraging community assets, cross-border collaboration, and developing regional plans for economic development.

The Lower Rio Grande Valley has long suffered from a regional "talent drain" in that many educated and skilled workers and professionals leave the region to build their careers and earn higher salaries. To help counter and reverse the talent drain, the Valley needs to emphasize the importance of grassroots development of entrepreneurial initiatives and celebrate homegrown entrepreneurial successes to:

- bring university and college graduates home to help build "smart infrastructure" in targeted industry clusters
- leverage the talent and networks of university and college alumni, where they are currently located for enhanced national and global exposure and access
- establish Binational Business Plan Competitions that feature Brownsville and Matamoros-based students and technologies
- develop cross-border entrepreneurial competitions for high schools focusing on technology, civic, and social entrepreneurship

INITIATIVE 4: FOSTERING CROSS-BORDER AND GLOBAL VALUE-ADDED PARTNERSHIPS FOR ACCELERATED DEVELOPMENT

On the one hand, the Lower Rio Grande Valley benefits from a strong binational economic and cultural heritage; on the other hand, the border region suffers from parochialism that fosters city-based and bilateral partnering rather than more regionally based, multilateral cooperative strategies and expanded cross-border cooperation. At one level the general objective is to target activities that foster regional and cross-border partnerships and alliances for targeted opportunities of collaboration that show business and quality of life results to the benefit of the Lower Rio Grande Valley and Matamoros academic, business, and government sectors. At another level the objective is to identify and eliminate relatively minor, but important, structural barriers to cross-border cooperation. Well-meaning rules and procedures designed for 20th-century business and society may inhibit value-added binational cooperation needed in the 21st century.

The South Texas/Northern Border region has a unique opportunity to build a binational model of wealth creation and prosperity sharing that other cross-border regions will seek to emulate. Technology capabilities of the 21st-century and assets and challenges of the Lower Rio Grande Valley can be linked. Improved telecommunications infrastructure is one key way to accelerate regional economic development and quality of life in knowledge-based economies. Community-based knowledge networks have the potential to enhance social capital that facilitates the exchange of knowledge among community leaders and organizations and the subsequent leveraging resources (Carbonara, 2002; CBIRD, 2002). The Internet brings new possibilities to the Cameron County/Matamoros region, from improved delivery of education and training, and health care services, to greater civic engagement and new economic opportunities. Cameron County/Matamoros community leaders can build new models for cross-border knowledge sharing, leveraging of resources, and problem solving. A willingness to tolerate multiple approaches needs to be accompanied by a willingness to apply lessons learned.

International initiatives include leveraging cross-border networks and entrepreneurial infrastructure of the South Texas Border Region and Northern Mexico with access to international markets with an emphasis on Latin America. However, it is also possible to build knowledge bridges globally to access world-class R&D and to develop and manufacture these technologies in the Cameron County/Matamoros region (Kozmetsky and Yue, 1997). R&D and technology expertise can be imported to the Valley to be developed locally for emerging clusters in health sciences, transportation and logistics, manufacturing and maquiladoras, and border security.

CONCLUSIONS

This chapter advocates four strategies for accelerated cross-border economic development (see Figure 15.16):

- the recruitment of key technology-based industries, service companies, and talent
- retaining and facilitating the growth of established and emerging industries and talent
- accelerating the growth of new technology-based industries and talent
- fostering academic-business-government partnerships binationally and globally

Figure 15.16
Four Interrelated Strategies for Accelerated Cross-Border Economic Development

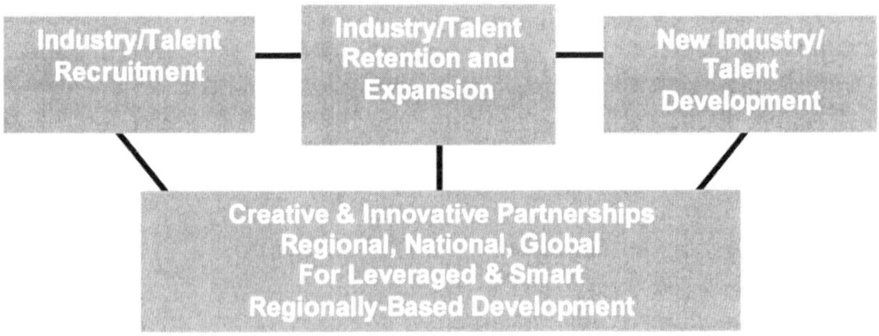

Industry/talent recruitment is an important and long-standing strategy for regional development. However, this strategy often leads to enhanced competition and win-lose scenarios as neighboring communities seek to outbid each other with tax breaks and other recruitment incentives. Such competition is prevalent in economic development environments that are challenged by domestic corporate downsizing and off-shore corporate growth, especially in labor-intensive and low value-added (non-knowledge-based) industries such as North Mexico and South Texas.

Retaining and growing regionally based industry and talent focuses public and private resources on firms and human resources that already have a presence in and a loyalty to the area. Facilitating the growth of talent and new company start-ups emphasizes the importance of regionally based entrepreneurship for diversifying the local economy for enhanced wealth and job creation. These two economic development strategies—industry/talent retention & expansion and new industry/talent development—have been key to the accelerated wealth and job creation of most successful and innovative technology/knowledge growth areas including Silicon Valley, CA; Boston, MA; and Austin, TX.[12]

Creative and innovative regionally based academic/business/government partnerships are key for knowledge-based economic development to "take-off" and "leap-frog" national and global competition. In the 21st century, geographic regions are the most appropriate unit of analysis for examining changes and opportunities for economic and social development. Defining an economically meaningful region is not so much a matter of national borders or geographic size, but rather of finding areas where educational, economic, cultural, and other linkages are strong or could be stronger relative to external linkages.

As emphasized, the Cameron County/Matamoros border region, as well as the Lower Rio Grande Valley, is at a crossroads in terms of (see Figure 15.2):

- business and industry development for accelerated wealth creation and the growth of globally competitive, career-oriented jobs
- "closing the gaps" in education in terms of participation, success, excellence, research, and workforce training for 21st-century jobs
- maintaining and enhancing an accessible quality of life for all the region's citizens and countering the regional "brain drain" of some of the most capable workers and professionals
- visionary leadership that fosters cross-border strategies for action

Institution-based excellence—whether it be academic, business, or government—is NOT sufficient. What is required is visionary public-private leadership and leveraging of cross-border assets to overcome challenges to accelerate cross-border development. How effectively Cameron County and Matamoros and other border business, academic, and government sectors collaborate, coordinate, and cooperate will, in large part, determine the region's ability to build and sustain requisite levels of education and training needed to accelerate economic growth and create high-value, career-oriented jobs while sustaining and enhancing an accessible quality of life for all: One Region–Un Futuro.

Key to the border region meeting these challenges is enhanced leveraging of cross-border industry assets, education and training, government and foundation resources, and community and cultural assets and knowledge to foster regional and binational development of talent, knowledge, capital, and know-how for accelerated development. Cross-border assets will be leveraged through business, civic, and social entrepreneurship that facilitates networking and the forming of binational visions and strategies for success. Committees are not collaboration. Collaboration is meaningful knowledge sharing and targeted leveraging of resources (i.e., financial, intellectual, and physical) to overcome specific challenges. For such collaboration to be realized it needs to be action oriented with metrics for success (short- and long-term).

NOTES

1. Special appreciation for this research project, data analyses, and report writing is extended to our collaborators Margaret Cotrofeld, Mark Gipson, and Richard Rodarte at IC^2 Institute, The University of Texas at Austin, and Oralia De Los Reyes and Ignacio Rodriguez at The Cross Border Institute for Regional Development (CBIRD), The University of Texas at Brownsville/Texas Southmost College. For the complete report that is the basis of this chapter, please see Gibson and Rhi-Perez (2003). A main objective of CBIRD UTB/TSC is to do cutting-edge border research and to incubate new attitudes and alliances needed for the border region to reach its full economic and social potential including regional, binational, biliterate, and multicultural principles: One Region – Un Futuro.

2. The observations listed here are based on interviews and focus group meetings in the Lower Rio Grande Valley during 2000-2002, including a Binational Knowledge-Base Benchmarking Workshop held at UTB/TSC, October 12-13, 2000. Special recognition is also due a group of community leaders that was convened on February 15, 2001, by Professor Pablo Rhi-Perez, UTB/TSC, and which included Dr. Tony Knopp, Colonel William H. Card, and Andres Cuellar among others.

3. Talent refers to special individuals or champions who make things happen and who facilitate the development and application of new technology products and processes. Capital, whether venture, angel, or grants, provides the fuel for development of the business idea. Business know-how includes management, legal, marketing, and sales and distribution.

4. An understood challenge of census-taking along the border region is the number of inhabitants who avoid being counted for a variety of reasons. Consequently, for both sides of the border, these are considered low population estimates.

5. In the State of Texas, the per capita income rose from $17,446 in 1990 to $27,752 in 2000: a net gain of over $10,000. Cameron County's per capita income rose from $9,946 to $14,906 in the same time period: a net gain of less than $5,000. In the year 2000, the per capita income difference between the State of Texas and Cameron County was nearly $13,000. Per capita income for Hidalgo County is below that of Cameron County, and for Matamoros in 2000, per capita income was $4,080 less than Cameron and Hidalgo Counties.

6. U.S. National Rankings published by the U.S. Census Bureau in November 2002, of counties in the United States with over 250,000 population.

7. The results of the UTB/TSC Cameron County survey need to be qualified as the response rate was about 5%. Indeed, very low survey response rates are common for the frequently surveyed border population. Cameron County's respondents' sample came from member lists of the county's chambers of commerce, economic development corporations, and a list of community leaders. Out of 245 returned surveys, 183 usable responses were received. A list of 350 Matamoros community leaders from the private sector, including *maquiladoras*, was created by ten long-term Matamoros residents. From this list a random sample of 100 was selected to be interviewed. Out of 100 interviews there were 81 usable surveys.

8. For Cameron County 27% of the respondents are company presidents, vice presidents or directors; over 50% are from the business sector and over 30% represent professional services and educational sectors. A high 85% of the respondents have lived more than five years in Cameron County, and 60% of the businesses represented have been operating more than 15 years in the region. Some 55% of the respondent's businesses are nonminority owned, 26% are Hispanic owned, and other minorities own 10% of the businesses included in the survey. Of the 81 usable surveys from Matamoros, 68% were company presidents, vice presidents, or directors and 19% were CEOs or managers. A very high 93% of the respondents lived more than five years in Matamoros.

9. These survey data were supported by Cameron County and Matamoros cluster and shift-share analyses and other data including wage analyses by cluster and income migration analyses. Please refer to Gibson and Rhi-Perez (2003). To assess the strength of manufacturing clusters in Cameron County and Matamoros, location quotients were calculated by comparing a cluster's share of total regional employment to the national share.

10. A critical index was calculated for each factor by multiplying the percentages of the "very important" responses by the percentages of the "not very effective" responses. A single point on the graph represents the intersection of each of the factors' importance and its regional effectiveness according to the respondents' opinion. The "critical index" was calculated for each factor by multiplying the percentages of the "very important" responses by the percentages of the "not very effective" responses.

11. These companies include Multimedia Production Center (MPC) in Brownsville that provides state-of-the-art Web page development and support services, South Texas Internet (STI) that provides advanced Web-based telecommunications support to facilitate cross-border business development, and Aventura Entertainment a Cameron County/Hidalgo County-based digital maquiladora that is building Spanish and Mexican video and film content for television and multimedia (Gipson, 2002).

12. Please refer to Florida (2002), and Smilor, Kozmetsky, and Gibson (1988).

REFERENCES

Aranda, J. (2002). "Texas Border Infrastructure Coalition: 2002-2003 Legislative Agenda Border Health," June.

Bell, E. C. (2002). "Barriers to Binational Cooperation in Public Health between Texas and Mexico." Senate Bill 1857, Texas 76th Legislature; Office of Border Health, Texas.

Brazier, G. D. and Gibson, D. (2001). *Assets and Challenges for Accelerated Technology-Based Growth in Hidalgo County: A Knowledge-Based Benchmarking.* A CBIRD Core Program, funded by the Economic Development Administration and University of Texas Pan American, October 31.

Carbonara, C. (2002). *Communities Networks.* Baylor University and Texas State Technical College.

CBIRD (2002). *Creating CBIRD 21st Century Communities: Networks, E-Learning and Technological Innovations*. CBIRD 1st International Conference, South Padre Island, Texas, May 8-9, see www.utbtsc-cbird.org.

Chapa, J. and Eaton, D. (1996). "Colonia Housing and Infrastructure: Current Characteristics and Future Needs." Draft, LBJ School of Public Affairs, The University of Texas at Austin.

De Los Reyes, O. and Rodriguez, I. E. (2002). "Manufacturing and Maquiladoras." CBIRD-University of Texas at Brownsville/Texas Southmost College, position paper, Summer.

Duca, J. V. and Yucel, M. K. (2002). "Biotech in Texas." Report on the "Science and Cents: Exploring the Economics of Biotechnology," Conference, the Federal Reserve Bank of Dallas, April.

Estrada, J. and Hardebeck, S. (2001). "An Analysis of Employment Trends in Cameron County, Texas: A Shift-Share Approach." Report from the Federal Reserve Bank of Dallas. *The Border Economy*, June.

Florida, R. (2002). *The Rise of the Creative Class: And How It's Transforming Work, Leisure, Community and Everyday Life*. New York: Basic Books.

Furino, A. (2001). "Health-Related Characteristics of the Regions of the South Texas AHEC." A statistical report prepared for the Area Health Education Center (AHEC), Center for Health Economics and Policy, University of Texas Health Science Center at San Antonio, August.

Furino, A. and Miller, D. (2002). "Changes in the Health Care Workforce: The Texas/Mexico Border Region 1996/7-2001." Regional Center for Health Workforce Studies at CHEP, The University of Texas Science Center at San Antonio, a preliminary report prepared for the 60th Annual Meeting of the US-Mexico Border Health Association, Chihuahua City, Mexico, June 4-7.

Patrick, M. J. and Garcia, B. (2000). "The Border: Texas' Roadway to the 21st Century." Report from the Texas Center for Border Economic and Enterprise Development, Texas A&M International University, June.

Gibson, D. and Rhi-Perez, P., with Cotrofeld, M., Gipson, M., De Los Reyes, O., Rodarte, R., and Rodriguez, I. (2003). *Cameron County/Matamoros At The Crossroads: Assets and Challenges for Accelerated Regional and Binational Development*, see www.utbtsc-cbird.org, and www.ic2.org.

Gipson, M. (2002). "Technology Infrastructure/Knowledge Base Benchmark." Position paper for CBIRD-TRAC, IC² Institute, The University of Texas at Austin, Summer.

Guzman, V. (2001a). "Director says Matamoros ISSSTE Hospital Suffers from Inadequate Funding," *El Mañana*, June 5.

Guzman, V. (2001b). "New Matamoros Neighborhoods Suffer Health Care Deficit," *El Bravo*, July 20.

Hartnett, D. (2002). "Port Gets a Powerhouse, New $2.5 Million Crane Expected to Boost Business at Port of Brownsville," *The Brownsville Herald*, May 30.

Henton, D. and Seline, R. (1998). "Index of the 1998 Texas Healthcare Technology Industry." Texas Healthcare & Bioscience Institute.

Kozmetsky, G. and Yue, P. (1997). *Global Economic Competition: Today's Warfare in Global Electronics Industries and Companies.* Boston: Kluwer Academic Publishers.

Lazcano, A., Leal, R., and Rhi-Perez, P. (2001). "Industria de Ensamble de Autopartes en Matamoros: Crecimiento, Impacto y Oportunidades." School of Business, University of Texas at Brownsville/Texas Southmost College, December.

Lower Rio Grande Valley Development Council, Texas State Technical College, and Texas Center for Border Economic and Enterprise Development at University of Texas at Brownsville/Texas Southmost College (2002). "Future of the Region, Regional Forum." Harlingen, Texas, June 25.

Murdock, S. (2000). Chief Demographer, Texas State Data Center, Texas A&M University.

Rodarte, R. J. (2002). "Transportation and Logistics: Assets and Challenges for Accelerated Technology-Based Growth in the Border Region of Cameron County, Texas and Matamoros, Mexico." A CBIRD Core Report and partnership between University of Texas at Brownsville/Texas Southmost College, IC^2 Institute Institute, The University of Texas at Austin, and CBIRD-TRAC, August.

Rodriguez, I. and De Los Reyes, O. (2002). "Retail Industries and Tourism" and "Regional Demographics and Educational Resources." CBIRD-University of Texas at Brownsville/Texas Southmost College, position papers, Summer.

Rylander, C. K. (2002). *Texas Regional Outlook: The South Texas Border Region.* Texas Comptroller, June.

Sharp, J. (1998). *Bordering the Future: Challenge and Opportunity in the Texas Border Region*, July.

Smilor, R. W., Kozmetsky, G., and Gibson, D. (eds.) (1988). *Creating the Technopolis: Linking Technology Commercialization and Economic Development.* Cambridge, MA: Ballinger Publishing.

Tech-Prep Texas (2001). *Closing the Gaps: How Tech-Prep Programs Have Increased Participation and Success in Texas Schools: A Five Year Study.* Statewide Tech-Prep Evaluation Region 5 Education Service Center, Beaumont, February.

Texas Department of Health, Bureau of Vital Statistics. www.tdh.state.tx.us/bvs/

Texas Department of Health, Selected Facts for Cameron County – 1997. http://www.tdh.state.tx.us/dpa/cfs95/cameron.pdf

Texas Education Agency, State Board of Education & Commissioner. www.tea.state.tx.us/

Texas Higher Education Coordinating Board and Texas Department of Health (2000). Texas-Mexico Border Health Education Needs. A report to the 77th Legislature. http://www.thecb.state.tx.us/reports/pdf/0295.pdf

Texas Higher Education Plan (2000). *Closing the Gaps, By 2015: Participation, Success, Excellence, and Research.* Texas Higher Education Coordinating Board, October.

Texas Workforce Network, The Texas Workforce Commission. www.twc.state.tx.us/

U.S. Census Bureau, City County Data Book 2001, Cameron County, Texas.

U.S. Census Bureau, United States Department of Commerce. www.census.gov/

U.S. Department of the Treasury, U.S. Treasury Advisory Committee on Commercial Operations of the United States Customs Service, Subcommittee on Border Security (2001). Technical Advisory Team Report on "Improving U.S. Border and Supply Chain Security," January 21.

Warner, D. and Hopewell, J. (1999). "NAFTA and the United States/Mexico Border Health: The Impact on HRSA-Sponsored Programs." Center for Health Economics Policy, The University of Texas Health Science Center San Antonio, December.

Wivagg, J. R. (2002). "Communications Technology and Community Research in the Community Network Planning Process: A Case Study of the Lower Rio Grande Valley." Ph.D. Dissertation, Baylor University, Waco, Texas, August.

Zavaleta, A. N. (2000). "Do Cultural Factors Affect Hispanic Health Status?" An article prepared for The University of Texas Health Sciences at Houston Health School of Public Health. Available at http://ntmain.utb.edu/vpea/elinino/newarticle.html.

16

Innovation and Knowledge Sharing Across Public and Private Sectors: The U.S.-Brazil Sustainability Consortium

John Motloch, Pliny Fisk, Rodolpho Ramina, and Pedro Pacheco

The shift to a sustainable future depends on innovation and knowledge sharing across public and private sectors at various levels, from global to local. This chapter reviews the U.S.-Brazil Sustainability Consortium as a case study for innovation, knowledge sharing, and partnering among public and private sectors to implement sustainability as a pathway for economic growth. It explores processes, tools, and techniques evolved by the consortium for global, regional, and local commercialization of sustainable technologies.[1]

INNOVATION, SYSTEMS, AND SUSTAINABILITY

Through the agricultural, industrial, and information ages, society has increasingly failed to identify and embed integrated innovations into contextual systems. Increasingly, advanced technology and market globalization have been pursued with inadequate understanding of ecological and social-system constraints. As a result, in nearly every country except for the poorest, development exceeds resource limits and regeneration rates; and people are encouraged to consume resources beyond the capacity of that country to live sustainably within its physical boundary. This resulting world is characterized by environmental degradation, gross disparities in wealth distribution, loss of cultural identity, and many other problems symptomatic of a metacrisis (Capra, 1982) of disconnect from context and a false belief that we can ignore natural laws and local limits (Quinn, 1995).

There is an urgent need for a shift to sustainability based on a systems view of the world and place-based innovation, where decisions are made in dialogue with the metabolic rates of human cultural understanding and environmental renewability. This shift needs to address social and ecological environments integratively with a commitment to sustaining the health and productivity of diverse systems (physical, ecological, human). It needs to include integrative management, planning, and design of people-environment relationships and interventions that address today's needs and sustain the ability to address future needs. It should apply information through innovation that integrates decisions to sustain resources and regenerate system capacity. It needs to address systems as integrated wholes (things and relationships) that function through interrelationships and that exhibit properties independent of their parts. Decisions to address short- and long-term human needs need to intervene in systems in ways that sustain health, productivity, and resource regeneration (Motloch, 2001) and address life-cycle flows in ways that sustain positive ecobalance (Fisk et al., 2000) where future productive potential equals or exceeds present potential.

William Rees and Mathis Wackernagel (1996) report on the development exceeding system capacity for 152 countries and the overall planet that is now 22% over capacity. The challenge is how we correct this trend and apply global talent, technology, capital, and know-how to promote a sustainable future and to enhance economic welfare, human health, education, and living standards. Unfortunately there is not a clear framework through which we can position ourselves and start to act responsibly.

INNOVATION IN THE BUSINESS KNOWLEDGE SOCIETY

Many people believe information technology draws the world together into a global community (based on the distance-canceling powers of Internet and Web-based communication), enabling greater access to global talent, technology, capital, and know-how; and promoting increased trade and cultural interaction. Others are concerned that despite their benefit to some people, the industrial revolution, knowledge revolution, and scientific and technical advances have increased the economic welfare, health, education, and general living standards of only a relatively small fraction of global population to record levels (Gibson and Conceição, 2003).

Technoapartheid in the Business Knowledge Society

Despite its empowering potential, access to the Internet is limited and exclusionary. According to the UNDP's Human Development Report of 1999, geographical barriers may have come down for communications but a new barrier—the so-called global *technoapartheid* barrier—has emerged.

- Southeast Asia, with 23% of global population, constitutes less than 1% of global Internet users.
- Globally, 30% of Internet users in 1998 had at least one university degree.
- Buying a computer would cost an average Bangladesh citizen more than eight years of salary, while it costs the average U.S. citizen less than a month's salary.
- Women comprise only 17% of Internet users in Japan and 7% in China.
- The majority of Internet users in China and the United Kingdom are younger than their 30s.
- English language prevails in 80% of Web pages, but is spoken by only 1 person in 10 globally.

Information technology (IT) also concentrates innovation diffusion in receiving countries, more educated social layers, and people with global lifestyles. While IT can accelerate knowledge-based industry growth, as it has for software companies in Ireland and computer services in India, Internet access is concentrated in a minority of people in the richest countries. For example, OECD countries, with 19% of world population, comprise 91% of Internet users.

While geographic proximity is becoming less important for business in the knowledge society (due to information network infrastructure development), sustaining over time the economic and social assets that may be generated locally depends on a stable network of local social relations and sense of community ownership. Stable social networks depend on symmetric flows of information, capital, costs, and benefits. Symmetry introduces complex diverse systems of values; and the global/local relationship is becoming a key issue in political agendas as the reaction against globalization of markets is being directed to international organizations including the WTO and IMF/World Bank (Ramina, 2003). Increasingly, therefore, global technology transfer, innovation, and commercialization must consider the triple bottom line (social, economic, and environmental dimensions) of sustainability.

Increasing Access to Technology in the Business Knowledge Society

The advantages of IT and the Internet include access to large amounts of information rapidly from Web sites or discussion groups. Its disadvantages include difficulty in determining information legitimacy and quality; and until recently, availability only to the well-educated and high-income people in developed and developing nations. The challenge now is to create ways for all people to benefit. Examples of low-income groups doing so include early 1990s cell phone introduction in rural villages of India to help local entrepreneurs and residents; and current success introducing Internet services in previously excluded communities and global regions. "These communication links have dramatically altered the way villages function and how they are connected to the rest of the ...world" (Prahalad and Hart, 2002: 11). Examples include the rapid rise of cybercafes and kiosks in Mexico that provide opportunities for people to interact and for TEC de Monterrey (ITESM) to implement its Virtual University (VU) to provide basic education and job-skill development to rural communities and low-income urban neighborhoods in Mexico and to Hispanic populations in other Latin American countries.

Postapartheid Business Innovation in Emergent Markets

In "The Fortune at the Bottom of the Pyramid," Prahalad and Hart (2002) maintain that emergent markets (former Soviet Union, China, India, Latin America) represent a great opportunity for business and innovation. They note these emerging markets are not the traditional middle-income groups targeted by large corporations in the 1990s, but the poor in nations that opened their doors to free enterprise. They contend that if this is true, "companies will be forced to transform their understanding of scale, from a 'bigger is better' ideal to an ideal of highly distributed small-scale operations married to world-scale capabilities" (p. 2). Whether Prahalad and Hart are correct or not, civil society, government, and private enterprise are aware that living conditions for the majority must improve if greater conflicts are to be avoided. They recognize that improvements must be done collaboratively. "Unless the private-public-union-scientific partnership takes shape and form and co-operates toward a common vision where we wish to be, there is little prospect that society as a whole can move forward" (Pauli, 1998: 198). With the growing role of collaboration, NGOs and citizen organizations around the world have gained importance as national and international alliances to address local conflicts within a global context.

PARTNERING FOR INNOVATION IN SOCIETY, BUSINESS, AND INFORMATION TECHNOLOGY

Partnering is a strategy whereby organizations around the world that deal with similar challenges take collective action and systemically organize to efficiently use resources and fulfill common interests. Groups of stakeholders join efforts toward common goals. Issues including scarcity of resources, maximization of potentials, and effective use of expertise motivate partnerships to address common challenges (Dean, Murk, and Del Prete, 2000). Individuals and groups work together to create partnerships with a life and culture of their own. As partnerships evolve, stakeholders develop social and psychological contracts to guide their work and improve the services and products they deliver.

In developed and developing nations, partnerships have become an important way to address social and environmental challenges. The not-for-profit sector has often taken the lead organizing and implementing collaborative initiatives for the common good. In recent years, and with support of international organizations, partnerships have been formed among local governments, NGOs, and private companies. In Mexico, for example, the World Bank, United Nations, Nature Conservancy, Habitat for Humanity, Kellogg Foundation, Walton Family Foundation, and others have partnered with communities, governments, and NGOs to alleviate poverty, provide public services, undertake productive projects, provide housing, and so on. In 2002, ITESM collaborated with local partners in Oaxaca, Mexico, federal agencies, and international donors including the Kellogg Foundation to create community learning centers that provide IT-based bilingual reading and writing programs to children and vocational training to youth of targeted ethnic groups. Similarly, in 1998, ITESM's (Monterrey campus) schools of architecture and civil engineering implemented 10X10 (Ten Houses for Ten Families), a housing and community program in low-income communities of urban Monterrey. This program was developed and implemented through a partnership of ITESM, local NGOs, municipal government, and a private enterprise.

Business Partnering for Innovation

Partnering to address a mutual agenda is also a common strategy in business. In the past, these partnerships have shared information and expertise to reduce costs, develop products, and so on. Internationally, Netscape and America Online have partnered to increase cyber links and improve Internet connections; and IBM, Apple Computer, and Motorola partnered to develop an operating system and microprocessor for a new generation of computers.

Innovation can emerge from this global trend of creating local entrepreneurships among corporations and local residents. By partnering globally, local companies can address the needs of the poor by increasing employment and securing financial and technical support needed by low-income groups to undertake context-based innovation. Local-global partnerships can support local groups and pursue vernacular technologies to enhance the lives of low-income people who for the most part do not now have access to formal financial and technical mechanisms available to other income groups.

Innovation in the Technology Sector

In most countries, institutional, political, and legal environments that promote technology innovation tend to have policies deeply connected with social structure. "States with transformative aspirations are, almost by definition, looking for ways to participate in 'leading' sectors and shed 'lagging' ones. ... These states are also hoping to generate the occupational and social structures associated with 'high-technology industry' ... to generate a multidimensional conspiracy in favor of development" (Evans, 1995: 10) As a result, most innovation policies are characterized by a focus on:

- creative shortcuts to produce superior outcomes (mostly profits) as compared to conventional productive activities already established
- global markets and global social class with international demands and tastes

- heavy investment in research and development, mainly with public funds, in highly competitive market "niches" in edge technologies, especially IT

Innovative entrepreneurs in the technology sector search for innovation that can produce higher growth rates than normal under traditional production cycles and that promote a *scale jump* in business and revenues. Classical economic theory defines two necessary conditions for gains in production scale: technological innovation, and new social production relations. These conditions are not independent. For each social arrangement a set of technologies applies. Also, innovative technologies that produce scalar growth imply new social structures. For example, Internet-based e-commerce depends on connection to growing numbers of potential buyers and suppliers with network access capabilities. This affects the educational sector and interpersonal relations; as bridging the "digital divide" strains deep-rooted social communication protocols. Progressive public policies often focus on new connections, with stress as a consequence. This occurs because policies give control to the state (investment in concentrated areas) and assure support from the private sector interested in high-tech markets; but do not invest adequate political effort and commitment (state structure and funds) to address complex social structure issues.

Technology is not inherently good or inherently bad.[2] It transfers modernization (old name for innovation) and globalization in ways that cause, aggravate, resolve, or mitigate local social problems.[3] Its effects occur through complex material and social relations that generate known and unknown consequences over a long time span. To achieve IT potential, fight concentration distortion, and improve access to IT, the UNDP Human Development Report (1999) proposes seven actions:

1. *More Connection*—with further development of telecommunication and IT infrastructure
2. *More Community*—enabling access to groups, not just individuals
3. *More Capacity*—building human qualifications for the knowledge society
4. *More Content*—increasing local perspectives into the Web
5. *More Creativity*—adapting technology to local opportunities and necessities
6. *More Collaboration*—developing IT to address local and global community goals
7. *More Money*—finding innovative pathways to finance the knowledge society

SUSTAINABILITY IN THE KNOWLEDGE SOCIETY

Historically, each culture evolved healthy, regenerative, supportive lifestyles. With globalization, cultures around the world lost this ability. The resulting environmental degradation, unequal wealth distribution, and loss of cultural values are now causing people to question their current lifestyles and the implication of these lifestyles to human survival. As a result, interest in sustainability has been growing since the 1990s, with both rich and poor societal groups feeling a need to integrate sustainable principles into their lives. By regaining the ability to evolve healthy, regenerative, supportive lifestyles, people can teach their culture and other cultures with similar life challenges.

Developed nations have made major progress toward implementing sustainable principles for development. Nevertheless, this application is only a portion of what needs to be done globally. It is estimated that the two-thirds of the global population living in developing countries continues to consume products and services in resource-intensive and excessively polluting ways. In many cases, products and services have been brought into local markets at the expense of cultural values and lifestyles. Until recently, it was thought that only developed nations could afford the luxury of thinking and acting sustainably. Now we understand sustainable development is urgently needed by, and available to, all people. The world needs alternatives that satisfy everyone's basic need for food, water, energy, health, and housing, and in ways that enhance local dynamics and respect nature.

Sustainable solutions tend to be fine-tuned to regional and local conditions; and sustainability in the knowledge society must address ecological and cultural diversity. *Ecologically*, there are 26 global biomes (14 terrestrial, 7 freshwater, 5 marine), at least 867 terrestrial ecoregions, and hundreds more marine and fresh water ecoregions (National Geographic Society, 2000). Each has its climate, physical features, and species that decisions must respect. In the past, limited technologies produced place-based vernacular construction that differed in mountain and lowland regions, as each responded to different conditions. Earth-based materials were widely used in dry and temperate regions but not in rain forests. Since regions also differ culturally, vernacular solutions also produce a rich global cultural palette. Historically, natural context influenced the construction system and culture influenced form and function (Rapoport, 1969). *Culturally,* diversity of people, norms, and interactions affect sustainability and the viability of development. For example, Amish home building in North America is a social activity that binds family members and strengthens community ties. Likewise, in Latin American urban neighborhoods, building is an important part of community life. In each case these interactions can produce sustainable solutions (environmentally healthy, socially responsible, and economically viable) that nurture and prepare individuals as responsible community members.

In the knowledge society, sustainable solutions emerge when people address four questions: "How can the universal concept of sustainability be translated into regional and local agenda?" "How can people learn from distant neighbors that differ culturally but live in the same ecoregion?" "How can IT and the Internet help people learn from distant neighbors who live in similar ecological or social contexts?" "How can people realize the potential of the Internet to increase dissemination of information about sustainability, ecoregional and cultural challenges, and sustainable solutions?"

GLOBAL PARTNERING, NETWORK INFRASTRUCTURE, AND SUSTAINABLE COMMUNITIES

The business sector has partnered globally since the early 1980s to increase profits and markets. In the process, corporations have benefited from the Internet, satellite broadcasting, and broadband IT systems. It is now imperative that society use IT to benefit individuals and communities around the world by raising awareness of local sustainable solutions. In so doing, global collaborative networks can help people learn from other cultures how to address local challenges and create new ideas, products, and services without repeating mistakes or replicating experiments. As stated by Gunter Pauli, "human beings can learn from nature, not by striving to be the strongest but by seeking collaboration across races and cultures, respecting the difference, and recognizing that only by co-operation will they succeed in converting limited resources into an abundance for all" (1998: 26).

IT System Innovation and Sustainability

Given the intricate set of economic and social relations, IT system innovations (product and process) can either solve or aggravate the problems they try to address. Considering the seven proposed actions above as design goals for innovation, two sets of design principles can be identified. The first addresses the sustainability of social relations. It reassesses global networks and issues of technology innovation, transfer, and commercialization in relation to social, economic, and environmental impacts at the local level. The second focuses on network infrastructure sustainability. For a knowledge society whose information infrastructure relies on the World Wide Web, sustainability depends on sustained flow of information in this network. This was central to the first studies and blueprints by the RAND Corporation in the 1950s and 1960s,[4] of what we now call the Internet. The challenge was to figure out what sort of connected communication system could keep working after major destruction caused by military attack. For a knowledge society, basic infrastructure network resilience was seen as necessary for continuous flow of information, even contradictory and ideologically conflicting information, to keep going under any kind of attack: physical, virtual, or ideological.

Sustainable Communities

In this new networked society, a structural schizophrenia emerges from the conflict between two spatial logics: the first based on human experience that still relates to the local level, and the second grounded in society's dominant functions and power organized in global flows of information, capital, and power (Castells, 1999). This local-global dynamic requires sustainable communities to be addressed at both local and global levels. Although community traditionally implied some sense of the local, emergence and evolution of the IT interconnected world may have produced a global community among people who are thousand of miles apart. This local-global dynamic requires that the concept of community be rethought. For some people, community means devoting time to common issues: neighborhoods, support groups, local politics, and so on. The social bonding and sense of community that results is probably the most valuable asset societies have, and is crucial to survival as a species. It brings feelings of continuity, safety, and familiarity; and a framework of commonly shared values to orient actions, perceptions, and reason to live. Sustainable communities are those groups of people who sustain themselves and their relations without impairing possibilities of future generations to sustain themselves. Community sustainability accrues from healthy people-people and people-environment relationships (Motloch, 2001).

A sustainable community relies upon sustained individual *commitment* to the community. In a free democratic society, people sustain their membership when this is seen as worthwhile. Freedom of choice implies that broad commitment is not sustainable without *symmetrical* relationships. This need for symmetry applies to IT systems. Symmetrical IT systems provide information that balance, in a given time span, all the relations each member has within a community. They balance the breadth, integratedness, and scale (spatial and temporal) through which the community makes decisions. They balance the flow of information in the community, and balance information flow in ways that regenerate the lifecycle flows of resources. They balance the extent of involvement of members of the community in decisions that affect their future and future availability of resources. Examples of symmetry in IT systems include the design of interactive voting systems that provide voters quick feedback on consequences of choices on local or regional issues. In industrial production, examples of symmetry include those that promote life-cycle design of products and processes and that build commitment to industrial ecology (Fisk et al., 2000).

As in sustainable local communities, commitment and symmetry are fundamental in constructing sustainable global communities. This applies, for example, to the design of well-informed global stock markets, where traded bonds reflect not only financial but also environmental and social bottom lines, including effects of investment on country economies. It also applies to the design of systems directed to assessing and calculating the value of trading carbon emissions at a global scale in real time, giving a new push for the Kyoto Protocol and Clean Development Mechanism.

NETWORK STRUCTURE AND INFORMATION FLOW

Many global infrastructure networks, including continental petroleum and natural gas distribution networks and national electricity systems (e.g., Brazil), show centralized structural patterns. This pattern type appears to associate with expanding systems, where economic efficiency produces networks with *treelike* patterns, few cycles, and a relatively small number of prominent central nodes (Ramina, 2000). Although cheap to construct and easy to control, these networks are vulnerable due to limited number of nodes and paths of flow. On the other hand, mature networks resemble forms that are more random, with higher connectivity, denser connections among nodes, and more cycles that distribute risk among larger number of elements. The sustainability of mature networks seems to be associated with rich connectivity and structural complexity, rather than the rationality of control and dominance.

Regarding one proposed action above—MORE MONEY—early-stage Web infrastructure construction is expensive, given the size of investments in IT innovation needed for implementation and the relatively small number of users ready to connect, especially in developing countries. This may be the cause of the Web's present centralization and small number of major hubs. But at the same time, this pattern is the footprint of a strong monopolistic environment.

Early in its formation (1964) the optimal structure for the Internet was analyzed. From three possible network architectures—centralized, decentralized, and distributed (see Figure 16.1)—the centralized and decentralized structures that dominated communication systems of the time were demonstrated to be too vulnerable. The conclusion that the Internet should be designed as a meshlike architecture was defeated by the military and AT&T, the communication monopoly of that time (Barabási, 2002).

Figure 16.1
Network Architectures

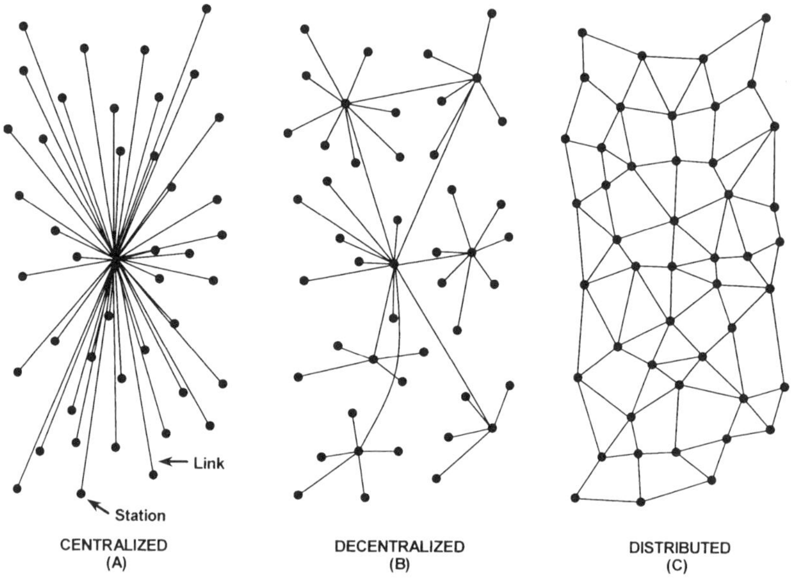

Source: Barabási (2002: 145).

Current Web Network

It is estimated that, starting from any page, one can reach only about 24% of documents on the Web. The rest are invisible, unreachable by surfing. It is estimated also that about 25% of all Web documents lie in *islands*, isolated groups of interlinked pages unreachable from the *central core*, the home of all major Web sites from Yahoo! to CNN.com. In a recent research report about Web topology and implications, Albert-Lázló Barabási remarked that there is a "*complete* absence of democracy, fairness, and egalitarian values on the Web ... the topology of the Web prevents us from seeing anything but a mere handful of the billion documents. ... the architecture of the World Wide Web is dominated by a few very highly connected nodes, or *hubs*. ... The hubs are the strongest argument against the utopian vision of an egalitarian cyberspace" (2002: 56-58).

Increasing Connectivity and Symmetry for Network and Community Sustainability

Applying connectivity as a design principle for World Wide Web innovation would increase the number of possible connections among members of a global community. This would not necessarily lead to less vulnerability unless connections are distributed among more nodes, making a dense network mesh. Such design could give credit to technological solutions like Internet radio, "echo-links," "packet-radio," and other interconnected technologies, some concentrated, some distributed (Ramina, 1993).

Increasing connectivity would also build local capacity and increase the match of IT development to local needs and conditions. Applying both connectivity and symmetry principles could result in a careful balance of local and external production of contents, addressing some of the seven actions, addressed above, proposed in the UNDP Human Development Report of 1999.

GLOBAL NETWORK OF SUSTAINABILITY CONSORTIA

University-community partnerships can lead the pursuit of innovation, knowledge sharing, and partnering among public and private sectors to implement sustainability as a pathway for economic growth. To do so, these university-community partnerships need to form a sustainability network, share information to translate the global need for sustainability into regional agenda, and advance understanding of regional sustainability (including indicators, baselines, and benchmarks). They need to identify value-adding ways to operate sustainably in specific regions, connect local with global funding to address these needs, and facilitate information flow that promotes behaviors that produce innovative streams of sustainable decisions (Motloch, 2003a). These partnerships need to include a global network of biome- or ecoregional-based institutes and landlabs that enhance information flow about sustainability and trigger behavioral change (see Figure 16.2). These institutes need to function as education-research-outreach-demonstration centers that connect people to sustainable relationships with local resources and cycles, translate national and global desire for sustainability into regional agendas, partner with others, and integrate regional agendas into a global sustainability information-flow network.

Figure 16.2
Global Sustainability Network of Ecoregional Centers/LandLabs

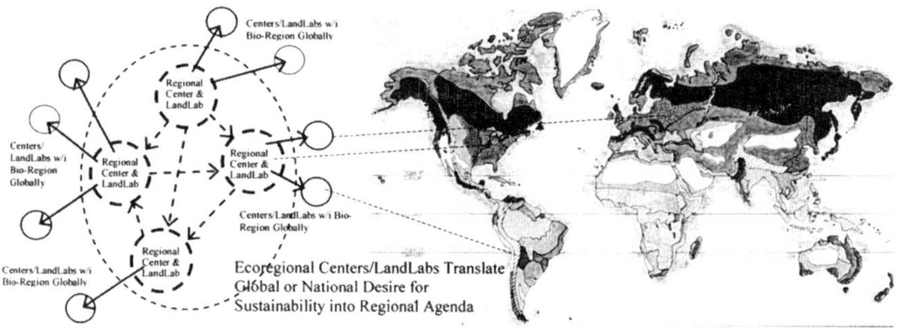

Source: Evolved by Motloch (2003b) from Motloch and Ferguson (1997).

The Land Design Institute (LDI) at Ball State University is currently building a network of sustainability consortia to help lead society to a sustainable future. These consortia pursue local-global partnering and symmetrical flows of sustainability information. They pursue *spatial symmetry* (global sustainability through local decisions that address people, economics, and environment at the local scale), *temporal symmetry* (addressing present and future needs), and *cultural symmetry* (balancing the developed world's environmental bias with the developing world's greater social and economic foci). LDI-facilitated consortia pursue a vision of life-cycle-based and ecobalancing land use, management, planning, and design. They pursue the *vision* of a global network of sustainability consortia that help lead society to sustainability through *innovation* that facilitates change. They facilitate community *partnerships* of diverse sectors and disciplines with adequate breadth of awareness to pursue *integration* and sustainability through value-adding processes and technologies that interconnect solutions with complex contextual systems. Consortia decisions seek to intervene in ecological and human systems and contexts to sustain and regenerate regional and local resources so downstream capacity (after decisions) equals or exceeds capacity prior to intervention. These consortia pursue sustainability through land management, planning, and design that integrates into complex regional and local contextual systems (physical, biological, cultural, economic) and life-cycle flows (energy, water, and so on). They pursue sustainability via value-adding solutions at increasing levels of integration into complex systems. They partner to identify appropriate levels of integration based on the complexity and nature of the systems into which decisions must integrate, information available to inform decisions, management structure sophistication, management resources available, and desired level of value to be added (see Figure 16.3).

Figure 16.3
Appropriate Value-Added Level

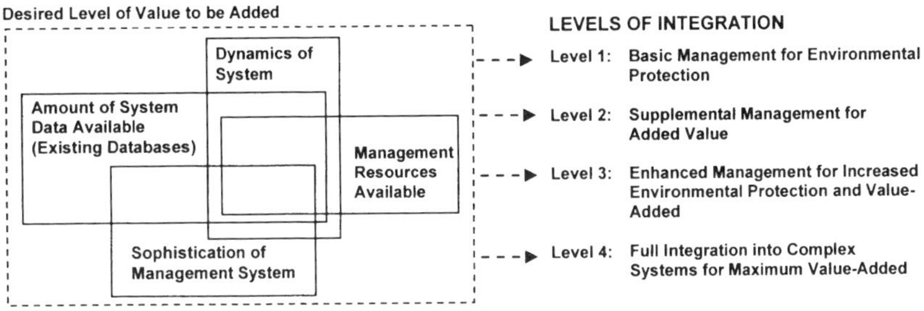

U.S. members of these consortia—the Land Design Institute at Ball State University, the IC^2 Institute at the University of Texas at Austin, and the Center for Maximum Potential Building Systems in Austin—partner with others internationally to share information and identify solutions to apply global and local knowledge to integrate decisions into local and regional life-cycle flows in environmentally responsible, socially equitable, and economically viable ways. They pursue innovation-intervention processes (Hatchuel, Agrell, and van Gigch, 1987) that bring people with different expertise into collective visioning. Participants transfer control of decisions within their expertise to the group to promote innovation through a collective vision that sees beyond the paradigmatic boundaries of specific disciplines. These sustainability consortia use these processes to discover complexity, potential, and needs; to identify and integrate value-adding sustainable innovations into complex local and regional systems; and to promote commercialization of sustainable technologies. They embrace partnering of universities, government, industry, and other societal sectors to access and integrate both academic and project funding streams into strategies for program seeding, short- and long-term implementations, and program sustenance.

Process of Building Sustainability Consortia

These sustainability consortia are being built via a four-stage process: emergence, seeding, implementation start-up, and sustained implementation. *Emergence* is usually facilitated by convening multisector partners who seek societal change to sustainability. This phase is usually funded by partnering institutions. Partners come together for short, intense visioning workshops that include a search for fit among partners, collective visioning, conceptualization of consortium agenda and potential projects, and commitments to build on this agenda and project list. *Seeding* usually includes institutional support-building for formal agreements to partner, and seed funding for interinstitutional and project visits (to identify sustainability initiatives, benchmark projects, and potential local projects that could benefit from consortium partnering). The team conceptualizes and embeds sustainability curricula in each institution, and meets with institutional administrators to build commitment. *Implementation Start-Up* usually includes multiyear initial academic and project funding and agendas for regional projects. Successful completion of a number of short-term projects builds the track record of partnering among consortium members essential for success in the sustained implementation phase of the consortium. *Sustained Implementation* usually includes a program of long-term academic and project funding to lead society to a sustainable future and to sustain the consortium. It includes institutionalizing the consortium through agency, foundation, project, or other funding.

U.S.-BRAZIL SUSTAINABILITY CONSORTIUM: A CASE STUDY IN "LEADING SOCIETY TO A SUSTAINABLE FUTURE"

The U.S.-Brazil Sustainability Consortium (USBSC) is included herein as the pilot project for this global network of sustainability consortia and as a case study for innovation, knowledge sharing, and partnering among public and private sectors to implement sustainability.

USBSC Process

The USBSC started with key individuals who had a desire to do projects together. It *emerged* in a two-day visioning workshop where shared views and values emerged. Partners committed to facilitating a paradigm shift by integrating sustainability into society and fostering sustainability education at all societal levels and audiences. They agreed to develop processes and tools that promote sustainability; and to engage in sustainability projects (as vehicles for education, research, outreach, and demonstration) and to build industry-government-education partnerships. Partners agreed to pursue parallel academic and project funding including seed, start-up, and sustained funding as an integrated academic-project pathway to societal change, sustainability, innovation, and green technology commercialization. They committed to pursue multiscale projects in resource-management, place-based decisions, community-building, energy systems, green-building, resource-balance, and ecobalance.

USBSC *seeding* (2002) included a 10-day visit of U.S. partners to Brazil. The team met with Brazil partner organizations, tested and upgraded distance-education delivery compatibilities; visited benchmark projects, *favelas* (slums), and computer game-development incubators; and identified potential projects. The USBSC entered its implementation *start-up* phase in 2003, with receipt of four-year consortium funding from FIPSE (U.S. Department of Education) and CAPES (Brazil Department of Education). The USBSC is at present pursuing its program for *long-term sustainability* including projects that can lead society to a sustainable future while sustaining the consortium. This includes major value-adding integrative technology programs, distributed energy projects, and resource-balancing projects. Consortium members are partnering to pursue project, agency, and international foundation funding through international, U.S., and Brazilian sources.

Proposals to Accelerate Sustainable Technology-Based Economic Development

The USBSC is seeking funding to accelerate sustainable economical development, including funding to develop the sustainable development dimension of IC^2's proposed Green Wetware (Skill) for Global Technology Commercialization Centers (TCCs); where IC^2 seeks to accelerate technology-based economic development through globally networked TCCs that foster growth of targeted small and medium-sized enterprises (Gibson and Conceição, 2003). The USBSC also proposed four-day "Toward Sustainable Development Workshops," where consortium members, IC^2 fellows from each partnering region, and IC^2 staff work together to translate the concept of sustainable development into an agenda, programs, and action items IC^2 can pursue to effectively realize the sustainable development dimension of its mission.

Sustainability Tools and Techniques

USBSC partners are committed to developing innovative tools and techniques to facilitate societal shifts to sustainability. These include networking and community-building tools and techniques that promote Ramina's (2003) symmetric flows of information, capital, costs, and benefits to sustain a stable network of local social relations, promote a sense of community ownership, build system complexity, and help balance global information, capital, and power flows with the innate need of people to connect to the immediate world around them. The USBSC is seeking to evolve and apply ecological footprinting, resource-balancing, and ecobalancing tools and techniques to make land use and master planning decisions based on Fisk and Armistead's (2003) Life Cycle Space™, regional resource dependencies, network analyses, and commercialization of value-adding sustainable technologies. The USBSC is pursuing tools and techniques to apply industrial ecology concepts to physical planning and design, to rethink land uses based on resource-balancing, and to increase efficiency through flexible manufacturing methods designed for renewable processes. The consortium is seeking to develop and apply tools and techniques that implement interagency collaboration models to create, evolve, and maintain linkages of people, places, and ideas for innovation (Pacheco 2003). These efforts will enable consortium partners from different countries to facilitate innovation and sustainability in community development through interagency collaboration. In these activities, the USBCS seeks to interconnect academic and project funding, and sustainability techniques, models, and tools, to make regionally and locally appropriate decisions that connect people, ideas, and resources, and that enhance sustainability knowledge flow and knowledge sharing to enhance economic and community development and regional and global commercialization of sustainable technologies (Motloch, 2003a).

TECHNIQUES AND MODELS

The USBSC is applying innovative techniques and models to make regionally and locally appropriate decisions that connect people into the dynamics of complex local and regional systems, to facilitate behaviors that produce sustainable decisions, trigger "streams" of sustainable decisions, and to help lead society and local communities to a sustainable future. To do so the USBSC is building collaborative partnerships that link international and local partners of diverse populations.

International Collaborative Partnerships

Organizations typically use different types of partnerships—monopoly, parallelism, competition, cooperation, coordination, and collaboration—to work together (Cervero, 1988). These types differ in the degree of interdependence among organizations. Collaboration is the most interdependent and widely used by organizations dealing with social and environmental programs. When partners collaborate, a sense of authorship develops around programs and group initiatives. Collaboration is embraced herein as the preferred partnership type, as is Donaldson and Kozoll's (1999) model for evolution in collaborative partnerships. This model's four components—emergence, evolution, implementation, and transformation—interact in a cyclical way (see Figure 16.4).

Figure 16.4
Developmental Stages of Collaborative Relationships

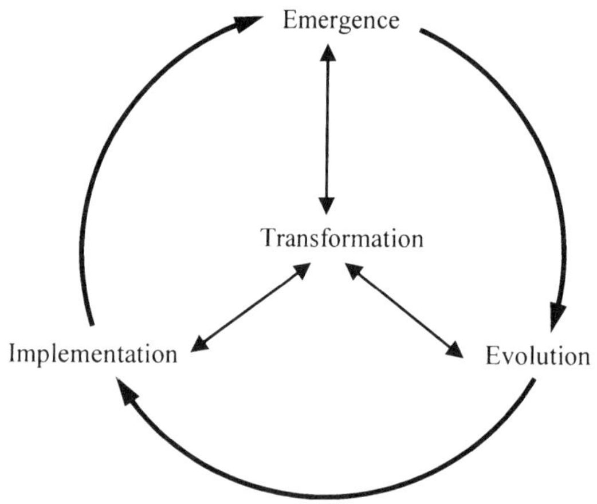

Source: Donaldson and Kozoll (1999).

Collaborative initiatives progress through the first three components in order. Transformation occurs throughout the relationship as partners assess their performance. In the emergence stage partners are selected, the decision to collaborate is established, and a common agenda is defined. In many cases, partners are selected based on commonality of interests, rapport developed among individuals, and compatibility of organizational values and principles (Kanter, 1994). The evolution stage establishes the direction of the effort and maintains the relationship. Implementation moves from planning to action and from general goals to specific tasks. It requires constant assessment to ensure that results respond to the interests of all stakeholders.

International Collaborative Partnerships for Housing and Community

The global sustainability network model (see Figure 16.2), is translated herein to the local scale via a global network of international collaborative partnerships for housing and community (ICPHC). Each ICPHC addresses local housing and community development issues, with each partner offering something important to, and benefiting significantly from, the partnership. In the ICPHC global network (see Figure 16.5), profit and not-for-profit organizations exchange information, expertise, and dollars to address similar issues. The network serves as a platform for learning and a laboratory for innovation. As stated by Chesbrough (2003), "Old-school R&D ... was strictly in-house. The new model for success requires collaboration with many innovators (and) draws on technologies from networks of universities, startups, suppliers, and even competitors." In each ICPHC, partners are identified based on their understanding of housing and community issues from a systemic and regenerative approach. Internet searches, printed information analysis, and on-site visits help determine compatible partners. Once identified, potential partners share ideas and explore opportunities for collaboration.

Figure 16.5
International Collaborative Partnerships for Housing and Community (ICPHC)

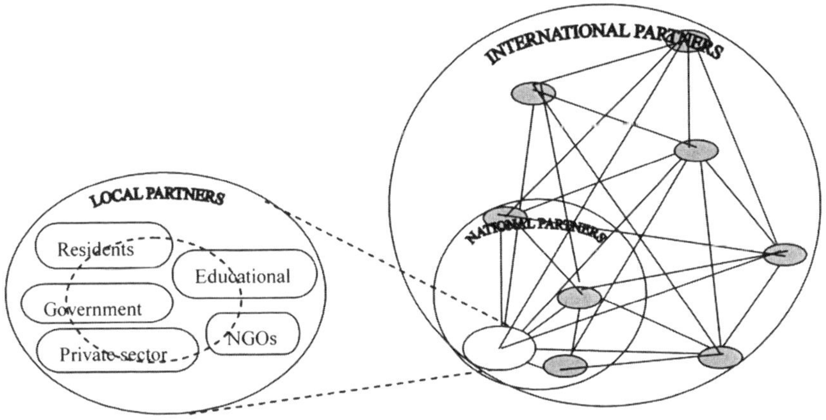

Principles of ICPHC Partnerships

As ICPHC partnerships emerge and evolve, stakeholders become aware of the context in which they may operate and the important contributions they can make. They also negotiate with other stakeholders to establish principles for collaboration that can help partners establish conditions for implementation (Donaldson and Kozoll, 1999; Kanter, 1994; Dean, Murk, and Del Prete, 2000). It is proposed herein that these principles for collaboration include:

- *Committing to sustainability*, where partners embrace sustainability as a regenerative paradigm.
- *Facilitating sense of ownership*, where partners see themselves as owners and promoters of the initiative.
- *Maintaining honest, constant communication*, where partners embrace open, continuous, and honest communication to help clarify ideas, move the initiative and projects forward, and help partners reach consensus to deal with conflicting interests in the partnership.
- *Facilitating empowerment*, where partners respect others for what they contribute to the relationship, and where partners protect the interests of other partners.
- *Sustaining commitment for action*, where partners commit to sustain a desire for change and action.
- *Promoting equity among partners*, where partners have equal rights and responsibilities in decision making.
- *Developing trust*, where partners cultivate trust toward other stakeholders and collaborative process.

Conditions for Implementation

To operate effectively, ICHPC partners can agree to conditions that define the rules under which they assume roles and responsibilities in an equitable and just manner; and that provide the general operating structure of the initiative. These conditions can include:

- *Obtaining legal representation*, where partners pursue legal status important for funding. This can include registering each organization in its corresponding country and registering the ICPHC in international organizations such as the Habitat International Coalition.
- *Establishing agreements*, where partners pursue the balance between formal and informal relationships crucial to sustaining momentum and good relationships among individuals. These can include formal institutional agreements to help establish broad agendas for action; and informal agreements necessary to build trust and communication among stakeholders.
- *Defining operational standards*, where partners establish baselines and benchmarks that help the ICPHC measure its level of success and improve its programs.
- *Establishing a sense of direction*, where partners define, orient, and scope ICPHC programs by collaboratively defining goals and objectives. In establishing this sense of direction, each partner can lead its own area and brings important resources to the group. A sense of shared leadership builds as goals and objectives can also be defined.

- *Defining a plan for action*, where, based on the established goals and objectives, partners can collaboratively define specific strategies and actions to be taken.
- *Defining roles and responsibilities*, where each stakeholder can play a specific role and protect particular interests. The collaborative initiative succeeds based on adoption of clear collaboratively defined roles and responsibilities based on partner expertise, economic capacity, and organizational values.

Implementing the ICPHC Initiative

In ICPHCs, individuals and institutions can collaborate to find tangible timely solutions (including organizational change) to difficult situations they could not solve as individuals. A crucial step in implementing ICPHCs and achieving organizational change is moving from planning to action and from individual to global responsibility (Anderson, 1998). Implementing this change, which is often crucial to sustaining participation and resources, requires that:

- *Action plans must be documented, monitored, and adjusted*, because documentation is an important tool for improvement, dissemination, and process management.
- *Information flow must be open and timely*, because quality information helps partners identify areas for improvement and prevent and resolve conflicts before they escalate. Open, honest communication helps partners reach consensus to deal with conflicting interests among partners; "achieving consensus requires flexibility and creativity of leaders and members alike in order to deal with differences constructively and reap the rewards of doing so" (Donaldson and Kozoll, 1999: 105).
- *Assessment must be conducted periodically*, because organizations must periodically assess their success in order to seek improvement and conduct midcourse corrections.
- *Success and lessons learned must be disseminated*, because dissemination of results must be part of the agenda to share outcomes and influence others.
- *Partners must celebrate their accomplishments*, because celebration is important in keeping partnerships alive and evolving. These informal interactions can also allow frustrations and potential solutions to emerge.

Links and Communication

State-of-the-art communication can help ICPHC members stay connected, interact, make decisions, plan, and implement programs. Innovative communication systems (two-way video, electronic mail, instant messaging, broadband communication) can facilitate partner interaction. Periodic on-site visits can reinforce relationships; and partners can get together at least annually in different places to become familiar with partner contexts and to celebrate accomplishments.

As part of the communication system, members can develop a network of Internet sites with an internal discussion room to exchange ideas. Additionally, to extend the network and its services, a discussion room can be opened to the public. This room can include information related to partnership innovations and projects. Educational institutions can be included in each ecoregion to help users take advantage of the research infrastructure and expertise of educational institutions.

Potential Projects

Potential projects likely to emerge as ICPHC projects include Internet courses for community residents to address sustainability, housing, and community development projects (and perhaps in the future other areas such as health and legal services). International online educational programs are likely to emerge in the areas of housing and community development for bachelor and master degree students in different parts of the world. International certificate programs oriented to practitioners and communities are also anticipated, as are videoconferences where partners interact with experts from different biomes (could be part of educational programs available to the public to increase awareness). International workshops to address members' projects or engage in products and services innovation are also anticipated, as are projects to develop Internet libraries and best-practice databases.

Benefits of International Collaborative Partnership Network

The proposed international partnership network can help partnering individuals and organizations, and the broader communities in different parts of the world, to discover traditional and contemporary vernacular lessons sensitive to natural and social contexts. This can help communities better understand and address their own local challenges. State-of-the-art technologies can enhance this exchange of ideas, expertise, and knowledge among cultures that address similar sustainability, housing, and community challenges.

In this effort, collaboration can be both a platform for education opportunities and a strategy for exploring innovation to address common concerns across cultures and communities. The international web of experts and learners can collaborate to maximize use of current technologies—satellite, teleconference, videoconference, and Internet—to interact, make decisions, and learn from each other. The stages, principles, and conditions of collaboration can enhance partnering among individuals and organizations around the world. Principles based on universal values can help partners establish social and psychological contracts to initiate, evolve, and sustain partnerships. An international partnership—carefully designed, implemented, and monitored—can link people, places, and ideas for innovation.

TOOLS FOR SUSTAINABILITY IN THE KNOWLEDGE SOCIETY

Growing awareness of systems, system dynamics, resource flows and the need to sustain ecobalance has motivated life cycle thinking. This thinking has progressed from Life Cycle Cost accounting (LCC) to Life Cycle Analysis and

Life Cycle Assessment (LCA) to Life Cycle Balancing (LCB) and Fisk and Armistead's (2003) introduction of Life Cycle Space™ (LCS) and Life Cycle Ratio™ (LCR). Each development depends on the previous; with emergence of Life Cycle Ratio™ being a direct result of establishing the spatial footprint protocols of Life Cycle Space™ (LCS).[5]

Establishment of Life Cycle Space™ as a planning and procedural tool involves acceptance of a series of conditions. These conditions provide metrics that enable us to survey geographically and otherwise the spectrum of existing and potential life cycle patterns needed to shift toward a more stable human-environment interface.[6] This section begins to provide the logic necessary to support the assertion that a region can best be sustained by regionalizing resource dependencies rather than maintaining national and international life cycle dependencies. The Life Cycle Space™ reduction approach herein should help shift our understanding of buildings, building systems and a wide range of sustainable human activity. LCS™ and LCR™ are based herein on the following 16 fundamental conditions that are codependant (i.e., for the most part not able to stand alone without recognition of the other conditions). It is assumed that all applications of these conditions are based on renewably based technological process where nature and the solar constant are the driving forces.

1. The life cycle is made up of links and nodes. Links contain four principle flows: *materials* (solids, liquids, gases), *energy* (all forms renewably derived), *money*, and *information*. Nodes are defined according to basic life-support needs (air, water, food, energy, and materials).
2. Land as a natural resource can be planned using long-term sustainable management practices once suitability analysis is accomplished to identify the lowest impact and highest use suitability areas for particular life cycle practices.
3. Incorporation of self-similarity and redundancy between life cycle topics at several scales (i.e., biomes, watersheds, building sites) must occur so as to reduce the possibility of system failure.
4. Efficiency of overall life cycle is often increased by reducing the number of transformation activities (nodes).
5. Reducing life cycle distances (links) between life cycle activities (nodes) relative to basic human physiological needs (nature-sourced air, water, food, energy, and materials) starts with the smallest scale and progresses only as necessary to larger scales of life cycle use.
6. All life cycle activities occur with definable boundaries either naturally or artificially derived in order that life cycle performance can be measured.
7. Increasing diversity within the life cycle (including all five kingdoms within the overall life cycle or within constituent phases) increases the health of the system by blocking disease throughput.
8. Reduce the need for larger life cycle scales by establishing multipurpose and/or highly integrated stages within the life cycle.
9. Reduce the complexity of the life cycle (as in #2) so that the quality of information needed by humans is manageable.
10. Extend the use phase of the life cycle by repair and maintenance necessitates accompanying increases in resource allocation for this activity by planning the life cycle accordingly.

11. Increase the adaptability (flexibility) of life cycle elements through separation of physical structure from function permits openness regarding how structures are placed on the land.
12. Establish place-based economic loops at all scales by purposely tying life cycle activity nodes to area resources and neighboring enterprises.
13. Do a life cycle balance of those sourced products supplied by nature through use of the necessary land area required for regenerative re-sourcing methods at each node or combination of nodes for the life cycle within a defined boundary scale.
14. Measure by ratio the sourcing and re-sourcing life cycles according to a set boundary scale.
15. Establish a pattern recognition procedure that codes possible life cycle patterns for purposes of measuring existing and potential balancing.
16. Establish points of entry into the existing system through network analysis whereby potential technological and sociological "triggers" can be identified.

The following sections describe several conditions more fully and in some cases link conditions to the National Institute of Standards and Technology (NIST) Incubator project at Montana State University. It is assumed that a second phase of work would enable more complete and measurable understanding of the Life Cycle Space™ process.

Conditions of Life Cycle Space™ Described

Several conditions are described more fully in the text below. For space-saving purposes other conditions are left out.

Condition #3: Incorporation of self-similarity and redundancy between life cycle topics at several scales (i.e., biomes, watersheds, building sites) must occur so as to reduce the possibility of system failure. This condition recognizes the state of duplication in structure and function at a variety of scales so the human-nature system remains robust and healthy. Sustaining levels of redundancy is the key to understanding performance because a certain amount of resource (energy, materials, information storage, etc.) must be put aside to guarantee system performance. The importance of understanding boundaries is the essence of Life Cycle thinking and a sustainably built environment.[7] A useful description of these boundaries is found in Di Castri and Hadley (1988) and in Figure 16.6.

Condition #4: Increase efficiency through miniaturization of the life cycle within a regional or site context. This condition states that priority should be placed on providing for the incorporation of all possible processes (or transformations) at the smallest possible scale, thus relieving the burden of impact necessitated by sole use of larger life-cycle systems. This condition requires the recognition of Life Cycle sequence as a fundamental planning tool (see Figure 16.7), and that the process within each life cycle overlaps and serves in a multifunctional manner into another life cycle as described in Condition #1. The efficiency of the life cycle process rises when fewer individual or separate transformations occur (see Figure 16.8.).

Figure 16.6
Condition of Systems Within Systems

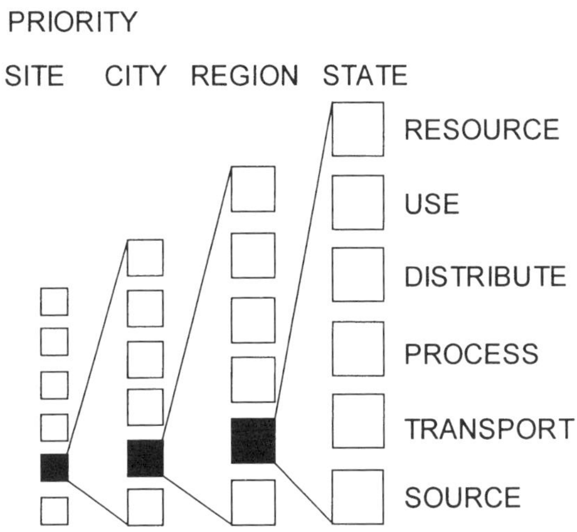

Source: Di Castri and Hadley (1988).

Figure 16.7
Efficiency Reduction According to Number of Life Cycle Phases (below in built environment terminology)

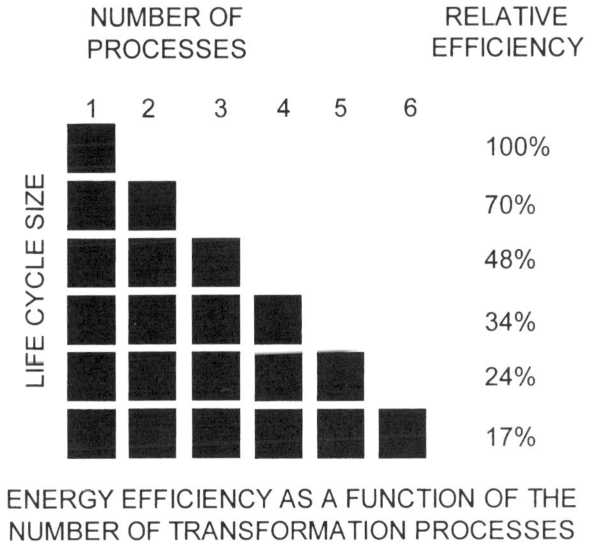

Source: Ayres (1989).

Figure 16.8
Examples of First-Level Reduction in Number of Life Cycle Phases

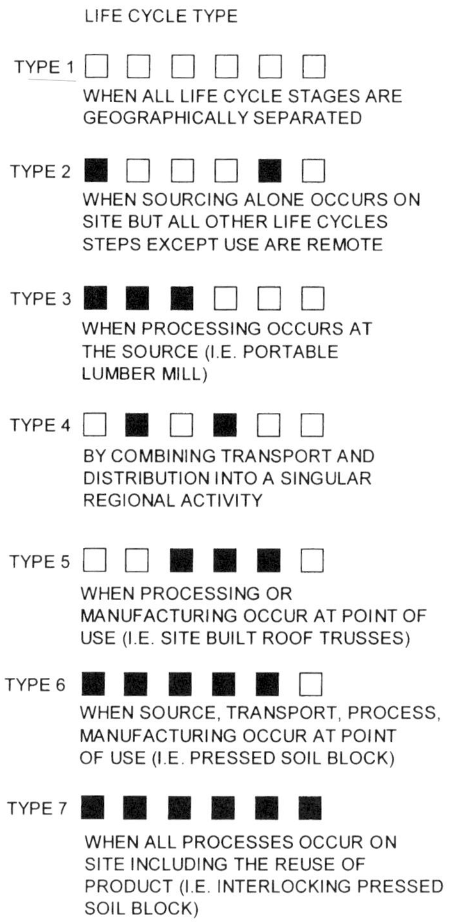

Condition #8: The technology of production and use at smaller scales can only compete with those at more centralized larger scales if they become multipurpose and highly integrated. There is a common belief that larger scale, centralized technologies are more efficient and environmentally superior to smaller scale operations due, for example, to effective centralized pollution control. However, trends show that with improved technology and enhanced integration between technologies, there is a greater possibility to achieve a balance in material and energy flows at all phases of the life cycle. Simply stated, integration is a more important concept in life cycle design than is conservation.

Condition #9: Reducing the complexity of the life cycle enables it to relate more directly to the amount of information processing by all actors involved, from design and engineering integration to users and environmental impact assessment. Working with simplified construction and mechanical systems aids both in information gathering and processing for environmental impact evaluation, and the ability to integrate one technology with another. Figure 16.9 summarizes the information and complexity issue.

Figure 16.9
Relationship of Scale, Complexity, Information, and Cost

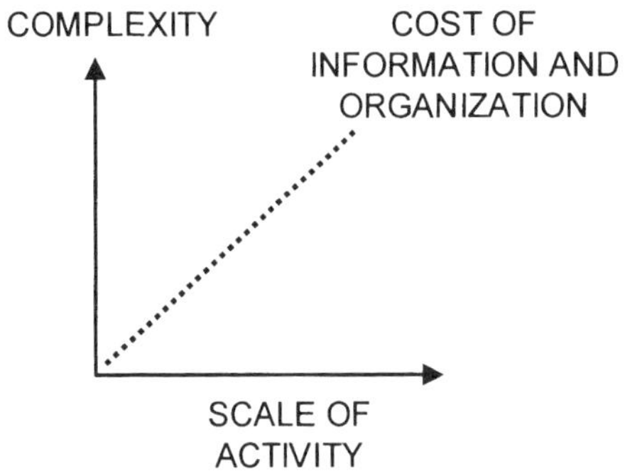

Source: Friedman (1973a, 1973b).

Condition #10: In architecture, plan for an extended use phase of the building's life cycle. This condition relates to length of time attributed to the use phase of a building, its environmental impacts, and the long-term economic investment a society places in the built environment. Design features such as flexibility, reuse, and material longevity lengthen a building's useful life, which, in turn, can affect the useful life of a building's materials. By building in an anticipatory manner, easily removable structures pay for themselves in terms of embodied energy and other resource uses several times over. This rule reflects the disproportionately large investments made for rebuilding and remanufacturing versus this money used for other investment practices that could reap greater social benefits.[8]

Condition #12: Support regionalized economic loops by respecting tight-knit life cycle integration. Each stage of the life cycle becomes a part of the region's economics. Life Cycle Economics promotes the close alignment of economic benefits with the benefits of designing highly integrated material and energy flows where wastes are considered as valuable as virgin resources. Linking economics and ecology, as practiced by industrial ecologists, develops the "tightness" necessary to achieve healthy, ecological, and economic facilities and regions.

Condition #13: Do a life-cycle balance of those sourced products supplied by nature through the use of the necessary land area required for regeneration re-sourcing methods at each node or combinations of nodes for the life cycle within a defined boundary scale; *Condition #14*: Measure by ratio the sourcing and re-sourcing life cycles according to a set boundary scale; and *Condition #15*: Establish a pattern recognition procedure that codes possible life cycle patterns for purposes of measuring existing and potential balancing. Taken together conditions 13, 14, and 15 are represented in Figure 16.10. Starting with the definition, any process being represented by the life cycle at whatever scale we are working, we then start dissecting this generalized pattern until we understand the missing elements of a particular pattern that needs to be balanced. The generalized pattern (there are 136 prime patterns and hundreds of secondary patterns) are then selectively described by number and explained using simple examples so the reader can understand the pattern recognition exercise.

Figure 16.10
Life Cycle Balance Ratios

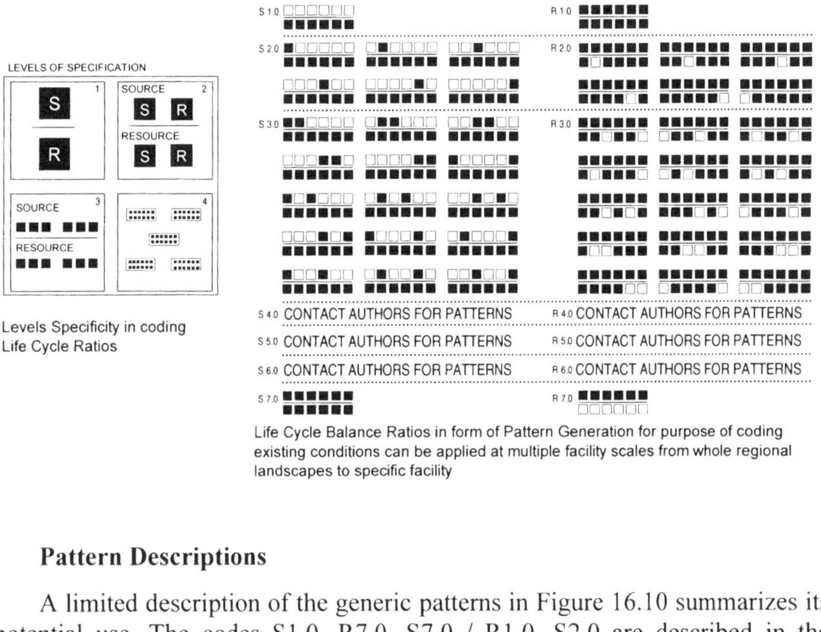

Levels Specificity in coding Life Cycle Ratios

Life Cycle Balance Ratios in form of Pattern Generation for purpose of coding existing conditions can be applied at multiple facility scales from whole regional landscapes to specific facility

Pattern Descriptions

A limited description of the generic patterns in Figure 16.10 summarizes its potential use. The codes S1.0, R7.0, S7.0 / R1.0, S2.0 are described in the following text. Then specific examples at different scales are diagrammed using the second-level specification. The latter demonstrates balanced energy, water, food, and sometimes material conditions relative to the specific subject topics. These examples have yet to be coded according to Figure 16.10.

S1.0 exemplifies a Life Cycle Space™ dominated by the re-sourcing part of the life cycle. The condition rarely occurs to create total balance of the re-source, but is reminiscent of Third World conditions where there is a large presence of garbage pickers who have derived hundreds of ways to recycle waste dumps, unfortunately in very unsafe manners.

R7.0 occurs when Life Cycle Space™ is dominated by the sourcing part of the life cycle. Representative of Third World extractive conditions that are exploitive, leaving no provision for balancing the materials or energy extracted.

S7.0 / R1.0 occurs when L.C.S.™ is in a near-balanced state and the ratio of source to re-source is 1. Examples of this occur primarily in highly integrated farming, industrial ecology passive solar architecture, water harvesting and treatment and depend on the defined topic areas being considered for balancing.

Life Cycle Space™ Conclusion

Buckminister Fuller (1982) stated that "A geodesic is the most economical relationship between any two events." The implication of Life Cycle Space™ is dependent on how nodes are configured in a manner that integrates processes and thus shortens linkages. The degree of balance within a given spatial context (boundary scale) must relate heavily on our understanding of a broad spectrum of ambient resources. The careful matching of nodal (processes) to these ambient resources in multiple fashions determines the Life Cycle Space™. At this time there are over 12.5 million businesses in the United States with the only procedural linkage being the Input/Output model of the U.S. economy. Spatial positioning on the geographic information system (GIS) of each business, its product and by-product condition, along with its relation to regional resources, could enable various degrees of Life Cycle Planning.

LOOKING FORWARD: SUSTAINABILITY FOR THE AMERICAS AND GLOBAL SUSTAINABILITY CONSORTIA

The USBSC is the pilot program of LDI's Sustainability for the Americas Consortia Initiative. Based on USBSC successful emergence, seeding, and implementation start-up, the Ball State University Land Design Institute recently facilitated emergence of the North American Sustainability, Housing, and Community Consortium (NASHCC) and Implementation Start-Up Phase. The NASHCC is likewise the case study for expanding the USBSC to address sustainability, housing, and community. It also serves as the model for other Sustainability for the Americas (SFTA) consortia. The SFTA network serves as model for the global network that pursues regional and global sustainability through innovation.

In these consortia, the U.S. members—the Land Design Institute, the IC^2 Institute, and the Center for Maximum Potential Building Systems—partner with other institutions and community groups to share information and identify solutions that integrate global and local knowledge and decisions into local and regional life cycle flows in sustainable ways. They apply innovative techniques and models to make regionally and locally appropriate decisions that connect people into the dynamics of complex local and regional systems. They help people shift to behaviors that produce sustainable decisions, trigger "streams" of sustainable decisions, and help lead communities to a sustainable future.

NOTES

1. This chapter reorganizes and integrates content from four papers—Ramina (2003); Pacheco (2003); Fisk and Armistead (2003); and Motloch (2003b)—presented at the "Innovation and Knowledge-Sharing Across Public and Private Sectors: The U.S.-Brazil Sustainability Consortium" roundtable at the 7th International Conference on Technology Policy and Innovation, Monterrey Institute of Technology, 2003. More comprehensive readings are available in these papers: business innovation and network structure aspects (Ramina, 2003), collaboration aspects (Pacheco, 2003), Life Cycle Space™ (Fisk and Armistead, 2003) and the U.S.-Brazil Consortium (Motloch, 2003b).
2. This affirmation is known as the "First Law of Kransberg," from the technology historicist Melvin Kranzberg, as cited in Castells (1999).
3. For a comprehensive study of undesired innovation outcomes where social, environmental, and economic relations are considered, see Poungsomlee and Ross (1992). For a collection of successful and not so successful cases of local social capital build-up based on information technology innovations in Brazil, India and Korea, see Evans (1995).
4. For a more appropriate report about the hardware and software evolution since RAND Corporation and ARPA in the 1960s and 1970s toward the present-day modems and TCP/IP protocols, see Horzepa (1989).
5. There are many reasons why the Center for Maximum Potential Building Systems (CMPBS) represents space that are beyond the scope of this chapter. They include incorporation of a geographic equal area projection system for the purpose of graphic visualization and to address the mathematical need for pattern finding around existing life cycle phases within the physical landscape.
6. Various projects of the CMPBS over the last 28 years have attributes that hint at these patterns and establish how one can work knowledgably with a region's resource base (virgin and by-products), labor skills, and enterprises to fit patterns that provide next step systemic conditions for directed change.
7. The intricacies of these scales, types, and relationships are addressed in other papers. If the reader wishes more familiarity with this concept, please contact P. Fisk (pfisk@cmpbs.org) and D. Armistead (darmistead@idmp.com).
8. "By increasing the durability of construction and renovation to 400 years would generate a 5 to 10 fold increase in the economic productivity of our resources and would reduce the economic cost of construction 80%-90%" (Bender, 1983). According to Duffy and Hanney (1989), capital invested into a structure over a 50-year period is overwhelmed by the cumulative financial consequences of three generations of services and ten generations of space. When combined, these end up costing approximately five times the cost of the structure.

REFERENCES

Anderson, R. (1998). *Mid-Course Correction: Toward a Sustainable Enterprise*. Atlanta: Peregrinzilla Press.

Ayres, R. U. (1989). "Industrial Metabolism." In J. Ausubel and H. Sladovich (eds.), *Technology and Environment*. Washington, DC: National Academy of Engineering, National Academies Press.

Barabási, A. L. (2002). *Linked: The New Science of Networks*. Cambridge: Perseus.

Bender, T. (1983). Winner, California's Affordable Housing Competition.

Capra, F. (1982). *The Turning Point: Science, Society, and the Rising Culture*. New York: Bantam Books.

Castells, M. (1999). *A Sociedade em Rede* (The Rise of the Network Society). São Paulo: Paz e Terra.

Cervero, R. (1988). *Effective Continuing Education for Professionals*. San Francisco: Jossey-Bass.

Chesbrough, H. (2003). "Reinventing R&D Through Open Innovation." *Strategy+Business*. Retrieved April 30, 2003 from the site http://www.strategy-business.com

Dean, G., Murk, P., and Del Prete, T. (2000). *Enhancing Organizational Effectiveness in Adult and Community Education*. Malabar, FL: Krieger.

Di Castri, F. and Hadley, M. (1988). "Enhancing the Credibility of Ecology: Interacting Along and Across Hierarchical Scales," *GeoJournal*, 17 (1): 5-35.

Donaldson, J. and Kozoll, C. (1999). *Collaborative Program Planning: Principles, Practices, and Strategies*. Malabar, FL: Krieger.

Duffy, F. and Hanney, A. (1989). *The Changing City*. London: Bullstrode.

Evans, P. (1995). *Embedded Autonomy: States and Industrial Transformation*. Princeton, NJ: Princeton University Press.

Fisk III, P. and Armistead, D. (2003). "Towards Resource Balanced Land Use Planning (The Emergence of Life Cycle Space™)." 7th International Conference on Technology Policy and Innovation, Monterrey Institute of Technology, Monterrey, Mexico, June 10-13.

Fisk III, P., MacMath, R., and Ramina, R. H. (2000). "Industrial Ecology as a Regional Planning Tool: A New Potential for Economic/Environmental Regional Planning." 4th International Conference on Technology Policy and Innovation, Curitiba, Brazil, August 28-31.

Friedman, Y. (1973a). "Critical Group Size," *Architectural Design*. London. January.

Friedman, Y. (1973b). "Towards a Poor World—How Scarcity Might Prevent Disaster," *Architectural Design*. London. October.

Fuller, B. (1982). *Explorations in the Geometry of Thinking Synergetics*. New York: Macmillan.

Gibson, D. V. and Conceição, P. (2003). "Incubating and Networking Technology Commercialization Centers among Emerging, Developing and Mature Technopoleis Worldwide." In L. V. Shavinina (ed.), *The International Handbook on Innovation*. Oxford: Pergamon.

Hatchuel, A., Agrell, P., and van Gigch, J. (1987). "Innovation as System Intervention," *Systems Research: The Official Journal of the International Federation of Systems Research*, 4 (1).

Horzepa, S. (1989). "Your Gateway to Packet Radio." The American Radio Relay League.

Kanter, R. (1994). "Collaborative Advantage: The art of alliances," *Harvard Business Review*, 72 (4): 96-108.

Miller, J.G. (1978). *Living Systems*. New York: McGraw-Hill.

Motloch, J. L. (2001). *Introduction to Landscape Design*, 2nd ed. New York: John Wiley.

Motloch, J. L. (2003a). "BSU Land Design Institute: Education for Sustainability." University of Massachusetts Lowell (UML) Conference on Education for Sustainable Development, Lowell, October 23-24.

Motloch, J.L. (2003b). "Partnering for Regional and Global Sustainability." 7th International Conference on Technology Policy and Innovation, Monterrey Institute of Technology, Monterrey, Mexico, June 10-13.

Motloch, J.L. and Ferguson, D.L. (1997). "The LandLab as a Hands-on Tool for Teaching Sustainable Concepts." Greening of the Campus II.

National Geographic Society (2000). Terrestrial Ecoregions of the World. From the site www.nationalgeographic.com/wildword.

Pacheco, P.D. (2003). "Partnering Globally: Connecting People, Places, and Ideas for Sustainable Development," 7th International Conference on Technology Policy and Innovation, Monterrey Institute of Technology, Monterrey, Mexico, June 10-13.

Pauli, G. (1998). *Upsizing: The Road to Zero Emissions, More Jobs, More Income, and No Pollution*. Sheffield, UK: Greenleaf Publishing.

Poungsomlee, A. and Ross, H. (1992). "Impacts of Modernization and Urbanization in Bangkok: An Integrative Ecological and Biosocial Study." UNESCO MAB, Institute for Population and Social Research in Mahidol University, Thailand and Centre for Resource and Environmental Studies at Australian National University.

Prahalad, C. and Hart S. (2002). "The Fortune at the Bottom of the Pyramid." *Strategy+Business*, First Quarter. Retrieved March 20, 2003 from the site http://www.strategy-business.com

Quinn, D. (1995). *Ishmael: [A Novel]*. New York: Bantam Turner.

Ramina, R. H. (1993). "CONNECTION—Some Thoughts for the Conceptualization and Design of an International Network Based on Amateur Packet-Radio." International Program for the Study of Communication for Sustainable Development, Ryerson Polytechnical Institute, Toronto, Canada.

Ramina, R. H. (2000). "Redes e Poder: o Processo de Metropolização e a Gestão dos Recursos Naturais" (Networks and Power: The Metropolization Process and the Management of Natural Resources). Ph.D. thesis, Universidade Federal do Paraná, Curitiba, Brazil.

Ramina, R.H. (2003). "Networks and Communities: Sustainability in the Knowledge Society." 7th International Conference on Technology Policy and Innovation, Monterrey Institute of Technology, Monterrey, Mexico, June 10-13.

Rapoport, A. (1969). *House Form and Culture*. Englewood Cliffs, NJ: Prentice-Hall.

Rees, W. and Wackernagel, M. (1996). *Our Ecological Footprint: Reducing Human Impact on the Earth*. Philadelphia, PA: New Society Publishers.

United Nations Development Program (1999). *Human Development Report 1999*. New York: United Nations.

PART V:
LEARNING FROM CASE STUDIES

17

Venture-Capital Investments in New Technology-Based Ventures in Mexico

Carlos A. Góngora-Caamal and Enrique Díaz de Léon López

INTRODUCTION

It is well known that developing countries are increasingly facing higher competitive challenges. Mexico is not an exception. There is a strong necessity for leveraging the integration of innovation, enterprise creation, and technology development, in order to overcome some of the competitive barriers imposed by the current knowledge-based economy. In these economies and in many others, bases of competition have changed from prices to value added, where ideas are the main source of innovation and value creation. A frequent challenge faced by enterprises is to transform ideas into products. Moreover, sometimes large corporations are not flexible enough to adapt to change and thus to commercialize their innovations without interfering with their current product line. On the other hand, small companies that are often agile and creative usually lack the financial resources needed to grow at a required fast pace. Yet, entrepreneurs frequently face the challenge of raising capital to commercialize their ideas. In general, there is no shortage of ideas and innovations that could transform into commercial products or services. Also, investors are challenged with the assessment of such innovations. Some of these ideas or innovations are typically based almost in intangibles, making the decision to invest a very hard challenge for the investor. Nevertheless, investors—and in particular venture capitalists—play a paramount job in the economy of a country. Most important, the benefits that a successful innovation could bring could be almost incalculable in terms of the creation of wealth to all shareholders and the society in general. This chapter is focused on gaining a better understanding of the current situation of investment in technology-based new ventures in Mexico.

Recent findings (Díaz de León, 2001; Timmons, 1999; Marcano, 1996) show that start-up firms create a substantial economic impact on most economies. However, the failure rate of start-up firms seems to remain high over time. Few researchers and authors have examined the influence of intellectual capital management on business performance. They realized that intellectual capital is a key factor for successful businesses development. For example, Peña (2002) suggests that the human capital of the entrepreneur (i.e., education, business experience, and level of motivation), the organizational capital of the new firm (i.e., firm capacity to adapt quickly to changes and the ability to implement successful strategies), and the relational capital of the new firm (i.e., development of productive business networks and an immediate access to critical stakeholders) are important intangible assets, which seem to be related positively to well-venture performance.

New technology-based firms (NTBFs) face some of the issues mentioned above. That is, NTBFs are enterprises with high rates of growth, providing social and economic benefits to their societies (Acs and Armington, 2003). Also, as part of their business model, these new ventures typically include aspects of innovation, entrepreneurship, and technology development. For example, they invest on average, more resources on research and development than other industrial sectors (PEUE, 2002; Timmons, 1999). This chapter aims to understand the current situation of the promotion and investment of these ventures in Mexico. Also, we are interested in comparing our findings with past experiences in countries with profitable NTBFs. One of these comparing factors is the development of venture-capital funding (CC, 2001; Marcano, 1996).

In the United States, venture capital has been a source of financing for more than 40 years. Furthermore, many of the world's largest technology companies started with venture capital as a way of financing their growth. Examples of these companies are Microsoft, Apple Computer, Intel, 3Com, Fairchild Semiconductors, Scientific Data Systems, Sun Microsystems, Netscape, Yahoo, and Digital Equipment Corporation among others (Cheesman, 2002; Timmons, 1999; Rock, 1987).

When the source of success for these ventures is carefully investigated, many complex factors are apparently involved. However, there seems to be at least two components in the formula: an extraordinary idea and someone to back it up with capital and management support. Moreover, we have experimented over the years and found that there is no scarcity of good ideas (Díaz de León, 2001). In particular, we can mention our university (ITESM) and its many programs established for many years promoting and giving support to entrepreneurship. However, one tends to wonder: What happens to those innovative and potentially attractive ideas? Are there options for entrepreneurs to obtain venture capital in Mexico? Who are the current investors in new technology-based ventures in our country? What are their criteria for investment? What is their preferred method to make investment decisions?

LITERATURE REVIEW

We define as new technology-based ventures those companies intending to commercialize a technology for the first time, expecting this technology to provide them with a significant source of competitive advantage. We focus on technology-based companies during their early stages of financing, which involve seed, start-up, and first-stage, at each of which the company has different requirements for financing: seed-stage financing, start-up financing and first-stage financing (Bachher and Guild, 1996).

The seed-stage financing is normally provided as an aid to develop a proof of concept and sometimes is used to build a prototype. Also, companies may obtain start-up financing to use during their product development and initial marketing stage. Companies may be in the process of being organized or may have been in business for a short while (usually a year or less), but have not yet sold their product. The first stage of financing is provided to companies that have used their initial capital and are starting to sell their products, but require additional funds to initiate full commercial production and sales (Bachher and Guild, 1996).

The focus of this chapter is on early-stage technology-based ventures that are within some of the following technology sectors: e-commerce, Internet services (Web hosting, Web services, outsourcing, Web content), electronics and computing hardware, software engineering, multimedia and entertainment, communication and wireless networks, biotechnology, health and life sciences (bionics, biomedicine, telerobotics), bioinformatics, and energy and environmental technologies.

These early-stage technology-based ventures have, in most cases, limited tangible assets. That is, when their entrepreneurs are seeking for financing, they typically face an interesting challenge communicating the value of such a new venture (Bachher and Guild, 1996). In other words, even though there are many successful cases of young technology-based ventures, investors frequently face some difficulties assessing their current asset value because it is mostly based on intangibles (Díaz de León, 2001; Zider, 1998).

Some studies analyze and describe the decision-making processes used by venture capitalists and business angels. Such studies show the importance to reveal intangible assets that investors frequently seek in entrepreneurs, and some of them show that business angels frequently decide using their intuition or gut feeling (Bachher and Guild, 1996; Diaz de Leon, 2001; Mason and Harrison, 2000; Nesheim, 1992; Rock, 1987; Van Osnabrugge and Robinson, 2000). Nevertheless, we found a lack of research directed to understand Mexico's current situation on venture capital investments.

There are several methods of evaluation that venture capitalists and business angels use in order to evaluate proposals and make a decision: real options, net present value using Monte Carlo simulation, comparables, adjusted present value, and other discounted cash flow approaches (Day et al., 2000; Lerner, 1999; Manigart et al., 1997). One objective of this chapter is to identify some of the most frequently used methods by investors in Mexico.

A business plan is the main communication tool used by entrepreneurs seeking venture-capital funding. A business plan can be structured in different ways. However, a typical business plan for technology-based ventures has at least some of the following sections: technology assessment, industry analysis, financial analysis, marketing plan, management, intellectual property, risk assessment, and project management (Centeno, 2001). Nevertheless, in this highly selective investment environment, technological entrepreneurs have to provide a better map for their strategies in order to increase their chances of raising funds. Similarly important is the proper and effective communication of their market plan, describing the overall market plan development.

In 1991, Roberts suggested that many technological entrepreneurs do not in fact write an appropriate business plan. Perhaps the number of incomplete or inadequate business plans that make their way to venture capitalists is an indication of the lack of knowledge in this area by entrepreneurs. Consequently, for those entrepreneurs who plan well, and present thorough, well-supported business plans, the opportunities of receiving financial support increase significantly. Venture capitalists differ in their assessment approach and have distinct prejudices when evaluating proposals. Nonetheless, a complete and well-written business plan presents higher probabilities of drawing the attention of an investor. Roberts (1991) suggests that a business plan should address four main functional areas: marketing, management, technology, and finance. What is more, there is scarce knowledge of the impact that a well-written business plan has within Mexican investors. Centeno (2001) obtained some evidence of the implications of effective business plans for investors in Mexico. In consequence, an objective of this chapter is to identify the perception of investors about business plans of technology-based ventures and entrepreneurs in Mexico as well as identify their expectations related to the content of a business plan.

RESEARCH METHOD

The study research was conducted in two steps. The first phase included a set of personal interviews with key players in the area of venture-capital investment in Mexico. Given the exploratory nature of the research question, we identified a group of individuals from both private and public institutions whose interests reside in the commercialization of new ideas or innovations. Following a snowball-sampling technique we were able to identify more people with the same characteristics. An objective of this phase of the research was to identify active investors in Mexico. Once we obtained their contact information, it was archived in a database to be used in the following phase. The second part of the research consisted of the design of a questionnaire to be applied to those investors from the list that would identify themselves as investors of technology-based ventures.

The questionnaire includes four main sections: (1) general information of the firm and investments, (2) business plan, (3) assessment process, and (4) Mexico's environment. We sent our questionnaire via electronic mail; answers were received via fax or electronic mail.

RESULTS

During the first stage of the research a total of 35 individuals were interviewed. The respondents were either investors themselves or representing investment institutions, individuals involved in enterprise incubators and technology development, and technological entrepreneurs. We found evidence that indicates the high degree of interest that this topic is currently raising in Mexico. Talking with them, we obtained first-hand information about Mexico's context from their perspective. The information elicited from this phase allowed us to obtain a better understanding of the current situation that investors, institutions, and the federal government are facing toward this issue. Moreover, by approaching these people we able to identify some of the main investors of technology-based ventures currently operating in Mexico.

For the second part of the research, a questionnaire was built based on some of the issues detected during the first stage. For example, we were able to detect some key criteria in decision-making processes as well as some evaluation methods used by venture capitalists and business angels, when assessing new ventures. Several of these key criteria are intangibles, and are based on perceptions, which in a way is consistent with some research results in this area (Bachher and Guild, 1996; Boocock and Woods, 1997; Díaz de León, 2001; Dixon, 1990; Fried and Hisrich, 1994; Lerner, 1999; Manigart et al., 1997; Mason and Harrison, 1996, 2000; Nesheim, 1992; Roberts, 1991; Van Osnabrugge and Robinson, 2000; Zider, 1998). We reviewed a study of business plans for technology projects made by Centeno (2001). Also, we considered some of the issues presented by the interviewees during the first phase related to technology development and enterprise creation. Some of these are important aspects that are currently limiting the venture-capital industry in Mexico. The survey was applied for a period of almost four months, from March to June of 2003.

Mexico has gathered a total amount of US$1,684 million in private equity placements from 1995 to 2001. At the end of 2000, Mexico's private equity placements represented 0.37% of total market capital of the main stock market (Schneider and Videgaray, 2001). A study made for Nacional Financiera (NAFIN) reports that there were only 40 to 50 active seed and venture funds (funds of risk capital for debt or capital investments named SINCAs and others), domiciled in Mexico and invested directly in Mexican firms. Moreover, these active funds accounted for only US$362 million committed and 256 investments as of December 1999 (Millenia, 2003).

In 1993 there were 58 SINCAs, but in 2001 there were only 28, and the expected number at the end of 2002 was between 20 and 25. The participation of Mexico in risk capital was considered between US$2,000 and $3,000 million, representing 0.47% of the world risk capital industry (NAFIN, 2002). A study made by Rozada and Alvear (2002) from Babson College also agrees that there is little venture finance through venture capital and private equity in Mexico.

From a total of 28 venture-capital funds, private equity firms, and independent investors that were identified by name, only 26 of them could be contacted. Some of these 26 firms had invested in early-stage technology-based ventures or are currently assessing their first ventures in Mexico. We found three firms that help entrepreneurs to find risk-capital funding, two of them are in Mexico City and the other is located in Monterrey.

A total of 26 invitations were sent to investing firms. From the 17 companies that accepted to participate in this study, only eight answered questionnaires were received. Such answers came from six venture-capital firms, one governmental office, and one business angel. They have been investing in early-stage technology-based ventures in Mexico for few years.

There is wide-ranging investing experience among these eight venture-capital funds. Four of them have been investing in early-stage technology-based ventures between one and three years. Another fund has less than one year of experience and one has more than seven years (two funds did not answer this question). These results show that most venture-capital funds supporting early-stage technology-based ventures have been operating in Mexico for only a few years.

General Information of the Current Mexican Funds

Figure 17.1 shows the main role currently performed by the person who responded to the questionnaire in the firm. From the companies interviewed, 50% of the venture capital operating in Mexico comes from Mexican funds (Figure 17.2).

Figure 17.1
Participant's Main Role in the VC Firm

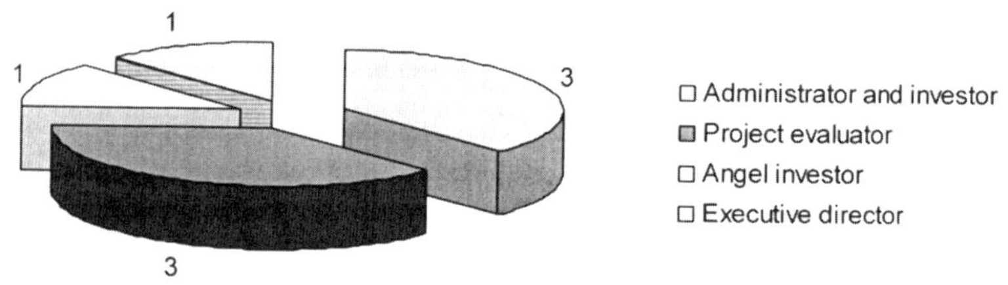

Figure 17.2
Where Venture-Capital Funds Come From

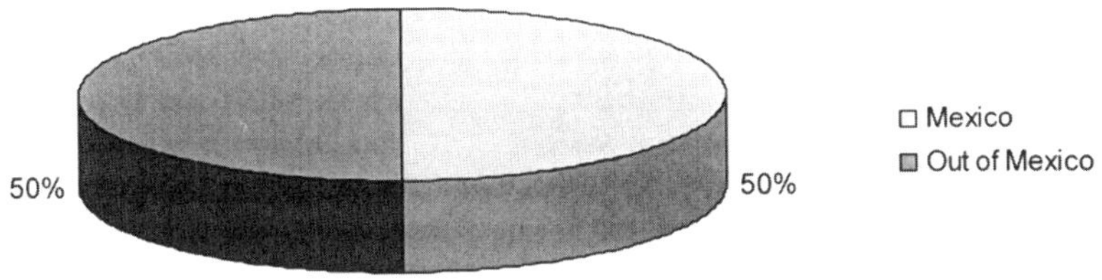

The 26 venture-capital fund offices are located mainly in Mexico City, Guadalajara, and Monterrey. Seven of eight participating firms have a total of eleven offices in Mexico: five in Mexico City, two in Jalisco (Guadalajara), two in Nuevo Leon (Monterrey), one in Estado de Mexico, and one in Tamaulipas. One firm has its office in the United States. These investors typically play the roles of both advisor and members of the venture's board. The sources of these eight funds come from corporate (three of them), private (two), pensions (one), government (two), and Inter American Development Bank (one), and one investor bets its own money (Figure 17.3). Four of them have a dedicated pool for financing technology-based ventures in any maturity stage between US$1 and 25 million; meanwhile one more has a pool between US$26 and 50, and another one between US$51 and 75.

Figure 17.3
Sources of Venture Capital Funds

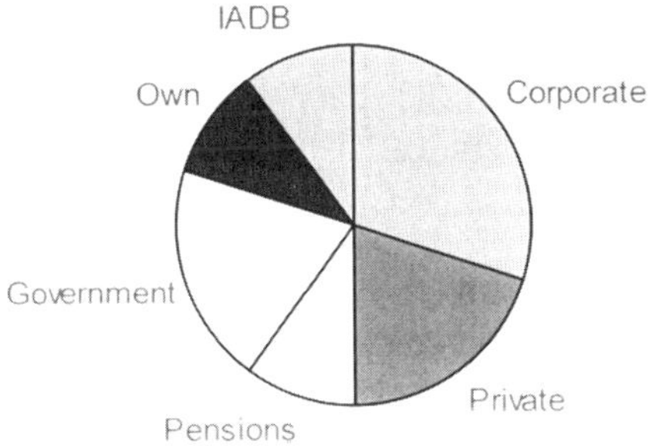

We asked for the number of investments in early-stage technology-based ventures made in the years 2000, 2001, and 2002, in order to identify whether they are increasing, decreasing, or without change. In 2000, two of these eight firms made one to three investments. In 2001, two firms made one to three investments, and one of them made four to seven. In 2002, six of them made one to three investments. Correspondingly, the number of investments of these firms is apparently increasing (Figure 17.4).

Figure 17.4
Number of Investments by Year

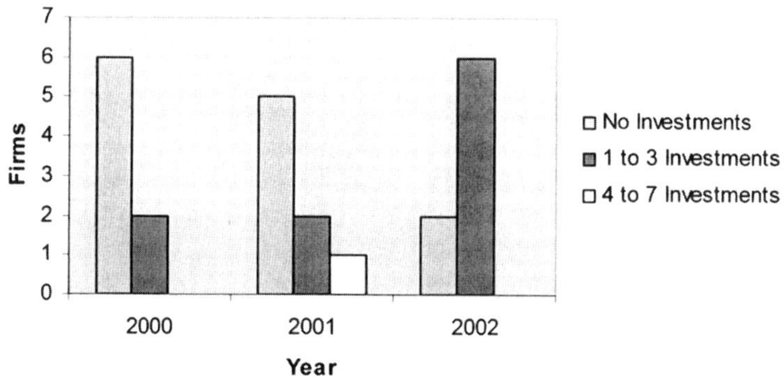

The technology sectors where these firms used to finance new ventures are described in Table 17.1. This table shows the number of firms investing by sector, that is, the venture-capital offer by sector. Software engineering, media and entertainment, communication and wireless networks are the three sectors that are of more interest to investors. There are four firms supporting ventures in each of these sectors. In electronics and computer hardware and biotechnology sectors we found minor venture-capital offers.

Table 17.1
VC Funding by Technology Sector

E-commerce	1
Internet-based services	1
Electronics and computer hardware	2
Software engineering	4
Media and entertainment	4
Communication and wireless networks	4
Biotechnology	2
Health and medicine	1
Energy and environment technologies	2
Bioinformatics	1
Security systems	1

Some preferences for investing by stage of venture's maturity are described in Table 17.2. Four of these venture-capital funds have a preference for investing in firms in later stages of maturity, when most of the risk of failure is practically nonexistent. Four funds support early-stage ventures as their first investing preference, and two of them in the seed stage.

Table 17.2
Preference by Stage

	Preference			
	First	Second	Third	Fourth
Seed	2			2
Start-up	1		**4**	
First stage	1	**3**		1
Later stage	**4**	2		

For each stage of maturity, ventures need different kinds of support and amounts of money. Only four answers were received; however, they illustrate important information. Figure 17.5 shows the amount of money typically invested in each of the three early stages of technology-based ventures.

Figure 17.5
Investments in Each of the Three Early Stages of Technology-based Ventures

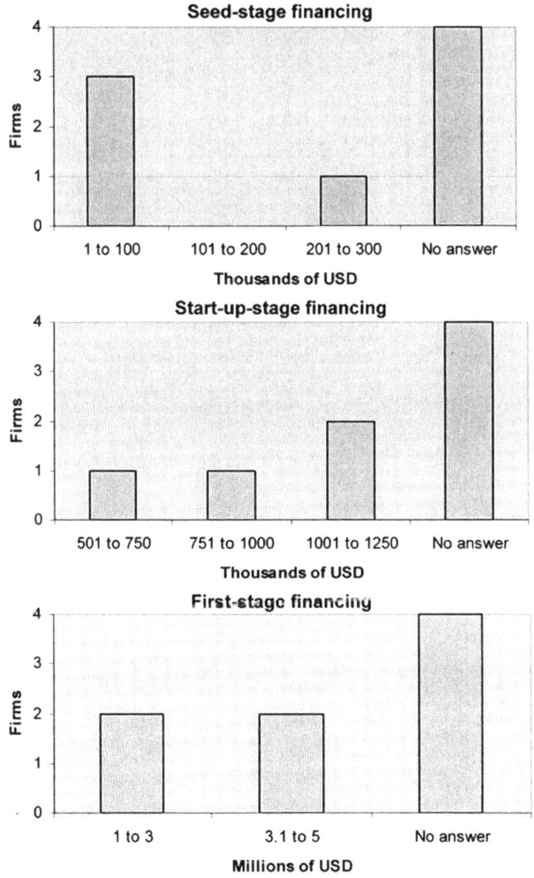

Venture-capital investments have active contributions of investors or people from venture-capital firms. The preference of the roles played by the people who responded to the questionnaire are shown in Figure 17.6.

Figure 17.6
Roles Played in NTBV

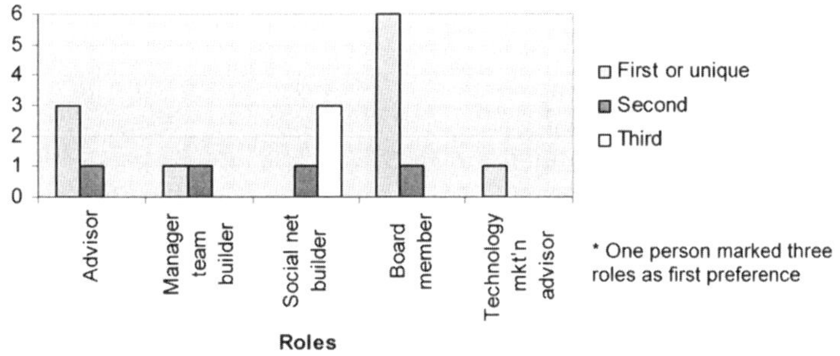

The preference by type of investments of these funds shown in Figure 17.7 ranks venture capital as first preferred. The preferences of coinvestment are with corporate and private funds.

Figure 17.7
Type of Investments preferred

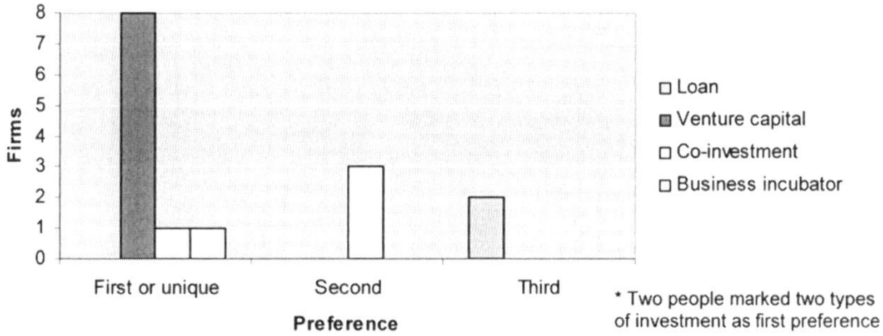

Exits

The time expected for exiting from investments is for five funds between three and five years and two funds responded that this time is between six to nine years, and one did not answer. The preference for exiting is described in Table 17.3. The most preferred exit is through merger and acquisition, and the second one is selling their shares to coinvestors. There is a special case for one fund, the government department, because its exit is directly related to the local government, where it is established. When funds exit from investments they normally sell all their shares. Five of these eight funds sell all their shares, one maintains between 31 and 45 percent of the venture's stock, and one keeps more than 45 percent of the venture's stock.

Table 17.3
Exit Strategies

	Preference			
	First	Second	Third	No clear
IPO				
Sell shares to entrepreneur	1		2	
Sell shares to coinvestors		**3**		
Merger/Acquisition	**4**			
Other	1			
Not exit yet				2

Business Plan

The business-plan sections expected for a technology-based project are shown in Figure 17.8. It shows that investors stress all categories: product and service description, industry analysis, financial sections, executive report, marketing plan, and risk assessment sections.

Figure 17.8
Business-Plan Sections

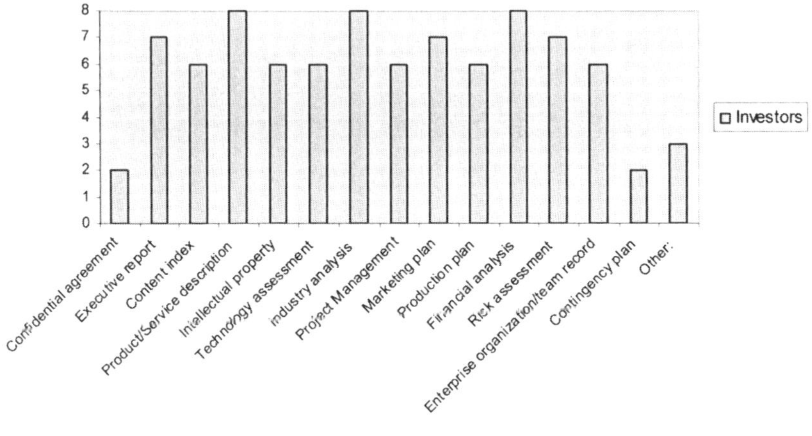

In addition to business-plan sections, a business plan should communicate the targeted market by the new venture. Typically, a decision to fund a venture could depend on the targeted market. Figure 17.9 shows the investors' preference in this aspect.

Figure 17.9
Market Targeted by Business Plan

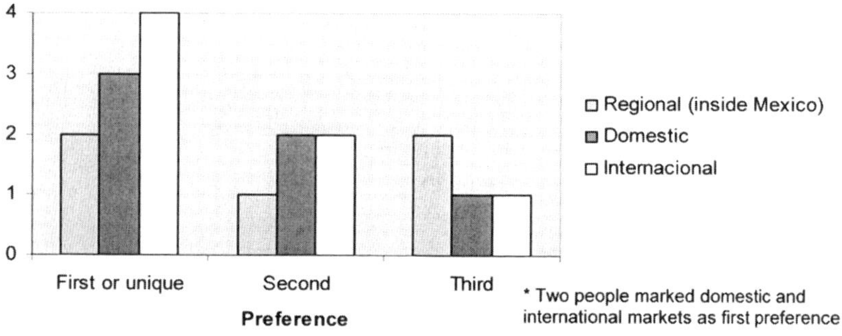

The number of business plans received during last year of the study by each fund varies significantly. Four of them received between 1 and 30 business plans, two of them received between 61 and 90 plans, and the last two received between 91 and 120 plans. However, only a few business plans overcame the screening stage. Six investors said that just between 1 and 20% of business plans that they receive overcame the screening stage, another one said between 21 and 40%, and the other said between 41 and 60% of business plans were used to overcome this stage. Other significant aspects are the perception of change of the number of business plans that they have received through the years, but their perception in terms of percentage of increment or decrement of the number of business plans is quite different. Figure 17.10 shows these results.

Figure 17.10
Perception of Change in the Number of Business Plans

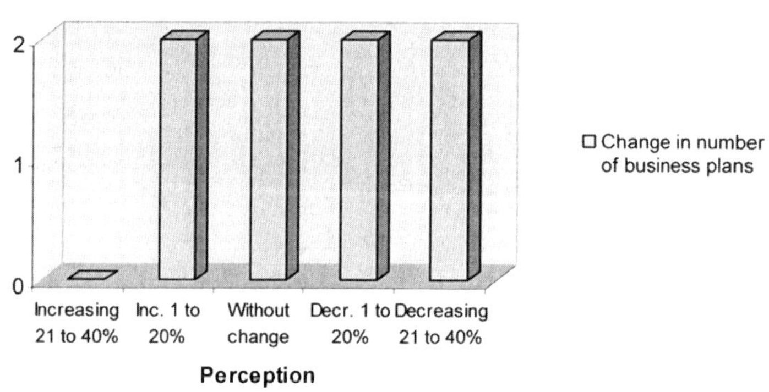

Assessment Process

The assessment process from the reception of the business plan to communicating an investment decision typically lasts between three to four months. During this period investors and/or evaluators have meetings with the management team. It is normal to have between one and five meetings, and sometimes between six and ten meetings, in order to know the proposal and the entrepreneur team.

The main investor requirements are shown in Figure 17.11, and the number of references and the type of them are shown in Figure 17.12. This figure illustrates that the number of references commonly asked is between four and seven, and also shows a clear preference for references coming from other investors.

Figure 17.11
Requirement for Being a Candidate for Evaluation

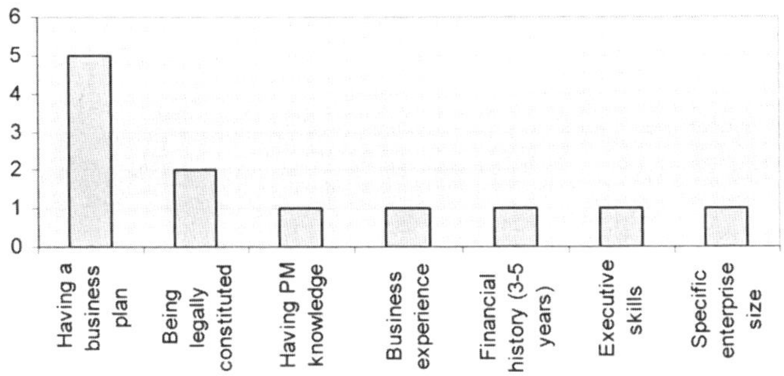

Figure 17.12
About References Asked

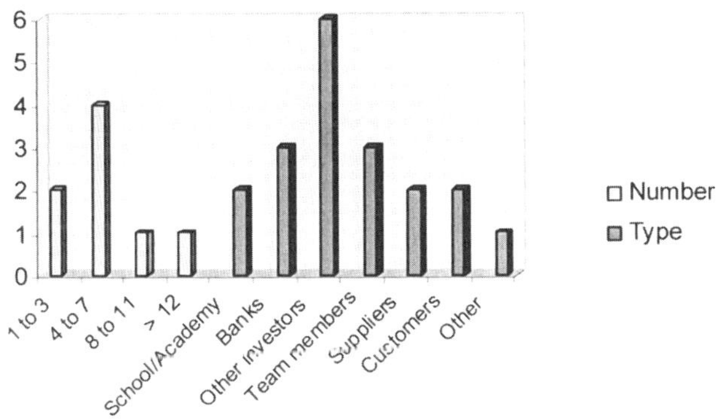

Sources of Proposals

The eight funds receive proposals all year long, the source being mainly friends and acquaintances, and other investors. Table 17.4 describes sources of proposals and the frequency.

Table 17.4
Sources of Proposals

	Frequency			
	First	Second	Third	No clear
Agencies for investors networks			1	1
Own agency (in our offices)	2			2
Friends and acquaintances	**4**			2
Other investors		**5**		2
Expositions, Conventions, or Contests			1	
Business incubators				
Internet site				

Evaluation Methods

The financial methods used most frequently are net present value, industry analysis (also known as fundamental analysis), and comparables. Some firms used more than one method for the proposal's evaluation (Table 17.5).

Table 17.5
Evaluation Methods

Industry analysis	6
Comparables	6
Real options	2
Net present value	7
DCF	4
Multiples	1

Decision-Making Criteria

When investors/evaluators are assessing entrepreneurs and ventures, there are key criteria for decision making that normally they consider by level of importance. Figure 17.13 illustrates criteria that investors consider as extremely important when making decisions to support a venture.

Mexico's Environment

All firms agree that Mexico offers good opportunities for developing new technology-based firms, and seven will continue investing in Mexico in technology firms during their early stages. Also seven expressed that Mexican laws help or facilitate the venture capital investments in early-stage technology-based ventures.

They also think that the taxes schema is an important topic to improve, because it limits the investments in technology-based ventures. These topics agree with results of a study of entrepreneurship in Mexico, status and opportunities of improvement (Fabre and Smith, 2003), and also some suggestions for taxes schema reforms (Morrison and Foerster, 2003) presented in a meeting at NAFIN's headquarters in 2003.

Figure 17.13
Decision-Making Criteria

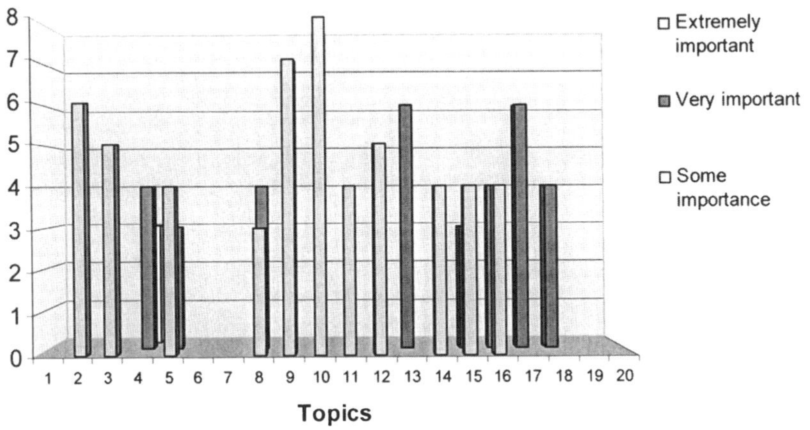

1. Enthusiasm of the entrepreneur(s)
2. Trustworthiness of the entrepreneur(s)
3. Expertise of the entrepreneur(s)
4. Investor liked entrepreneur(s) upon meeting
5. Track record of the entrepreneur(s)
6. "Chemistry" with entrepreneur(s)
7. Well clothing and cleanness
8. Clear idea or proposal presentation
9. Sales potential of the product
10. Growth potential of the market
11. Quality of product
12. Market niche
13. Informal competitive protection on the product (know-how)
14. Perceived financial rewards (for the investor)
15. Expected rate of return
16. High margins of the business
17. Investor's involvement possible (contribute skills)
18. Investor's strengths filling gaps in business
19. Feel good intuition from proposal and team
20. Near to firms' offices

There are other important aspects that could be improved. Figure 17.14 shows the importance that investors give to 11 aspects in order to leverage there venture capital investments in technology-based ventures (TBVs) in Mexico. The maximum agreement was with four investors.

Other important aspects to improve in order to leverage the development of technology-based firms in Mexico from the investors' perspective are shown in Figure 17.15. The maximum agreement was with three investors, about the number of requirements and government transactions.

Figure 17.14
Aspects to Improve in Order to Leverage VC Investments in TBVs

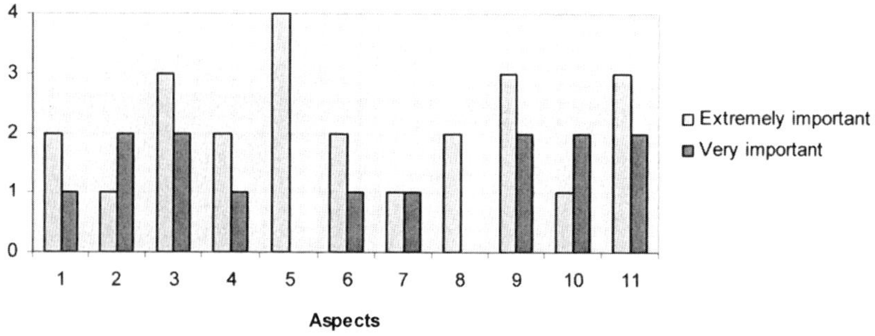

1. Venture-capital understanding
2. Venture-capital Investors' directory
3. Competitive entrepreneurs
4. Entrepreneurial promotion in universities
5. Venture capital funds
6. Government funds
7. Pension funds
8. Better science & technology laws
9. Critical mass of entrepreneurs
10. Taxes & related laws reforms
11. Specific regulation for venture capital

Figure 17.15
Aspects to Improve in Order to Leverage TBV Development

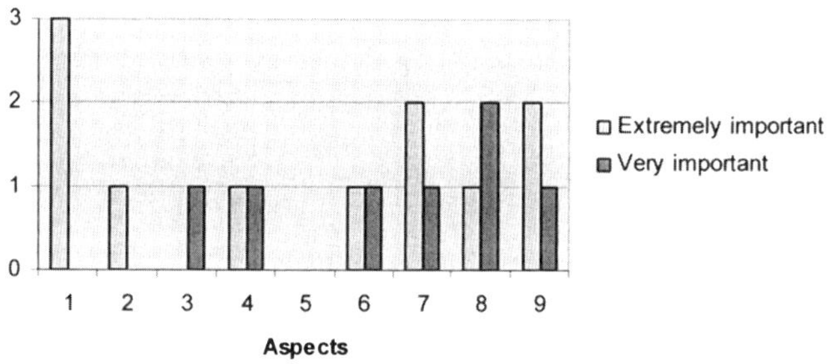

1. Number of requirements and government transactions
2. Time spent in accomplishing government transactions
3. Public safety
4. Confident legal framework
5. Country's economic growth
6. Production costs (taxes, energy, communications)
7. Government policies
8. Potential ideas for doing business
9. Corruption (transactions, permissions, licenses)

In addition to the last two sets of results, investors also pointed out as key opportunity areas to Mexican technological entrepreneurs some subjects—knowledge, skills, and competencies—shown in Figure 17.16, in order to obtain financing to their proposals and be able to deal with the day-to-day management of the young firm. The maximum agreement in importance was with four investors, the subjects are marketing and technology assessment.

Figure 17.16
Subjects That Mexican Technological Entrepreneurs Must Improve

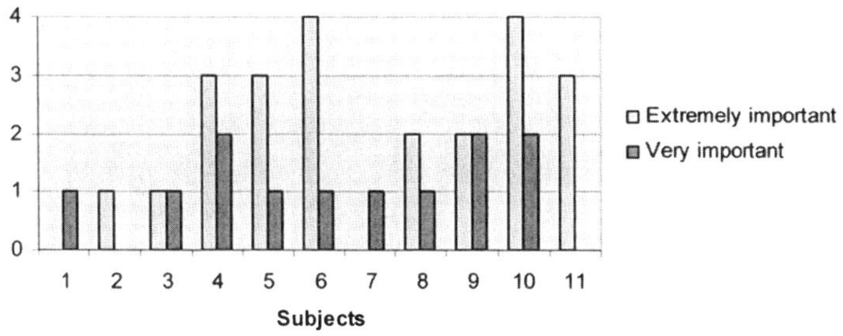

1. Knowledge management
2. Project management
3. Management of technology & innovation
4. Strategic management
5. Finance
6. Marketing
7. Production, manufacturing
8. Technology forecasting
9. Intellectual property, patents
10. Technology assessment
11. Creative ideas

Apparently there are only a few number of funds, but is possible that these funds are close to the population of funds that support early-stage technology-based ventures in Mexico. Table 17.6 shows the answers to the question "how many venture capital firms or investors do you think are financing early-stage technology-based ventures in Mexico?" This question has the objective to get information from the population.

Table 17.6
How Many VC Firms or Investors Do You Think Are Financing Early-Stage TBV in Mexico?

1 to 3	2
4 to 7	3
8 to 11	2
12 to 15	0
More than 15	0
No answer	1

CONCLUSIONS

Venture-capital financing for early-stage technology-based ventures is scarce, not only in terms of the number of funds currently operating in Mexico, but also in the number of investments that these funds make. However, some results show that this activity is increasing. Also, most of the funds dedicated to early-stage technology-based ventures have only a few years operating in Mexico, and are very difficult to contact. There is not an association of venture-capital firms and investors, and they do not know each other or set a limit to give references and to reach other investors through them. Eleven funds that finance technology ventures in Mexico were identified; nevertheless only eight responded to our survey instrument. Although these eight funds support early-stage technology-based ventures, most of them have preference for later stages of ventures' maturity.

The technology sectors where there are more opportunities to find support are those related to software and communication applications and services. It is remarkable that there is financial support for sectors with intensive use of technology, such as bioinformatics and biotechnology.

The business plans for technology-based projects that these firms receive per year are few, and from them, only a small percentage manage to overcome the screening stage; even fewer manage to obtain financing, thus competition is very tough. Although Mexico is ranked eighth in its total entrepreneurial activity by the Global Entrepreneurship Monitor, most entrepreneurs are not working in technology sectors, as Fabre and Smith (2003) found, and most entrepreneurs are in low-value-added and low-risk ventures. This situation shows areas of improvement to entrepreneurs, because it is possible to write a business plan that contains the information expected by investors, it is possible to learn and to develop the knowledge, skills, and capabilities that demand technology enterprises.

In fact, the current situation of venture-capital investments on new technology-based ventures depicts a venture-capital industry that is starting in Mexico. This means that there are many things to do in order to leverage and to facilitate venture-capital investments, including technology-based ventures. Mexico has good opportunities to build a technology entrepreneurship culture, and an important reason to do it, as all investors, who have money, experience, and are willing to support technology-based firms, perceive that Mexico offers good opportunities to create and to develop technology-based ventures.

The current situation also brings issues for future research that might help in the development of technology-based ventures during their early stages. Some of these issues include:

- Identify reasons of scarce early-stage preferences.
- Investigate why there is high entrepreneurial activity in Mexico and few new technology-based ventures.
- Identify areas of improvement for investors and entrepreneurs in order to facilitate the interaction between them, and to help them to develop skills to have better

evaluation approaches (investors) and better opportunities to obtain financing (entrepreneurs).

REFERENCES

Acs, Z. J. and Armington, C. (2003). "Entrepreneurial Activity and Economic Growth." http://www.babson.edu/entrep/fer/Babson2002/XI/XI_P1/XI_P1.htm.

Bachher, J. and Guild, P. (1996). "Financing Early Stage Technology Based Companies: Investment Criteria Used by Investors." Frontiers of entrepreneurship. www.babson.edu/entrep/fer/papers96/bachher/.

Boocock, J.G. and Woods, M. (1997). "The Evaluation Criteria Used by Venture Capitalists: Evidence from a UK Venture Fund," *International Small Business Journal*, 16 (1), October-December: 36-57.

CC (2001). "Creación y Gestión de Nuevas Empresas de Base Tecnológica: Reflexiones y Propuestas." Cluster Conocimiento, Ed. Cluster del Conocimiento.

Centeno, C. (2001). "Business Plan Development Method for Technology-based Projects to Seek Venture Capital." Masters degree in Administration of Information Technologies. ITESM Campus Monterrey, Mexico.

Cheesman, D. (2002). "Nothing Venture..." *IEE Review*, 48 (2), March: 21-25.

Day, G. S., Schoemaker, P. J., and Gunther, R. E. (2000). *Wharton on Managing Emerging Technologies*. New York: John Wiley.

Díaz de León L., E. (2001). "Toward an Expert Assessment of Intangibles in Technology-Based New Ventures." Doctor of Philosophy in Management Sciences Thesis. University of Waterloo, Ontario, Canada.

Dixon, R. (1990). "What Do Venture Capitalists Look For?" *Management Accounting*, 68 (2), February: 36-37.

Fabre, F. and Smith, R. (2003). "Building an Entrepreneurial Culture in Mexico." Venture Finance Institute of Mexico and Venture Finance Institute, Mexico.

Fried, V.H. and Hisrich, R.D. (1994). "Toward a Model of Venture Capital Investment Decision Making," *Financial Management*, 23 (3), Autumn: 28-37.

Lerner, J. (1999). *Venture Capital and Private Equity: A Case Book*. New York: John Wiley.

Manigart, S., Wright, M., Robbie, K., Desbrières, P., and De Waele, K. (1997). "Venture Capitalists' Appraisal of Investment Projects: An Empirical European Study," *Entrepreneurship Theory and Practice*, 21 (4), Summer: 29-43.

Marcano G. L. F. (1996). "Empresas de Base Tecnológica." V World Conference on Science Parks, Rio de Janeiro, Brazil.

Mason, C.M. and Harrison, R.T. (1996). "Why Business Angels Say No: A Case Study of Opportunities Rejected by an Informal Investor Syndicate," *International Small Business Journal*, 14 (2), January-March: 35-51.

Mason, C.M. and Harrison, R.T. (2000). "Investing in Technology Ventures: What Do Business Angels Look for at the Initial Screening Stage?" http://www.babson.edu/entrep/fer/XV/XVA/XVA.htm.

Millenia Consulting (2003). "Consolidated Paper on Venture Capital Development in Mexico." Christina Kappaz and Kenneth O'Hare of Millennia Consulting and John B. McNeece III of Luce, Forward, Hamilton and Scripps, LLP, with input from Robert Heard of Edge Development Capital.

Morrison and Foerster LLP (2003). "Comparative Review of Legal and Regulatory Frameworks Supporting Venture Capital." Rafael Hernandez Mayoral, Tom Eldert and Gustavo Struck.

NAFIN (2002). "Construyendo la Industria de Capital Semilla y de Riesgo en México." April.

Nesheim, J.L. (1992). *High Tech Start Up: The Complete How-to Handbook for Creating Successful New High Tech Companies*. Saratoga, CA: Electronic Trend Publications.

Peña, I. (2002). "Intellectual Capital and Business Start-up Success," *Journal of Intellectual Capital*, 3 (2): 180-198.

PEUE (2002). "Las Claves del Crecimiento Futuro: Innovación y Empresa." Presidencia Española de la Unión Europea, Spain.

Roberts, E.B. (1991). "High Stakes for High-Tech Entrepreneurs: Understanding Venture Capital Decision Making," *Sloan Management Review*, 32 (2), Winter: 9-20.

Rock, A. (1987). "Strategy vs. Tactics from a Venture Capitalist," *Harvard Business Review*, 65 (6), November-December: 63-67.

Rozada, R. and Alvear, J. (2002). "Financing the Entrepreneurial Venture in Mexico." Final Project, Professor Les Charm. December, Babson College.

Schneider, F. and Videgaray, L. (2001). "Private Equity in Mexico." Conference of Financial Markets in Mexico organized by the Center for Research on Economic Development and Policy Reform at Stanford University, October.

Timmons, J. A. (1999). *New Venture Creation: Entrepreneurship for the 21st Century*, 5th Ed. New York: McGraw Hill.

Van Osnabrugge, M. and Robinson, R.J. (2000). *Angel Investing*. San Francisco: Jossey-Bass.

Zider, B. (1998). "How Venture Capital Works: Before You Can Understand the Industry, You Must Separate Myth from Reality," *Harvard Business Review*, 76 (6), November-December: 131-139.

18

Technological Capability Accumulation in the "*Maquila* Industry" in Mexico

Gabriela Dutrénit and Alejandro O. Vera-Cruz

INTRODUCTION

A group of countries of East and South East Asia achieved important successes in their processes of industrial and technological progress from the development of local suppliers of the manufacturing industry.[1] Based on processes of learning and accumulation of technological capability, these local suppliers could advance rapidly from simple assembly activities in the 1960s and 1970s, toward product design in the late 1980s, and finally to introducing their own brands in the international markets and carrying out R&D activities for new products in the 1990s.[2]

The industrial relocation process toward the northern border of Mexico began in the mid-1960s. On one side, the Mexican model was different from the East and South East Asian one: Transnational firms established their own assembly plants on the northern border, which were denominated *maquilas*.[3] On the other side, the evolution in Mexico has been less successful in terms of national industrial and technological development.

As a consequence of the conditions and restrictions under which these plants were established, and of the poor attempt by the *maquilas* to form links with different Mexican industrial and governmental organizations, the stereotype that they were technologically poor establishments, where workers were submitted to repetitive and inhuman exploitation processes, was consolidated.

During the decade of the 1990s the *maquila* industry consolidated its role as employment generator in the manufacturing industry. Between 1990 and 2000, it created employments at an annual rate of over 11%. In 2000, the 1,285,007 workers employed in the 3,590 plants of the *maquila* industry represented 31% of the total personnel working in the manufacturing industry.[4]

However, the evolution of the *maquila* industry was not limited to the growth of the number of establishments and employees. As a result of internal learning processes and changes in the strategies of global firms, various *maquilas* in Mexico have experienced important qualitative changes. Although there are no precise data to evaluate the depth and magnitude of such transformation, recent studies confirm that during the 1990s a change occurred in the nature of the productive and technological activities of a group of *maquilas* toward more complex products and more sophisticated technical activities.[5] But there are also dimensions that have evolved more slowly, like the linking of national firms to their suppliers chains, particularly of components.

The aim of this chapter is to present an analytical framework to help study the technological capability accumulation in the *maquila* industry in Mexico, based on that to analyze the levels of technological capability accumulation of three *maquilas*, and to bring to light some stylized facts of the accumulation process in this industry.

The analytical framework proposed draws on the taxonomy of technological capabilities proposed by Bell and Pavitt (1995) for the manufacturing industry in developing countries, and the adaptations carried out by Figueiredo (2001) and Ariffin and Figueiredo (2003). This chapter adapts it to the particularities of the *maquila* industry in Mexico, in this sense the new taxonomy adds technical functions that are relevant to this industry and redefines activities that correspond to various levels of accumulation.

It is important to highlight that the framework focuses on intrafirm accumulation processes; it reveals the paths, processes, and strategies of accumulation. However, it is less suitable to explain the links between these internal processes and the external context. Therefore, it is important to consider that the characteristics of the evolutionary paths of the firms depend on internal and external factors. Among the internal factors stand out the particularities of the firms' foundation, their organizational and technological culture, and the business and technological strategies. These factors affect the building of organizational routines that shape the path of technological capability buildup. The most relevant external factors are associated with the economic and social environment in which firms operate, and with the characteristics of the local and national innovation systems.

The analysis of the accumulation processes of the *maquilas* in Mexico is placed in a national environment that has been characterized over decades by a macroeconomic instability and by the existence of an immature national innovation system, with a fragile structure of linkage among the different agents. Additionally, the *maquilas* were initially placed in localities in the northern border with little manufacturing tradition, a young educational system, nonexistent R&D centers, an immature local institutional structure, and so on. In these localities it was impossible to talk about a local innovation system, because such a system was barely established. This affects both the accumulation of the *maquilas* and the development of local suppliers. This local environment has evolved slowly because the *maquilas* have established very few external links. In fact, the local environment has been transformed into a binational regional environment, which increases the number of actors and opens spaces for further links. The locality of Ciudad Juarez is a representative case for the analysis of the *maquila* industry. It concentrates approximately 8% of the plants and 20% of the employment.

This chapter is based on a case-study methodology of the business lines of three *maquilas* in Ciudad Juarez: Thomson Multimedia, Philips Corp., and Delphi Corp. Two business lines correspond to the consumer electronics industry and the third is specialized in electronic products for the auto-parts industry. The evidence was collected between April 2001 and October 2002 in Ciudad Juarez. The main sources of information are interviews conducted with the personnel of different hierarchical positions within each *maquila*. The next section presents the analytical framework to evaluate the levels of technological capability accumulation in the *maquila* industry. Next is an analysis of the trajectory of technological capability accumulation of the three cases, followed by a comparison of the trajectories of accumulation of the three *maquilas*. The chapter ends with some stylized facts of the technological capability accumulation processes in the *maquila* industry.

A TAXONOMY OF TECHNOLOGICAL CAPABILITIES OF THE *MAQUILA* INDUSTRY

Since the beginning of the 1980s, a group of authors has contributed to the gradual buildup of an analytical framework to help analyze the processes of accumulation of technological capabilities by firms in developing countries.[6] The basic idea is that capabilities represent abilities to do things, and technological capabilities reflect the dominion of technological activities. Based on empirical research at the firm level, this literature has elaborated taxonomies to help describe the gradual processes of accumulation, from a stage that reflects minimum levels of knowledge (needed for routine operation) to the stage of advanced innovative capabilities. In this section we present a new version of the taxonomy of technological capabilities adapted to the particularities of the *maquila* industry in Mexico. An index of technological capabilities that measures the results of the accumulation processes is also presented.

The design of the taxonomy of technological capabilities for the *maquila* industry draws on the analytical framework proposed by Bell and Pavitt (1995), which gathers up the advances of knowledge in this area,[7] and on the adaptations carried out by Figueiredo (2001) and Ariffin and Figueiredo (2003). The philosophy of the taxonomy is kept but, based on the evidence of the characteristics of the accumulation processes at the maquila industry, new technical functions are added and some of the activities corresponding to each level of accumulation are redefined.

The files group out the main technological capabilities according to the degree of innovativeness, including four levels of accumulation: a level of routine production technological capabilities, and three levels of innovative technological capabilities—basic, intermediate, and advanced. By columns, the taxonomy distinguishes the technical functions in which firms can develop technological capabilities. There are three groups of technical functions: (1) investment functions that refer to the creation of technical change and the administration of its implementation during large investment projects; (2) production functions that refer to the generation and management of technical change in the processes, the production organization, and the products; and (3) supporting functions that consist of the development of links and interactions necessary for innovative activity.

The taxonomy of technological capabilities in the *maquila* industry has the following particularities:

1. It is defined for the activities of the *maquilas* in Mexico, so it shows the accumulation of technological capabilities in each plant, business line, or the whole operation in Mexico, independent of the technological capabilities of the global firm.
2. There are three technical supporting functions where defined: internal linkages, external linkages, and equipment modification.
3. The difference between activities of internal and external linkages was due to the fact that they reflect two relevant aspects of the relationships of the *maquilas*: intrafirm links and links with the context. These dimensions have evolved differently.
4. Following Figueiredo (2001), the technical function of equipment modification was added because it is relevant in many firms in developing countries.

Table 18.1 presents the taxonomy of technological capabilities for the *maquila* industry. In each stage of accumulation of each technical function, the most characteristic activities of that level are listed.

ACCUMULATION OF TECHNOLOGICAL CAPABILITIES IN THE *MAQUILA*: THE EMPIRICAL EVIDENCE

Based on the framework described above, this section analyzes the trajectory of technological capability accumulation of three *maquilas*: Delphi Corp., Philips Corp., and Thomson Multimedia.

Because there exist differences in the technological capability accumulation processes per each *maquila*'s business line, in each case the most representative line of the process of technological capability accumulation in Mexico was chosen. In the case of Delphi we analyze the business line of sensors and actuators, as it is the line that has had the most advanced trajectory of technological capability accumulation in Mexico. In the case of Philips the business line of televisions is analyzed, with which it started activities in Mexico and which has worked like a seed for the rest of the activities in this country.

Table 18.1
Taxonomy of Technological Capabilities for the *Maquila* Industry

	Investment functions		Production functions		Supporting functions		
Capability level	**Decision making and control**	**Project preparation and implementation**	**Processes and production organization**	**Product centered**	**Developing external linkages**	**Developing internal linkages**	**Equipment modification**
Basic Operative Capabilities	• Engaging primary contractor • Payment estimation	• Preparation of initial project outline • Construction of basic civil works • Simple plant erection	• Replication of process specifications • Routine operation of simple and/or more complex assembly process • Improvement in the workstations based on supervision systems and/or quality control • Basic process engineering	• Replication of product specifications • Routine quality control based on quality control processes	• Relationship with suppliers, clients and institutions through the headquarters	• Relationship with the headquarters to receive authorizations for inputs, technical specifications of products and processes, and investment projects	• Routine maintenance of equipment • Simple replication of plant specifications and simple machinery parts • Basic maintenance without planning

Table 18.1 (continued)

Basic Innovative Capabilities	• Active monitoring and control of: -feasibility studies -technology choice/sourcing -project scheduling	• Feasibility studies • Standard equipment procurement • Simple ancillary engineering	• Minor adaptations to assembly process based on times and movement studies • Shaining and Taguchi methodologies • Implementation of Poka-yokes in critical stations • Forming of work groups • Layout improvement, scheduling • Total productive maintenance • Scaling of assembly process and/or manufacture of pieces of different size	• Minor adaptations to market needs • Increasing improvements in product quality	• Relationship with clients through the product specifications • Searching and negotiating with suppliers of indirect materials • Search of links with local institutions for personnel training	• Establishing work groups to help build links between plants, design centers, divisions and the headquarters	• Copy and minor adaptations of specifications of existing test equipment • Reconstruction of small equipment without technical assistance • Basic programmed maintenance

Table 18.1 (continued)

Intermediate Innovative Capabilities	• Search, evaluation and selection of technology/ sources • Tenders negotiations • Overall project management	• Detailed engineering • Equipment acquisition • Environment assessment • Project scheduling and management • Designation of the work group • Training and recruitment • Starting up	• Redesign and/or design of parts of the assembly process and/or manufacture • Validation of processes according to the product • Stretching capacity based on line balancing • Slim manufacture, quality systems and continuous improvement	• Incremental product design	• Technology transfer to local suppliers to increase efficiency, quality and local supply • Attract suppliers of direct material to the region • Joint projects with universities for professional training	• Delegation on behalf of the headquarters in the decision making about process designs, clients, suppliers and institutions	• Adaptations of large equipment • Reverse engineering • Engineering and building of test equipment • Preventive maintenance
Advanced Innovative Capabilities	• Developing new production systems and components	• Basic process design and related R&D	• Innovation in processes and related R&D	• Design of basic characteristics for new products • Product innovation and related R&D	• Links with universities and R&D centers for technological developments • Collaboration in technological development with suppliers, clients and partners	• Autonomy in the decision making related to production, supply of components and indirect materials, new products	• Design and building of equipment and components • R&D for new components

In the case of Thomson we analyze the business of televisions, decoders, and cable modems, where the four plants of Ciudad Juarez are integrated.

Delphi Corp.: Business Line of Sensors and Actuators

Delphi Corp is an auto-part producer, specializing in mobile electronics and transportation components and systems technology, organized in six divisions. It is a firm oriented toward global integrated production that makes decisions in different parts of the world.[8] In 2001 it had approximately 192,000 employees and was operating in 42 countries. Its headquarters are located in Troy, Michigan, and it has regional headquarters in Paris, Tokyo, and São Paulo.[9]

Delphi was first established in Mexico as Delco Remy, under the *maquila* regime in 1979. In 2001 it had 72,000 employees, 50 productive plants in 14 states, 8 coinvestments, 3 technology licenses, and a Technical Center. In that year the production in Mexico represented 14.9% of the total sales of the group.

The Mexican Technical Center (MTC) was established in Ciudad Juarez in 1995 with 714 employees who came from the plants in Mexico, other facilities of Delphi in the United States, and the hiring of personnel in both Mexico and the United States.[10] It is Delphi's biggest Component Engineering Center. Initially Delphi transferred to the MTC the engineering area of sensors and actuators. Gradually all the divisions have established areas of engineering and different laboratories in the MTC. In 2002 there were 2,097 employees, of which approximately 1,100 were engineers.

Most of the activities at the MTC are oriented to make developments for production. Only three of the six divisions operating at the MTC carry out product design activities. The most advanced division is Delphi Energy & Chassis Systems,[11] which carries out product design and advanced engineering. Inside this division, the largest capabilities in terms of R&D are in the business of sensors and actuators.

The sensors and actuators are produced in six plants all over the world: 1 in Ciudad Juarez, 2 in Chihuahua, 1 in Brazil, 1 in Portugal, and 1 in China. It participates in the sensors and actuators market at the international level with 8.8% of the total. The engineering, design, and development activities are located at the MTC. Here there is a group of advanced engineering of sensors and actuators—six with doctorate degrees, 13 with masters degrees, and 1 engineer. This group is in charge of developing the technology in the business line of sensors and actuators at the world level. The activities carried out at the MTC are the following: (1) part of the necessary applied research, (2) all the advanced engineering, and (3) the strategic and technological planning.[12] To carry out development projects they interact with Delphi Technology, Inc., which carries out part of the basic and applied research required, and with universities, mainly American, who provide basic research for the projects.[13]

From the analysis of the productive and technological history and of the structure of its links, three stages of evolution of Delphi's business line of sensors and actuators were identified. The definition of the beginning of a new stage is associated with a jump in the process of technological capability accumulation.[14] Tables 18.2 and 18.3 summarize the main features of the accumulation at each stage and the accumulation levels in the technical functions according to the taxonomy of technological capabilities.

Table 18.2
Main Features of the Accumulation Process of Delphi's Business of Sensors and Actuators

Stage I. Simple assembly of few components, 1978-1988	**Stage II. Complex assembly of product families, 1989-1994**	**Stage III. Product design, 1995-2002**
• Simple assembly of components and manufacturing processes • Few products • 1979: first plant in Ciudad Juarez (SEC-plant 35) • 1986: second plant in Chihuahua-Chihuahua • Basic engineering of processes • System of conventional drive manufacture • American managersForeign inputs and other components • Minimum links with the local and regional context	• Complex assembly of components and automatized manufacturing processes • Various families of products • 1990: third plant in Chihuahua-Chihuahua • Engineering of assembly processes • System of synchronized manufacturing with the client, multifunctional work cells in U form, 1 engineer every 2-3 cells, subplants per families of products • Statistical controls • Development of Mexican managers in subplants • Global suppliers • Late 1980s: transfer of the indirect material buying area to the MTC	• Complex assembly of components and subsystems, complex manufacturing processes and manufacture of pieces for test and production equipment • Creation of a technical center (MTC) • From engineering of sensors and actuators toward advanced engineering and R&D activities • Lean manufacturing • Equipment improved through the 6 sigma, documented quality control (PIBAB) • MTC makes some decisions locally • 90% of direct material is bought in United States, 10% locally • Various Mexican managers • Global suppliers and some national • Major links with the local and regional context: agreements with regional universities for professional formation, incipient links with Mexican research centers

Philips Corp: Business Line of Televisions

Royal Philips Electronics was founded in 1891 to produce incandescent lamps and other electric products. Its headquarters is located in Amsterdam, the Netherlands. It has seven business sectors: lighting, consumer electronics, electro-domestic equipment and personal care, medical systems, components, semiconductors, and miscellaneous. It is oriented toward globally integrated production. In 2001 it had approximately 229,000 employees and operated in over 60 countries. It has regional management offices in Europe; Asia, Middle East and Africa; Latin America (Brazil); and North America.

Table 18.3
Accumulation Levels of Delphi's Business of Sensors and Actuators

	Technical investment functions		Technical investment functions		Technical supporting function		
Stages	**Decision making and control**	**Project preparation and implementation**	**Processes and production organization**	**Product centered**	**Developing external linkages**	**Developing internal linkages**	**Equipment modification**
Stage I	Operatives	Operatives	Operatives	Operatives	Operatives	Operatives	Operatives
Stage II	Basic Innov.	Basic Innov.	Intermed. Innov.	Basic Innov.	Basic Innov.	Operatives	Basic Innov.
Stage III	Intermed. Innov.	Intermed. Innov.	Advanced Innov.	Advanced Innov.	Advanced Innov.	Intermed. Innov.	Intermed. Innov.

Philips began its operations in Mexico in 1939 as a wholesaler of imported products from Europe with the name Philips Mexicana, S.A. de C.V. In 1973 the first plant of Royal Philips Electronics was established in Ciudad Juarez under the regime of a *maquila*. In 2002 Philips Mexico had a total of 15 plants all over the country, of which 12 were *maquila* plants of different products, and had 11,500 employees. At present Philips Mexico belongs to the Latin American region with headquarters in São Paulo, Brazil.

The business of televisions belongs to the sector of consumer electronics. Philips assembles and manufactures televisions in different parts of the world. The activities of engineering, design, and development of this line are located in Singapore; Bruges, Belgium; and Knoxville, Tennessee.

As in the previous case, three stages of evolution of the business of televisions in Philips were identified.[15] Tables 18.4 and 18.5 summarize the main features of the accumulation in each stage.

Table 18.4
Main Features of the Accumulation Process of Philips's Business of Televisions

Stage I. Simple assembly of components (1973-1983)	**Stage II. Assembly of chassis (1984-1986)**	**Stage III. Assembly of televisions (1987-2002)**
• Simple assembly of components and manufacturing processes • Few products • 1973: first plant (SESA) • 1974: first line of assembly of chassis • Creation of several plants of subassembly and components • Basic engineering of processes • 1973: machining workshop at the plant SESA • Foreign managers • Minimum links with the local and regional context	• Assembly of more complex products and manufacturing processes • Creation of several plants of subassembly and components • Engineering of assembly processes • Design engineering: original proposal of the manufacturing process for the televisions • Plant engineering: development of capabilities in large investment projects • Reconstruction of small equipment • Some few Mexican managers • Components bought from Knoxville (US) with global suppliers • 1980-97: "Program for the Development of Local Suppliers" of indirect material, integration of a limited group of local suppliers	• Assembly of televisions and complex manufacturing processes • 1987: plant 5 for final assembly of televisions (previously SESA) • 2000: plant 10 for assembly of chassis for all kinds of televisions • 2001: line of televisions PTV • Evolution of the capabilities in equipment modification: technical support for all the plants in Ciudad Juarez (1991), new business line (1997), and new plant (1998)* • Various Mexican managers • Global suppliers and some national suppliers • Major links with the local and regional context: creation of a public-private training center—CENALTEC—oriented to the training of technicians in machine tools

Note: * In 2001 Philips decided to sell Enabling Technologies Group (where this business was located) because it was not considered a central business. This led to reorienting the activity of this plant toward the business of plastic molds for injection, stopping the process of accumulation in equipment modification (Interview at Philips).

Table 18.5
Accumulation Levels in Each Stage of Accumulation of Philips's Business of Televisions

	Technical investment functions		Technical investment functions		Technical supporting function		
Stages	**Decision making and control**	**Project preparation and implementation**	**Processes and production organization**	**Product centered**	**Developing external linkages**	**Developing internal linkages**	**Equipment modification**
Stage I	Operatives	Operatives	Operatives	Operatives	Operatives	Operatives	Operatives
Stage II	Basic Innov.	Basic Innov.	Basic Innov.	Operatives	Operatives	Operatives	Basic Innov.
Stage III	Intermed. Innov.	Intermed. Innov.	Intermed. Innov.	Basic Innov.	Basic Innov.	Basic Innov.	Intermed. Innov.

Thomson Multimedia: The Business of Televisions, Digital Decoders, and Cable Modems

Thomson Multimedia is a firm of consumer electronics products whose headquarters is located in Boulogne, France. It was founded in 1879 under the name Compagnie Française Thomson-Houston. It has five business sectors, and the consumer products sector represents 62.7% of its sales. It manages three large brands: Thomson, RCA, and Technicolor. It is oriented toward global integrated production. In 2001 it had approximately 73,000 employees, operated in over 30 countries, and had 31 production plants. In 2001 Asia represented 33% of the sales, America 40% (18% corresponding to the United States), and Europe 27% (America includes the United States, Canada, Mexico, and Brazil[16]).

The history of the beginning of Thomson's operations in Mexico is linked with RCA. In 1952, RCA established an assembly plant for radios, kinescopes, and television beams in Mexico City. In 1969 an assembly plant was established in Ciudad Juarez (Chihuahua) to carry out assembly processes of electronic components for radios and televisions produced in the United States under the regime of the *maquila*. At present, Thomson Multimedia's activities in Ciudad Juarez correspond to Thomson Consumer Electronics: it has four plants that assemble components and final products and one plant for the rebuilding of products of the RCA brand. From 1998 to 2002 it had a Support Center. Additionally Thomson has a plant in Torreón (Coahuila) and a plant in Mexicali (Baja California).

As in the previous two cases, three stages of evolution of the business of televisions, decoders, and cable modems were recognized.[17] Tables 18.6 and 18.7 summarize the main features of accumulation in each stage.

COMPARISON OF ACCUMULATION LEVELS

The analysis of the technological capability accumulation trajectories of Delphi, Philips, and Thomson put forward some common features as well as certain differences. In this section the trajectories of technological capability accumulation of the three *maquilas* are compared and the similarities and differences extracted.

Tables 18.3, 18.5 and 18.7 present the levels of technological capability accumulation of Delphi, Philips, and Thomson in Mexico. In each case three stages of accumulation were identified, which were defined in function of jumps observed in the evolution. Stage I starts in the first years in Mexico, Stage II is one of transition, and Stage III reflects the present profile.

In the three cases there was a gradual accumulation of technological capabilities, the three *maquilas* evolved from having basic operative technological capabilities to having more and more innovative technological capabilities.

Table 18.6
Main Features of the Accumulation Process of Thomson's Business of Televisions, Decoders, and Cable Modems

Stage I. Simple assembly of components (1969-1980)	Stage II. Complex assembly and equipment modification (1981-1992)	Stage III. Final assembly and design (1993-2002)
• Simple assembly of components and manufacturing processes • 1969: plant RCA • Basic engineering of processes • Test technology was introduced, ATE (automatic test equipment) • Development of capabilities to modify test and assembly equipment • American managers • Minimum links with the regional context	• Assembly of more complex products and manufacturing processes • Engineering of manufacturing • Redesign of processes and assembly lines • Increasing improvements to the basic product design according to the production needs and clients' requirements • 1981: the small, more sophisticated manuals and the semiautomatic ATEs were built in Ciudad Juárez • Plant engineering: development of capabilities in large investment projects • Incorporation of Mexican managers • Global suppliers	• Final assembly of televisions, decoders and cable modems, and complex manufacturing processes • 1993: plant TTM to begin the final assembly of televisions, decoders, and cable modems • 1998: plant MASA to increase the assembly of digital televisions • Division of work between the three plants but they all have flexibility to change their assembly lines • 1996-1998: a support center is created to turn it into a fourth global design center for basic televisions of 19" and 27",* in 2002 the global strategy changes, the support center is closed and some activities are decentralized to the plants • 2000 a team of software design for decoders GLA is created and located in a plant • Design and manufacture of test equipment, design of tools to adjust processes, and export of this equipment to Brazil • Global suppliers and some national suppliers of machining pieces, packaging and simple welding • Various Mexican managers • Major links with the local and regional contexts: professional formation and training

Note: * In 2002 there was a change in the strategy of the global firm, when it decided to turn Thomson into a global organization, thus it was no longer profitable to have another engineering group in design development in Ciudad Juarez.

Table 18.7
Accumulation Levels in Each Stage of Accumulation of Thomson's Business of Televisions, Decoders and Cable Modems

	Technical investment functions		Technical investment functions		Technical supporting function		
Stages	**Decision making and control**	**Project preparation and implementation**	**Processes and production organization**	**Product centered**	**Developing external linkages**	**Developing internal linkages**	**Equipment modification**
Stage I	Operatives	Operatives	Operatives	Operatives	Operatives	Operatives	Operatives
Stage II	Basic Innov.	Basic Innov.	Intermed. Innov.	Basic Innov.	Basic Innov.	Basic Innov.	Intermed. Innov.
Stage III	Intermed. Innov.	Intermed. Innov.	Intermed. Innov.	Intermed. Innov.	Basic Innov.	Basic Innov.	Advanced Innov.

In Stage I they all acquired basic operative technological capabilities, necessary for efficient production, and in Stage II the development of basic innovative technological capabilities predominates. In Stage III Philips and Thomson advanced toward intermediate innovative technological capabilities while Delphi obtained advanced innovative technological capabilities in most of its technical functions.

Differences in the level of accumulation can be observed in each stage of accumulation. In Stage III Delphi reached a highest level of accumulation in most of its technical functions because it transferred the product design of the sensors and actuators business line to Mexico, and the group of advanced engineering at the MTC also began to carry out R&D activities. Philips presents the lowest level of accumulation; in fact Philips's present stage corresponds to Stage II of Delphi and Thomson, because Philips did not advance toward product design and also deaccumulated in the equipment modification activities.

The differences observed in the evolution are associated with both the specificities of the process of internal accumulation in each *maquila* and their corporate strategy. In this sense, the transfer of the product design activities of Delphi's sensors and actuators line to Mexico is the result of a corporate decision based on the accumulation of local technological capabilities in this business line. The slow evolution of Thomson toward design activities is more associated with a corporate decision to concentrate the design in the three existing global centers, than with a scarce internal accumulation of technological capabilities.

Aside from the differences in the specific years in which the jumps in the accumulation processes took place, it can be established that the stages are as follows: Stage I: early 1970s–early 1980s; Stage II: early 1980s–early 1990s; and Stage III: early 1990s–2002.

Although there has been an advance in the accumulation of innovative technological capabilities locally, the three cases show that the evolution has been slow, particularly in relation to East and South East Asia. It took Delphi 19 years to advance toward product design; the first plant was established in 1979 and the group of advanced engineering was not established until 1997. Thomson took 31 years to advance toward product and software design, from the creation of the RCA plant in 1969 until the constitution of the first product design group in 2000; these activities are still incipient. Philips has accumulated engineering capabilities, but has accumulated neither design capabilities nor R&D locally.

The accumulation was gradual, but the technical functions evolved differently. In fact in some functions the accumulation was faster than in others, thus the level of innovativeness reached was different. As shown in Table 18.8, there are certain similarities in the characteristics of the accumulation per technical function.

The technical functions where the accumulation was faster in the three *maquilas* were: (1) centered in processes and production organization, (2) equipment modification, and (3) decision making and control, and project preparation and implementation. In the three *maquilas* these functions reached intermediate innovative technological capabilities in Stage II of accumulation.

These technical functions are based mainly on accumulation processes at the plant level, needed to assure efficient assembly processes given the local specificities. In contrast, the functions where there was less accumulation, or the accumulation was slower were: (1) product centered, (2) internal linkages, and (3) external linkages. These technical functions depend on decisions that transcend the plants and are made at the firm level. In this sense, the activities of product design and the buying of key components are made globally. Also, as the plant level is transcended, other kinds of interactions are established between different units of the firms.

Important differences can be observed in the accumulation of the three *maquilas* in the function centered on the product. Delphi and Thomson progressed toward intermediate or advanced innovative technological capabilities in Stage III, this because Delphi began to carry out product design activities, and Thomson software design activities. The slow evolution of the technical function of external linkages reflects the existence of few local links for innovation, the stronger links are for professional formation and training. Delphi's higher accumulation shows that, although it is a global firm, it is difficult to prevent a disintegration of the basic activities to ensure the constant renovation of the competitive advantages. In addition to that, carrying out activities of design and R&D raises the need of some links with local, regional, and international institutions for innovation. It is important to point out that external links are internalized in the region, where out-of-region firms set main business policies.

Finally, Delphi accumulated more in all its technical functions, which is related to the transfer of a global line, and not just a plant, to Ciudad Juarez. Thus, a more important part of the decision center is in Ciudad Juarez.

FINAL REMARKS

The analysis carried out allowed us to identify some common features in the evolutive process of technological capability accumulation of the three *maquilas* in Mexico. These common features in the accumulation processes suggest the existence of some stylized facts.

- The accumulation process differs in each business line, given the specificities of the process of internal accumulation and the corporate strategy of the global firm.
- As plants learn, they spread in technical activities with a higher grade of innovativeness, and develop innovative technological capabilities.
- The learning processes in the plants lead to the accumulation of technological capabilities locally and to shorten the distance between the production and technology functions. This generates pressure on the headquarters to acknowledge the technological capabilities accumulated and allow them to develop technical activities of higher innovativeness.
- Local accumulation is a necessary condition, though not sufficient, for global firms to transfer technical activities to Mexico; the global logic governs over the internal accumulation of technological capabilities.

Table 18.8
Comparison of the Technological Capabilities Accumulation of the Three *Maquilas*

	Technical investment functions		Technical investment functions		Technical supporting function		
Maquila	**Decision making and control**	**Project preparation and implement**	**Processes and production organization**	**Product centered**	**Developing external linkages**	**Developing internal linkages**	**Equipment modification**
Delphi 1995-2002	Intermed. Innov.	Intermed. Innov.	Advanced Innov.	Advanced Innov.	Advanced Innov.	Intermed. Innov.	Intermed. Innov.
Philips 1987-2002	Intermed. Innov.	Intermed. Innov.	Intermed. Innov.	Basic Innov.	Basic Innov.	Basic Innov.	Intermed. Innov.
Thomson 1993-2002	Intermed. Innov.	Intermed. Innov.	Intermed. Innov.	Intermed. Innov.	Basic Innov.	Basic Innov.	Advanced Innov.

- The *maquilas* in Mexico are not firms, initially they were simply plants, so that they learned and accumulated the technical functions related with plants. Thus there has been a faster process of accumulation in the technical function centered in processes, in the modification of test equipment, and in the investment functions in large projects.
- The headquarters maintain the power of decision on the technical functions centered on the products (design and R&D), internal linkages, and in the links with the suppliers of components that correspond to the function of external linkages.
- Some *maquilas* evolved in the sense of attracting global business lines, which allowed accumulating in technical functions related to product innovation, internal linkages, and external linkages.
- The development of managerial abilities among Mexicans has been slow because of the lack of opportunities to assume positions of high level locally. As they assume positions of higher responsibility, they look to strengthen the development of more innovative technological activities and integrate Mexican suppliers.
- The *maquilas* have followed an evolutive dynamic related with the pressures of international competition. Beyond the efforts to develop national suppliers, the logic of businesses is radically different; moreover the gap between the types of firms has accentuated.
- The limited external links are associated with three factors: (1) the decision center outside the locality, (2) the profile of the activities at the plant—more productive than technical—that has demanded mostly links for training, and (3) the weaknesses of the local production and innovation systems, which do not have the capacity to respond to links for innovation.
- In general, the fragility of the production and innovation systems, and of the social interaction, has not facilitated the linking processes of the local agents with the *maquila*.

NOTES

1. This chapter is part of a research project, "Technological learning and industrial upgrading: The generation of innovation capabilities in the maquila industry in México," COLEF/FLACSO/UAM (Project CONACYT, no. 35947-s).
2. See, for instance, Hobday (1995) and Kim (1997).
3. From the mid-1960s, the Mexican government established a border industrialization program (Programa de Industrialización Fronteriza) with the purpose of cutting down the high unemployment rates in the northern border of the country. This program had the purpose of attracting foreign investment, mainly from the United States, to establish a 10-mile strip from the northern border. The plants created under this scheme were denominated *maquiladoras* (shortened to *maquilas* in this chapter). See Lowe and Kenney (1999), Buitelaar (2000), Barajas et al. (2002).
4. In 1990 the number of employees was 446,436 and the number of establishments was 1,703.
5. Carrillo and Hualde (1997), Lara (2000), Dutrénit and Vera-Cruz (2002).
6. Dahlman and Westphal (1982), Katz (1984), Dahlman, Ross-Larson, and Westphal (1987), and Lall (1992).
7. A critical description of this taxonomy is presented in Dutrénit, Vera-Cruz, and Arias (2003).

8. Delphi was a part of General Motors Company. In 1999 it separated and became an independent industrial group, Delphi Automotive System Corp. In 2002 it changed its name to Delphi Corp.

9. Delphi concentrates its activities in North America (United States, Mexico, and Canada), where 78% of the sales are generated. Europe represents 18.4%, South America 1.5%, and other areas, which include China, 2.3% (Delphi, 2002).

10. The establishment of the MTC in Ciudad Juarez responded to different motives: (1) closeness with the plants to reduce the costs of line installation, response time and communications cost; (2) restrictions to grow in the United States by the "head counter"; (3) search to reduce operation costs, through reducing the cost of labor (Interview at the MTC).

11. Delphi-E&C belongs to the dynamic and propulsion business sector and has 14 business lines—sensors and actuators is one of them. It has two lines of products, sensors and actuators, and these combine in a subsystem that allows creation of intelligent modules of control. The modules combine sensors, actuators, electronics, and software.

12. Additionally the planning of sales, marketing, and investment are carried out (MTC, 2000; and interviews at the MTC).

13. One of the most important results of the innovative activity is registering the intellectual property. In 2001 the advanced engineering group obtained 88 records of investments, presented 55 applications of patents, was granted 15 patents, and realized 7 defensive publications and 3 industrial secrets (Interview at the MTC).

14. Arias (2002) presents a description of the three stages.

15. Urióstegui (2002) presents a description of the three stages.

16. www.thomson-multimedia.com

17. Sampedro (2003) presents a description of the three stages.

REFERENCES

Arias, A. (2002). "Capacidades Tecnológicas en I+D y Diseño en la Industria Maquiladora Mexicana: El caso de Delphi Corp.," working paper, Doctorado en Ciencias Sociales, UAM-X.

Ariffin, N. and Figueiredo, P. (2003). *Internacionalizacão de competências tecnológicas*. Rio de Janeiro: Editora FGV.

Barajas et al. (2002). "Industria maquiladora en México: perspectivas del aprendizaje tecnológico-organizacional and escalamiento industrial," Monografía No. 3 del proyecto "Aprendizaje tecnológico y escalamiento industrial: Generación de capacidades de innovación en la industria maquiladora de México," COLEF/FLACSO/UAM.

Bell, M. and Pavitt, K. (1995). "The Development of Technological Capabilities." In I.u. Haque (ed.), *Trade, Technology and International Competitiveness*, pp. 69-101. Washington, DC: The World Bank.

Buitelaar, R. (2000). "Maquila, Economic Reform and Corporate Strategies," *World Development*, 28 (9): 1627-1642.

Carrillo, J. and Hualde, A. (1997). "Maquiladoras de tercera generación. El caso de Delphi-General Motors," *Comercio Exterior*, 47 (9): 747-758.

Dahlman, C. and Westphal, L.E. (1982). "Technological Effort in Industrial Development. An Interpretative Survey of Recent Research." In F. Stewart and J. James (eds.), *The Economics of New Technology in Developing Countries*, pp. 105-137. London: Frances Pinter.

Dahlman, C., Ross-Larsen, B., and Westphal, L.E. (1987). "Managing Technological Development," *World Development*, 15 (6): 759-75.

Delphi Corp., Annual Report, several years.

Dutrénit, G. and Vera-Cruz, A.O. (2002). "Rompiendo paradigmas: acumulación de capacidades tecnológicas en la maquila de exportación," *Innovación y Competitividad*, Publicación trimestral de ADIAT, 2 (6): 11-15.

Dutrénit, G., Vera-Cruz, A. O., and Arias, A. (2003). "Diferencias en los perfiles de acumulación de capacidades tecnológicas en tres empresas mexicanas," *El Trimestre Económico*, No. 277, January-March.

Figueiredo, P.N. (2001). *Technological Learning and Competitive Performance*. Cheltenham: Edward Elgar.

Hobday, M. (1995). *Innovation in East Asia: The Challenge to Japan*. Aldershot: Edward Elgar.

Katz, J. (1984). "Domestic Technological Innovations and Dynamic Comparative Advantage: Further Reflexions on a Comparative Case-Study Program," *Journal of Development Studies*, 16 (1-2): 13-38.

Kim, L. (1997). *From Imitation to Innovation. The Dynamics of Korea's Technological Learning*. Boston: Harvard Business School Press.

Lall, S. (1992). "Technological Capabilities and Industrialization," *World Development*, 20 (2): 165-186.

Lara, R. A. (2000). "El nacimiento de las maquiladoras de tercera generación: El caso Delphi-Juárez," *Comercio Exterior*, 50 (9): 771-779.

Lowe, N. and Kenney, M. (1999). "Foreign Investment and the Global Geography of Production: Why the Mexican Consumer Electronics Industry Failed," *World Development*, 27 (8): 1427-1443.

MTC (2000). "Qué pasa MTC," September.

Royal Philips Electronics, *Annual Report*, several years.

Sampedro, J. L. (2003). "Aprendizaje y Acumulación de Capacidades Tecnológicas en la IME: Thomson-Multimedia de México," thesis, Maestría en Economía y Gestión del Cambio Tecnológico, UAM-X.

Urióstegui, A. (2002). "Del ensamble de componentes al producto final: el caso de Philips México," thesis, Maestría en Economía y Gestión del Cambio Tecnológico, UAM-X.

19

A Successful Experience of Innovation and Technological Learning in the Automobile Industry: The Tremec-Chrysler Case

Salvador Padilla Hernández and María de la Luz Martín

INTRODUCTION

This chapter examines the economic and technological relationship between two companies, Chrysler and Tremec. The first organization is an American corporation, while the second belongs to the Mexican industrial group DESK, as we can see in Table 19.1.

Beginning with studies of the economic and technological relationships of the two firms, it's convenient to give a brief description of the structure and strategies of production in the automobile industry in Mexico. The participation of the automobile industry (spare parts and finished products) as a proportion of the manufacturing sector is 10%, and 2.3% in relation to the GNP. Regarding employment, this industry occupies 6.2% of the work force from the manufacturing sector and 0.7% of the total employment (JICA, 1997).

The industry is integrated by two sectors: one refers to the production and assembling of vehicles (subcompacts, compacts, luxuries, and sport cars), trucks (heavy and light) and buses; and the other sector refers to the spare-parts manufacturing firms. The car industrial sector is made up of nine transnational organizations: the Ford Motor Company, General Motors, Daimler-Chrysler, Nissan, Renault, Volkswagen, Mercedes Benz, BMW, and Honda.[1] Vans and pick-up trucks are produced by Chrysler, Ford, General Motors, Nissan, and Volkswagen; while the large trucks are produced mainly by Camiones y Motores Internacional de México, Consorcio G, Grupo Dina, Kenworth Mexicana, Mercedes Benz México, Motor Coach Industries México, Scania de México, and Volvo Bus de México (INEGI, 2000).

Table 19.1
User-Producer Relationship Between Chrysler and Tremec

Chrysler of Mexico	Tremec
Year of establishment in Mexico: 1938	Year of foundation: 1964
Facilities in Mexico, D.F, Toluca, and Coahuila	Localization: Queretaro, Queretaro.
Industrial activity: car assembly, trucks, and small trucks	Industrial activity: spare parts production (mechanical and automatic gearboxes)
Capital: 100% American	Capital: 100% Mexican, acquired by Spicer of UNIK corporation, which at time belonged to the Mexican DESK group, in 1995.
Number of employees: 10,500 (including D.F., Toluca, and Saltillo).	Number of UNIK employees: 3,500

The spare-parts sector includes more than 500 companies (*Expansión*, April 2002). In this sector, the big companies are internationally recognized and they produce for both national and international markets. They focus on the production of mechanical components—systems with a higher grade of technical complexity and expense that in many cases requires higher production scales. These components include the motor, the gearbox and the drive shaft steering, brakes, alignment, and suspension—in other words, all of the functional automobile parts.

In Mexico, the big spare-parts companies are regulated by international competition, and for that reason production adopts norms that the assembly plants impose. So, instead of supplying the market based on prices, the purveyor's system is founded on factors such as quality system application as the global sourcing, or just-in-time and product development, factors that reflect the close relationship between purveyor and client.

In order to fulfill the quality standards imposed by the assemblers, some of the big Mexican spare-parts manufacturing companies—such as Spicer, Condumex, Vitro, and Cifunsa—have made investments in research and development, allowing them to participate successfully in the international market and to maintain their competitive advantage.

In addition, to participate in this highly competitive industry on a worldwide scale, the big national producers like Spicer and Condumex have established strategic alliances with foreign companies through different forms of cooperation, such as licenses, covenants, or direct technology transference. These alliances have had successful results because foreign support adds knowledge to the national market of Tremec in aspects such as client preferences, relationship with assemblers, and governmental regulations (NAFIN-IMEF, 1995: 176).

TECHNOLOGICAL EVOLUTION

As we will see, the technological evolution of Chrysler and Tremec is complementary, as much in their technological capacities and their ability to establish contracts and infrastructure as for their production costs reduction and dynamic economies and learning opportunities. In the next two sections the technological evolution of both companies is explored.

The Chrysler Technological Evolution

Chrysler of Mexico started operation in 1938 with the assembly plant installation in Mexico City. In 1954 the general offices of the corporation were inaugurated there and were extended in 1962. Two years later the Chrysler productive capacity expanded activities to facilities in Toluca. In 1965 this organization began to export motors, and in 1995 it sold 30,000 vehicles each year to the external market.

Today Chrysler has an assembly plant in Mexico City and an automobile complex in Toluca where cars are assembled, engines are manufactured, gearboxes are produced, and metallic stamps are fabricated. Furthermore, the company has industrial facilities in Saltillo that contain an engine plant, another stamping plant, and a third for assembly. In these facilities there are approximately 10,500 employees: executives, technicians, administrators, plant workers, and engineers dedicated principally to design.[2]

Technology process is endogenous. The support of the R&D laboratory is the direct responsibility of the corporation. In fact, Chrysler has in Mexico a product engineering department and a purveying department. This department is responsible for setting, together with the producers, the changes in the technical specifications according to the innovations to be introduced in new cars, interchanging information and technological knowledge, and solving technological problems.

At the same time, in relation to the production and market characteristics, in 1996 Chrysler of Mexico sold $3,500 million worth of products; 80% corresponded to exportations to Canada, Europe, and South America, and the rest was traded in the domestic market.

Vehicle distribution is realized by agencies and in-house advertising or is contracted mainly through mass media. In addition, the corporation does market research to ascertain the demand and the requirements of the clients.

Technological Evolution of Tremec

Tremec is one of Chrysler's principal supplier companies, and specializes in gearbox manufacturing for automobiles in Mexico. The company belongs, together with TSP (Figure 19.1), to the division of Spicer's gearboxes and to the industrial DESK group. Tremec is a metal-mechanical company, funded by 100% Mexican capital, and is located in Queretaro, Queretaro. In the spare-parts industry this company is dedicated to the design, development, production, sales, and service of products such as gearboxes for the automobile industry and for the military, components for automobile gearboxes, timing gears for motors, gearboxes and components for the agriculture industry, and also replacement parts. In 1997 the employees and workers numbered 1,809.

Figure 19.1
UNIK Group

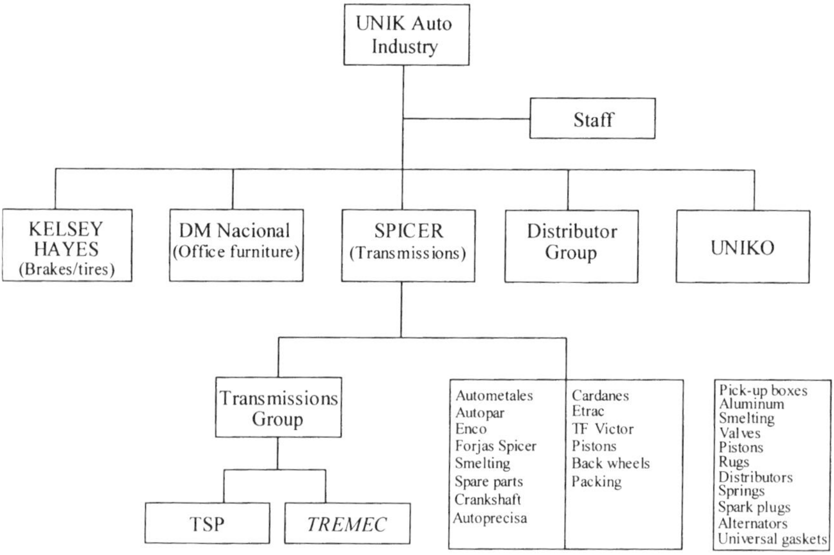

In the period 1964-1968, Tremec made investments in three types of development: (1) marketing and technological innovations for distribution; (2) introducing methods of learning for innovating process and products; and (3) administration and organization for planning competitive strategies in order to face the world market demands. In this way, the company grew thanks to exportation abilities, the improvement of marketing systems, distribution, client services, and spare parts, with an emphasis on the knowledge of the Mexican and foreign clients' needs. Since 1960 the main customers are described in Table 19.2.

Table 19.2
Principal Tremec Clients

Years	Clients
1960s	General Motors, Chrysler, Ford, Vamsa
1970s	General Motors, Ford, Vamsa, Dina, Ford USA, Chrysler in the United States, and General Motors USA in Mexico
1980s	General Motors, Chrysler, Ford, Vamsa, Mercedes, Dina, Ford USA, Getrag, Cummins
1990-1995	General Motors, Chrysler, Ford, AMG, MB, Dina, Getrag, Cummins, Rockwell ZF, DSA, Autopar
1996-1998	General Motors, Chrysler, Ford, ANG, MB, Dina, Getrag, Cummins, Rockwell ZF, DSA, John Deere, BTR, Sang Yong, Izusu, Autopar, DSA/TTC, ETREC, Troler

Source: Tremec.

In 1980 the world market strategies of the automobile industry changed radically in terms of recognition of engineering and quality, and Europe and Japan were probing alternatives for the American markets. Gearbox manufacturing wasn't the exception, inasmuch as the design conception changed, the permissible noise was reduced, and an effort was made to change the velocity lever. These changes in the design of the automobile were a result of the interactive relationships and new organizational and productive routines established between Tremec and their most important users like Ford, Chrysler, and General Motors, among others.

In addition, during the 1981 to 1985 period the users' attention was redirected in the search for gearbox manufacturers that would take the responsibility for engineering, product development, and as a consequence, new production processes development. So, in 1981, Tremec started to design, develop, and produce its own products. The result was that in 1991, the first gearbox with a 100% Mexican design was presented on the market.

From 1987 on, Tremec developed innovations in processes and products that can be observed in the company's manufacturing technology trajectory evolution (Table 19.3). At the same time, the R&D in Tremec's laboratories, besides solving problems in gearboxes, made applied research that allowed the company to create the technology for new developments.[3]

On some occasions, in order to make a new product, a new process is required even though it would be desirable for the new products to be manufactured with the existent equipment. Frequently, a new process that doesn't exist might be necessary, and during the initial production phase a process for manufacturing must be created as well.

Table 19.3
Tremec: Evolution of the Technological Trajectory of Manufacturing

Years	Technological Manufacturing Evolution
1998-2000	Teleconferences with the clients. Changed probe simulation in laboratory. Application of dynamic Adams design. Gearbox with five automated velocities. Gearbox with six velocities. Gearboxes TR 3440/50, T-45 Y, TR 4040/50.
1996-1997	Electronic direct communication with the client. Gearboxes FSM-5005, 100% Tremec. Gearbox T-5. Gearbox TR-3450 for GM with Tremec innovations.
1994-1995	Acquiring information system for laboratory. Use of lithography for prototypes by CAD-CAM-CAE.
1992-1993	Laboratory items automating. Electronic communication of CAD achieved. Gearbox manufacturing TR-3340, TR-3350, and TR-3550 completely designed by Tremec.
1990-1991	CAD application in three dimensions; laboratory probe automation; FEA applications in work stations; software developments for engineering calculus CAE, vibration, and noise analysis; better technology for gearbox 199F; gearbox manufacturing TR-2250, the first completely designed by Tremec.
1987-1989	CAD application in two dimensions, FEA applications in PCs; effort for experimental analysis.

Source: Tremec.

TECHNOLOGICAL BEHAVIOR

Within technological behavior there are elements by which the interfaces between Chrysler and Tremec are informal accords between companies with solid established prestige and where the product quality is guaranteed.

Technological innovation in the automobile industry is endless and covers all the parts and components of the automobile. Even when the dominant model of the automobile has been defined since the beginnings of the 20th century, the innovations in the industry are the basis for competing globally. In this way, the greatest growth in this industry has been degrees of technological complexity of product and process innovation, where the design is the most important difference between producers.

The process of innovation in industry consists of the use of advanced technological electronic software (CAD, CAM, CIM, among others) and it has been imposed by the Japanese model in the organization of production that includes Kanban, just-in-time, inventory reduction, work teams, organization for joint technical problem-solving, and work rotation in the assembly processes.

Also, product innovations include motors, stamping, gearboxes, and all the varieties of parts and ensembles that a car contains and that make up transport equipment. In this sense, Chrysler has instrumented a system called "Cap-Forward" for the innovation of their cars. Within this framework, one tries to increase interior space and the automobile handling. This innovation was realized by Chrysler at the R&D laboratory located in Detroit, along with their purveyors (Carrillo and Hualde, 1997).[4] Chrysler acquired from Mexican or foreign purveyors between 70% and 75% of the 20,000 parts that integrate a car, or between 30% and 25% of the total car.[5]

In Detroit, Chrysler works with its purveyors to introduce the technological innovations to the cars that will be on the market in the years ahead. In addition, the Chrysler purveyors contribute considerably to the innovations and they are dedicated to the design area. This working arrangement joins the corporation purveyors and the technicians and engineers of the company (Chrysler, interview in November 1997).

For a company to be accepted as a Chrysler purveyor requires the technological capacity to participate in the prototype designs that the corporation is planning to introduce on the market five years ahead of time. The Chrysler purveyors must be efficient in respect to manufacturing: They must abide by the rules that the company has established; they must be at the level of service required by the transnational company; they must be capable of making and proposing their own designs; and they must grow and export. Only in this way can the spare-parts manufacturers be partners of Chrysler and establish technological relationships and commercial and long-term business agreements.

In the new business philosophy, the big transnational firms see their purveyors as an extension of the company, what Chrysler calls "Extended Enterprise." In other words, this is equal to a "vertical integration," but in which each company keeps its autonomy, independence, and responsibilities at the time of making decisions. This kind of system implanted by the company implies that the buyers, as much as the purveyors, learn and contribute knowledge to the technological development. In the relationship there exists a benefit for both kinds of companies, so they interchange knowledge and technologic information that make it possible to make innovations, elevate productivity, and confront the challenges of worldwide competition in the industry.

Chrysler in Mexico does not have a research and development laboratory, but has a design engineering department or product engineering division. According to one of the interviewed executives, the company has one of the most advanced information systems that the automobile industry possesses. This is used for the efficient handling of the different production areas of the company: From manufacturing process control to communication systems to cells and production programs. This system is also used for interchanging information with the central offices of the corporation in Detroit and with the other plants of the company located in different parts of the world.

Tremec is one of the 223 vehicle parts suppliers that Chrysler has in Mexico. It has services purveyors that make designs for turbocharged motors, plant installation, and offer other kinds of advice. According to the information provided by the personnel of the engineering and design department, since 1995 Tremec has one patent registered and two in the process of registration.

Patenting is not a high priority of the company because, according to Tremec's manufacturing manager, patenting of products is not required. He says:

> One of the principal factors that impedes approval and makes patenting unnecessary are the temporary technological barriers implied in the development of a new product, so in the spare-parts industry copying a design and developing it takes at least three years. Because of this lag, the company that copies the design cannot actualize it in time to compete. Also, the investment in the specific activities necessary to realize the copy would be considerable, and recuperation wouldn't be guaranteed.

Otherwise, the technological linkage between Chrysler and Tremec, such as organizational innovation, is defined by the introduction of organizational production systems, which in the case of the second company has the objective of satisfying the clients' needs. These organizational methods include just-in-time, Kanban, continual bettering processes, statistics control processes, quality control, integration degree, and purveying agreements in order to obtain certified international quality or national prizes.

At the same time, the mentioned organizational methods indicate the changes that the purveyors have resorted to for improving the relationship with their clients with the objective of meeting international quality standards, and maintaining themselves as first-level purveyors.

In fact, from 1993 Tremec transformed its organization from a hierarchical system to a cellular organization.[6] This organizational method allows the company to accomplish global goals, and implicates participative leadership, direct communication, horizontal organization, teamwork, commitment, and a policy with a common objective.

For example, there is the manufacturing cell that includes the gear cell, a department dedicated to the internal manufacture of that component of the gearbox. The cellular organization provides flexibility because an operator in a cell can be independent enough to operate the machines and tools in his department. Besides that, the manufacturing cell has to service the quality systems, maintenance, and supply.

With respect to the most important manufacturing of gearbox parts, Tremec is integrated horizontally. For example, the companies of the Spicer group manufacture gears, wheel rods, chassis manufacturing equipment, or the chassis itself. Other components of less technological complexity, like ball bearings or screws, are acquired outside of the company, but most spare parts are produced inside the company or in companies of the same industrial group. This means that within this industrial group the production of spare parts is between 40% and 50%; 30% is acquired from foreign suppliers, and of the rest, 20% to 30% of the standard spare parts are acquired on the national market.

It is important to mention that Tremec's technological sources are endogenous and exogenous. In relation to endogenous sources, Tremec counts on an R&D laboratory and production engineering and design department. With respect to exogenous sources, the technical relationship with the users not only is the most important technological source for Tremec, but also has established covenants that contribute to Tremec's technological progress.[7]

According to an engineer interviewed at Tremec's facilities, a prototype to be manufactured becomes a product at the moment that it passes all the valuation tests and then it is ready to manufacture. This means that the moment all the details have been resolved, Tremec has taken all the necessary steps described or codified in the Advanced Quality Plan. Then come all the product manufacturing phases, from the feasibility analysis, to the client's requirements, to prototypes, to the production phase, and so on.

In the automobile industry, new methods of production introduced for increasing productivity and competitiveness and adopting the best practice ("benchmarking") have modified the relation between firms, because the competition between them today is principally associated with the product's quality. The competition is presented now as the pressure for reducing inventories and just-in-time deliveries, particularly by the purveyors, under punishment of paying substantial penalties to the final assembly companies, or losing the purveying contract.

On the other hand, the behavior between Tremec and Chrysler is not only influenced by the way in which both companies decide to learn, buy, or interchange technology, but also by the routines that have been established on the interface between them. For instance, there is the need to carry out periodic reciprocal visits and/or offer help when either company needs advice and/or supervision from the other.

The previously mentioned issues are explained by the fact that in this industry the user-producer relationships are permanently modifying product technology. Even more, the technological complexity of the new production model creates the need for establishing interfaces that are created through cooperative and intense relations, and promotes information technological knowledge interchanges between the different specialists and engineers of companies, beginning with the first designs of the prototype that must be produced. In this way, knowledge of technological process is traded through interactive learning.

With respect to the construction of the routines between Chrysler and Tremec, both firms are able to interchange information and technological knowledge, and periodically the engineers and technicians of the two firms meet to solve technical problems. There is constant interchange of information and technological knowledge, and both firms have the capacity to change their routines and adjust to the necessities and specific exigencies of the counterpart to accommodate functional technical changes (Figure 19.2).

In relation to the technological learning, when we talk about a market that is regulated by quality and prices, the competitiveness is achieved when the producing company satisfies the transnational companies' specific demands. In the automobile industry there are some quality rules that go through the channels between organizations wherein the purveyor has to produce under the quality rules that the clients demand. It is common that the car assemblers carry out quality control for the purveyors, preparing them in their concept of quality, and the efforts of the purveying firm are challenged to obtain a quality certification by the assembler.

In that sense, in 1994 Tremec obtained the quality certified ISO-9001 and in 1995 the QS-9000 (quality rule created by Ford, General Motors, and Chrysler) along with the Underwriters Laboratory (UL) certification. Having these international quality systems obliges the producing company to carry out organizational and technological innovations. In fact, at the end of five years the clients establish new specifications for new products. In this sense, when setting a new specification, it has to improve the old product to develop a new one. This constitutes an option for interactive technological learning; in this case an interface is made between the producers and users to create or design the new prototype.

Figure 19.2
Chrysler-Tremec: User-Producer Relationship

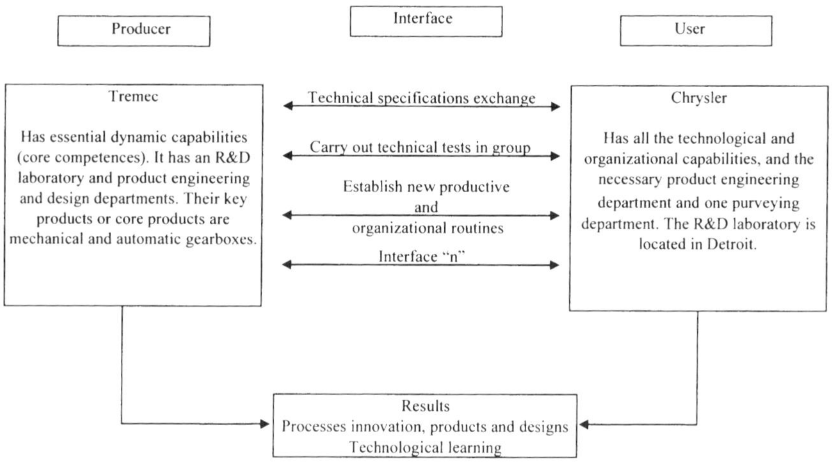

Another form of technological interactive learning that results in incremental innovations is the engineering design that is accorded in the purveyor-user interface. Sometimes, according to an interviewed engineer,

Tremec acquires the gearboxes that the competitors manufacture and observes the composition of those devices. In most of those cases, the details that make the difference in gearboxes are minimum, so in this way the gearbox design doesn't change. Following this procedure the company obtains some of the technical characteristics of the competition's gearboxes and applies them to their products. But it cannot use all the gearbox concepts because the technological capacities and the production equipment of Tremec and the competing companies are not equal.

In fact, given that Tremec manufactures components for companies like ZF and Rockwell, among others, the advantage for Tremec is that when manufacturing different products for distinct organizations, they can use some mechanisms or pieces and adapt them to their designs, costs are lessened, and they obtain competitive advantages.

Tremec often fulfills visits to clients in which the user companies establish specifications of the new product. Besides this kind of routine, the organization has established other interfaces, buying electronic systems and production control with the idea of being in daily contact with the clients' requirements. This means that at Tremec there exists an interface that allows electronic communication with the client about products that will be delivered just-in-time.

Also, the interface that allows interchange communication and the specifications with the clients obligates the company to carry out innovations of material substitution for the parts manufactured or designed. For example, in gearbox manufacturing there exists the tendency to reduce noise, and to make changes in gearbox forces and the size of the artifact. Since motor sizes were reduced, the gearbox had to be reduced considerably. Currently they are working at the Tremec R&D labs to reduce the noise in the gearboxes. It is hoped that the manufacturing alloy used will increase the material durability because it can resist a higher number of changes.

Each time the clients are supplied with a new model, they establish specific necessities for the improvement of the motor's performance. Therefore, Tremec has to improve its products. For example, according to the Tremec R&D laboratory boss, the problem means reducing noises. Some improvements have to be made in the gears because the new designs must be noiseless. To reduce gear-changing forces, the synchronizers' size has to be increased to increase the capacity, or double or triple cones must be used, because the change force becomes slower. The size reduction requires added processes for selecting the material, and in doing that, the production period becomes shorter.

In that sense Tremec is in a process of searching for new materials and new manufacturing processes to make more resistant gearboxes—for instance, manufacturing more resistant gears or the substitution of iron materials for aluminium.

Tremec works continually on the design of prototypes, which involves the necessary tests of the gearboxes, and making the plant and equipment adequate in such way that the specifications of gearboxes required by the clients is fulfilled.

With respect to the instrumentation of new organization, the just-in-time system in the relation of Tremec to Chrysler implies that provisions required for the gearboxes be made by electronic direct communication and teleconferences with the client. The company uses Kanban system sense of the moment of interchange of technological specifications and establishing of work "cells" to impart the clients' technical specifications, for the organization of work groups for solving specific problems.

During one of the visits to Tremec one of the organizational routines was observed that related to the periodic meetings that are carried out by the work cells for discussing different technical problems, interchanging information, planning, and studying suggestions for the solution of technical problems. The working team meetings are for analyzing specific problems. There is a leader in each work group, which is six or ten people. These work meetings promote equality—that is, the hierarchy is dissolved momentarily and the qualified employee operations are evaluated by the majority with the same seriousness of the technicians and most experienced engineers of the company.

Of course in Tremec there also exists the opportunity to transform part of the tacit knowledge to codified knowledge. Tremec is looking at this process through the integration of the work cells; nevertheless there doesn't exist an ideal organization model such as that proposed by Nonaka and Kono (1998) that makes that labor easier and increases the creation, reproduction, and diffusion of new technological knowledge to the other areas of the company.

Another important aspect related to the analyzed organizations' behavior refers to purchase or technology transference. In this case, for example, Tremec and ProDrive (an English company) in 1997 prepared an agreement that would start at the beginning of 1998 with the objective of designing and developing an automated gearbox prototype. That agreement included three phases, the first of which was the concept phase that implies the basic engineering for the prototypes. For the second phase, a detailed program for specific design, development, and calibration of gearboxes was established. The third phase included the calibration and test development for the prototype vehicle. In this case the technological relationship established between two companies of two different countries was examined. ProDrive is the technology producer and Tremec the user or buyer.

In the contract or interface between the companies, technical specifications were exchanged that would be applied to five different kinds of gearbox. These specifications were based on the fact that the mechanical components would be common to those joint gearboxes as much as possible, except for the minor modifications that are required for adapting the components of each gearbox, which depended on the vehicle in which the artifact would be installed. Through such agreement, ProDrive committed to effecting the change of calibration that requires its kind of gearbox for its vehicles.

On the other hand, the technical specifications established in the interface included the design's characteristics, which implied an intense relation of technical knowledge interchange between R&D and engineering product in Tremec's departments and the provider company. Also, in the contract the cost and the delivery time of the prototypes were determined, and also the basic information exchange between the companies about the materials that should be delivered by Tremec.

In addition, the gearboxes' technical specifications exchange in this kind of project included: pictures, gearbox operational methods, synchronization details, depth, specifications, mechanism liberation, and strength of the engagement, among others. According to the information obtained in this case study, the technological agreements were established between the provider companies of the first level, but not with the assembly companies.

In fact, there is a high degree of competition between producers that compels them to establish this kind of technological collaboration agreement. On the other hand the cooperative assemblers obtained contracts of no more than five years, and on occasion these are of informal character, usually when the client and the purveyor have a prestige that redounds in minor transaction costs. For instance, when a product is offered, the assembly organization evaluates it according to purveyor politics, and if it signs a contract with the purveyor, protection stipulations, investment costs, and provided volumes are included, among other aspects. The supplying parts and components contracts are established then for maintaining or increasing the market participation by the purveyor.

Tremec signed another agreement with Columbia University in 1996. The university founded a dynamic gear laboratory that is sponsored by many companies and Tremec. The activities of that laboratory are oriented to research related to the gears, software development for the vibration analysis of such pieces, projection of gearbox mistake analysis for predicting the noise level, and technology for manufacturing continuous-variable gearboxes (or without gears).

The agreement of technological assistance was signed at the beginning of 1998 and the first prototype for evaluation was obtained. When this contract was signed, the company started the automatic gearbox production, and for that reason a company that had the necessary technological experience was searched for.

On the other hand, Tremec could work on an experimental scale in automated gearbox manufacturing. Nevertheless, because of the innovations' speed, the organization would spend more time in reaching the competitors. In this way, the foreign technology purveyor's contract conforms to the fact that in other ways the technology is able to catch up to offer competitive products with a higher degree of technological complexity.

According to the interviewed engineer at Tremec's R&D laboratories, the most important part of the technological collaboration agreement between Tremec and ProDrive is that Tremec took into account that ProDrive has been dedicated to the automobile motor line for a long time. In this way the negotiation resulted in being more productive and obtaining satisfactory results, given that normally when working with institutions or national organizations, success is not quick because those organizations do not have the infrastructure or the necessary experience for obtaining concrete products. The former constituted a problem for the national purveyor development in the car industry.

In relation to the innovations of novelty products, Tremec works together with a Mexican supplier of electronic modules for integrating them to a gearbox of 18 gears for heavy trucks. The innovation means installation of a semiautomatic system for gearboxes, which works through buttons and sensors in order to make the trucks easier to drive.

Another innovation was the use of aluminum instead steel for manufacturing specific gearbox pieces. The substitution of one material for another will allow increasing the pieces' resistance and thereby increasing the number of changes in the gearbox. For example, the steel parts endured 50,000 changes, while the aluminum ones supported at least 500,000.

In the visits carried out at Tremec's facilities, we observed that in the work cells there was a transformation process of tacit to codified knowledge through the exchange of experiences and searching for the solution of concrete technical problems that emerged in the social work process.

QUALITY STANDARDS AND PRODUCTION CONTROL

In Mexico many products and process innovations come from intrafirm quality rules (Micheli, 1996). In this sense Tremec has the QS-9000 certification that is a combination of the systems ISO-9000 and QS-9002 established by Ford, Chrysler, and General Motors, which constitute the quality requirements system of those companies.

Tremec was awarded the Ford's Q1 prize in 1986. In this year the former company started the manufacturing of components for their competitors, and so this firm got technological competitive advantages.

With the development of its own technology, during 1991-1992 Tremec introduced its first gearbox in vehicles like the General Motors Blazer, and for the spare-parts market (aftermarket), and sells them for sports cars like the Mustang and Camaro.

From 1993 to 1995, the spare-parts exports were 60% of Tremec's production. In 1995 the economic crisis diminished sales in the Mexican market, so foreign trade became a vital strategy for company survival. In that year Tremec was acquired by Spicer.[8] In that way Tremec became stronger with technological and financial support.

With products and technology development, the company produced the best gearbox prototype qualified by Ford in 1996. In that year Tremec gained major market penetration with the Spicer's acquisition of TSP, an American company, manufacturer of gearboxes of categories T-5, T-45, and T-56. Moreover in December of that year Tremec obtained the ISO-9001/QS-9000 distinction.

In 1997 the company was honored with the recognition from General Motors Corporation as the best supplier of 1996 in the world (QSP Award) (Table 19.4). This prize is given by the corporation to only 173 of the 30,000 worldwide purveyors. With this award Tremec was able to sign new contracts and invest in specific technological assets dedicated to produce and design equipment with high technology for the models that General Motors launched on the market in 1999. In other words, the producer's organizational behavior stimulated the interactive technological learning and organizational innovations of processes and products. On the other hand, the fact stands out that Tremec signed an agreement of technology development with a British company in order to gain skills to innovate spare parts and gearboxes.

Table 19.4
Technological Evolution of Tremec (1986-1997)

Year and/or period	Technological events	Results
1986	Tremec obtained the Ford prize Q1	Interactive technological learning
1994-1995	Tremec got the ISO-9001, QS-9002, and UL certifications	Interactive and organizational technological learning
1996	Tremec established direct electronic communication with users	Technological innovation and organizational learning
1999	Tremec signed a technology development agreement with ProDrive	Interactive technological learning, investment in technological activities

AUTO INDUSTRY USER-PRODUCER RELATIONSHIP

The strength and length of technological cooperation between Chrysler and Tremec depends on user requirements. In other words, it transforms the user-producer interface (Andersen, 1991; Klein, 1996)[9] and the life cycle of the products on the market.

In this way, Chrysler makes specific demands to Tremec, which transforms the information to technical specifications for manufacturing new products. This means that given the requirements that Chrysler asks of Tremec, the technological linkages between both companies becomes more intense for working together to establish routines for solving the client's requirements.

In that sense, for the development of new products there is an intense relationship between both companies. An assembly plant has different specialized areas, such as the pre-assembly of motor area, composed of different segments: personnel who are in charge of gearboxes, clutches, pneumatics, and so on, and these segments are jointly dedicated to the assembly a of vehicle. Tremec has direct communication with those areas and helps the company to make new developments, and to carry out research and analysis for determining if a product completely covers the quality requirements.

In other words, written documents do not exist that establish technological cooperation agreements. This relationship is more informal. For instance, if Chrysler has some technological problem, then this company asks Tremec to carry out specific research. This means avoiding the transaction costs of a formal technological cooperation agreement.

Besides the technical specifications interchanged for the new product development, Tremec makes tests and does analysis for Chrysler. For example, the clients send to Tremec's R&D laboratory vehicles for analysis and solving noise, resonance, and comfort problems. The first visit to Tremec facilities (July 1997) was realized by Chrysler when they sent to Tremec the 1998 model in order to get a series of tests. For example, the interior was an original design and Tremec's work consisted of making a complete analysis of the motor, clutch, gearbox, differential, suspension, wheel rods; this was a full study for detecting if there were problems in the Chrysler vehicles.

The technological information interchange with clients is carried out mainly through purchase and sale, product engineering, and service departments. In general, those departments have a special program of periodic visits to the clients, in which are shown the product tendencies—for example, how the design of the vehicles is going to change in a period of three years, if there will exist changes in the carriage or the gearboxes, or other requirements.

In this aspect, Tremec started to design or acquire the equipment and the specific assets that are necessary for making a gearbox according to the Chrysler requirements for a determined time. Those requirements were met in the development area, which is in charge of doing technological monitoring, and this department makes recommendations if the existent gearbox is adequate to the client's necessities or if it is necessary to design a new one.

According to JICA (1997: 37), the technological cooperation between assemblers and spare parts providers in Mexico is more dynamic when the final assembler decides to introduce innovations into new models. This is a factor that forces the first-level suppliers to carry out innovations in a constant mode and maintains purveyor contracts for a five-year period. The long term of these contracts is related to the fact that the assemblers maintain constant innovation in their sourcing processes and quality control.

In general, as has been mentioned, the assemblers look for a policy of global sourcing, through which they acquire spare parts through international competition. This allows the establishment of contracts and long-term alliances with spare-parts purveyor companies in an attempt to reduce transaction costs.

Nevertheless, even in Mexico there are specific regulations about the parts that the automobiles must integrate. The spare parts that the national purveyors supply is submitted constantly to competitions of the "global sourcing" scheme. Even though that scheme makes the constant adequation of capacities and qualities in spare parts necessary, national companies can limit the long-term contracts (NAFIN-IMEF, 1995).

In the global sourcing policy of the assemblers and the innovator profile of Tremec, the linkages between this company and its suppliers would be rich in interaction and therefore can develop product and process innovations that allow Tremec to maintain its status as a first-level purveyor.

At the same time, Tremec as a client transfers codified information to its foreign purveyors. In the company the technical specifications are generated—for instance, the composition of aluminum according to the characteristics of the spare parts that will be submitted to be checked out. Later the development department, together with the purchasing and engineering departments, supplies the production department with the requirements for manufacturing the requested gearboxes.

Once the development department finishes the gearboxes, the product is sent to the production division, then it returns to the R&D laboratory where they make the final tests in order to verify that the products meet all the specifications that Tremec pointed out.

Tremec has purveyors that supply some of the most important spare parts of the gearbox, so between Tremec and those suppliers there is exchange of technological information. This means that the organization uses the purveyors' technological capacities from other countries in order to develop the gearbox systems and can then use the applied knowledge to manufacture gearboxes.

In that sense, Tremec works jointly with two American companies that manufacture friction materials for synchronizers (Performance Friction Products and Select). This working method allows Tremec cost advantages in R&D, and as a result the total manufacture of those materials depends on the purveyors, which, at the same time, suggest the kinds of materials that are satisfactory for gearbox manufacturing.

In the same way, referring to other gearbox spare-parts manufacturing, Tremec works with foreign companies that supply these spare parts. This working form allows the company to reduce costs so all development processes of the materials are the charge of the purveyors.

On the other hand, one of the innovations that Tremec wishes to incorporate into its products is the automatic gearbox. The innovation is based on the introduction of electronic modules that move sensors in such a way that these parts automatically execute velocity changes. The electronic module that has a direct communication with the motor for monitoring the number of revolutions per minute will be connected to the generator for verifying speeds. These variables are taken into account for determining the moment of clutch-pedal liberation when speeds change most.

So those companies are developing a new product. The selected company to sign a technological agreement with Tremec needs data about the destiny or application of the new technological development.

From the beginning of the technological agreement, a meeting schedule and an organizational routine was established and accepted for defining the agenda of problem solving between two companies. Also, all kinds of electronic communications are kept open between the two companies.

Summing up this work it has been discovered that between Chrysler and Tremec positive conditions have been established for innovation and technological learning. This relationship is archetypical of favorable technological conditions, information and knowledge exchange, purveyor selection, establishment of network collaboration, and investment in specific assets and organizational innovations necessary for building a successful user-producer relationship.

The successful relationship is a result of three factors. First, Tremec is a Mexican company specializing in manual and automatic gearbox manufacturing.

It is a functional automobile subassembly company with a certain degree of technological complexity. It is also a large company that has the necessary technological capabilities for being a first-level national and international purveyor in the automobile industry.

Second, the success of the purveyor relationship between Tremec and Chrysler is the outcome of the establishment of interfaces that allows the exchange of information and complex technological knowledge. Also, the accomplishment depends on the implementation of certain organizational routines for confronting and determining technical problems jointly.

Third, the success of Tremec as a Chrysler purveyor depends on a technological and organizational permanent innovation process. This phenomenon is explained given that the large automotive firms like Chrysler are subjected to the incessant pressure of world competition.

Even it wasn't the purpose of this research, one of the problems detected at the Tremec facilities visits was the fault of connection of productive linkages between this organization and the local automotive parts and components manufacturers. It was clear that Tremec lacks basic information with respect to the available offer of the national small or medium-sized producers,[10] or because these producers do not have technological and organizational conditions for becoming Tremec's trusted purveyors. At the same time, the national producers of the second or third levels lack enough information about which automotive components and spare parts are demanded by Tremec and therefore they lack the basic knowledge in relation to the requirements, conditions, technical specifications, and technological capabilities that Tremec requires for selection of its potentials purveyors.

Finally, this relationship has resulted to be technologically complementary for both companies as can be observed in Table 19.1. The interface between Tremec and Chrysler is shown to be rich in information and technological knowledge exchanges that transform in innovation and technological processes in both companies.

NOTES

1. BMW built an industrial assembly facility in Mexico in 1995; Honda began the manufacturing of its Accord in Jalisco in 1995; and Porsche has introduced vehicles to Mexico since 1995. Also Peugeot, Volvo, and Audi arrived in Mexico with new automobile models and Renault came back to Mexico in 2000.
2. Information revealed by Chrysler executive in 1997.
3. An example is the gear noise in the gearbox, which was analyzed in the laboratory by eliminating or diminishing the noise, working with different geometric types and distinct kinds of gear until getting the optimums results.
4. In that sense, an interesting case is General Motors. The firm relocated the R&D design center to Ciudad Juárez, Chihuahua, in 1995.
5. In contrast, Ford purchases 50% of parts from suppliers and the firm produces the remaining 50%, whereas General Motors is more vertically integrated. This firm internally produces 70% of its parts, assembly, and so on. The remaining components are purchased from suppliers.

6. Until 1992 the company was organized hierarchically. That means that the production was obtained through departmental goals, and there existed communication problems, individualized work, and too many organizational levels.

7. Tremec signed two technological collaboration agreements: one with a British corporation, ProDrive, and another with Columbia University.

8. This firm is integrated by a group of firms of the automobile industry.

9. To see a historical, economic, and technological experience about the user-producer relationship and the vertical integration as an organization property, look to the General Motors-Fisher Body case analyzed by Klein (1996).

10. Nevertheless, the Secretary of Economy publishes monthly the "Directorio Electrónico de la Industria Maquiladora de Exportación" (Carrillo, 2001) and the Industria Nacional de Autopartes publishes the "Catálogo de Fabricantes de Autopartes." This catalog counts on information about 900 firms associated with the Industria Nacional Automotriz (INA), as well as firms not affiliated.

REFERENCES

Andersen, E.S. (1991). "Techno-economic Paradigms as Typical Interfaces Between Producers and Users," *Journal of Evolutionary Economics*, 1 (2).

Carrillo, J. (2001). "Maquiladoras de exportación y la formación de empresas mexicanas exitosas." In E. Dussel Peters (ed.), *Claroscuros: Integración Exitosa de las Pequeñas y Medianas Empresas en México*. Santiago de Chile: CANACINTRA-CEPAL-JUS.

Carrillo, J. and Hualde, A. (1997). "Maquiladoras de tercera generación. El caso de Delphi-General Motors," *Comercio Exterior*, 47 (9), September.

Expansión (2002), April.

INEGI (2000). *La Industria Automotriz en México*. Mexico: INEGI.

Japan International Cooperation Agency (JICA) (1997). "The Study of Master Plan for the Promotion of the Supporting Industries in the United Mexican Status," mimeo.

Kein, B. (1996). "La Integración Vertical como Propiedad Organizativa: Una Revisión de la Relación Fischer-General Motors." In O. Williamson and S. Winter (eds.), *La Naturaleza de la Empresa: Orígenes, Evolución y Desarrollo*. Mexico: FCE.

Micheli, J. (1996). "Industria, Calidad y Poder (A Propósito de la Industria de Autopartes en México.)" In J. L. Calva, M. Capdevielle, and C. Pérez (eds.), *Industria Manufacturera, Situación Actual y Desarrollo bajo un Modelo Alternativo*. Mexico: Universidad Autónoma Metropolitana.

NAFIN-IMEF (1995). *La Competitividad de la Empresa Mexicana*, Biblioteca NAFIN, no. 9, Mexico.

Nonaka, I. and Kono, N. (1998). "The Concept of 'ba'. Building a Foundation from Knowledge Creation," *California Management Review*, 40 (3).

20

The Geography of Innovation in the Pharmaceutical Industry: Assessing Implications for Developing Countries

Beatriz C. Fialho, Lia Hasenclever, and José M. C. Mello

INTRODUCTION

The pharmaceutical sector is an innovation-intensive industry, highly internationalized, dominated by large multinational firms, and with a high degree of market concentration. This sector is also characterized by a clear international division of labor in terms of production and innovation, with concentration of production, trade, and R&D in few developed countries. This geography of production and innovation reflects the evolution dynamics of knowledge production, and its interrelation with other factors reflects the organization of economic activities, especially the industry organization, at international, national, and subnational levels. In the case of the pharmaceutical sector, such features have severe implications for developing countries, where access to pharmaceuticals is a problem, and where companies do not invest in research and development, especially in respect to public health problems, like tropical diseases.

In the following sections, we discuss, through an account of the geography of innovation in the pharmaceutical industry, the implications for developing countries of concentration of R&D and innovation as well as of production of active principles in few developed countries. Special attention is given to an ongoing research concerning the Brazilian case. In Brazil there is technological and industrial capability in the final production stage (formulation of active principles), although there are few firms with R&D capabilities, and industrial and technological capabilities in production of active principles.[1] In addition, among developing countries, there are in Brazil important nuclei of knowledge production in the life sciences. Nevertheless, the Brazilian pharmaceutical industry has been characterized by a high dependence upon importation of active principles and upon knowledge produced in developed countries, especially knowledge economically useful for the purpose of innovation. Furthermore, access to medicines and treatment for endemic diseases like tropical diseases, are still important public health problems.

The next section discusses the current theoretical approaches dealing with the dynamics of knowledge production and innovation as a basis for the explanation of the differences among countries. Next the general features of the pharmaceutical sector, emphasizing the evolution of knowledge production and innovation, and its current dynamics are presented. Then the Brazilian case is discussed, highlighting the evolution of the Brazilian pharmaceutical industry and the possibility for the country to benefit from a window of opportunity for the location of knowledge production. The chapter ends with conclusions and the contribution of the ongoing research.

GEOGRAPHY OF INNOVATION: THE THEORETICAL STANDPOINT

The dynamics of knowledge production have proven unequal among countries. This uneven distribution also reflects the differences of research efforts accomplished among firms and among countries and in divergent industrial and technological development trajectories.

From the theoretical standpoint, since the end of the 1980s, several authors studying the dynamics of innovation have proposed different ways to explain such differences. From these approaches, two have attracted more attention. The first approach is the "national system of innovation," whose main references are the works of Freeman (1987), Lundvall (1992), and Nelson (1993). The second is the *triple helix* model that has been developed based on the works of Etzkowitz and Leydesdorff (1997 and 2000).

In the case of the national systems of innovation it is argued that the nature of this system, in which knowledge economically useful is generated, absorbed, and transformed, establishing a foundation to create and sustain competitive advantage and leadership, lies in the heart of the explanation of differences among countries. These authors emphasize the role of learning and technological accumulation; the role of R&D, with firms as the main locus; the influence of appropriability mechanisms in the generation and diffusion of innovations; the role of geographical and organizational proximity in knowledge production; the role of sectoral differences; and the differences in the institutional environment.

In the case of the *triple helix* model the interest is upon the understanding of the transformations in the dynamics of knowledge production with emphasis upon the role of universities in the innovation process, and the interaction among university, industry, and government (UIG). The focus is placed upon the roles attributed to those involved in knowledge production and the fact that it has become more difficult to delineate the borders among them once knowledge production involves recursive interaction. In this model, depending on the nature of the dynamics of UIG interaction, a given environment can be more or less favorable to innovation (Etzkowitz and Leydesdorff, 1997 and 2000).

These two approaches serve as a departure for a better understanding of the relation between innovation and development, and if some of its prerequisites are relaxed, they may also serve to analyze determined situations in which systems requirements are not filled. Notwithstanding, these models are limited to understand the differences between developed and developing countries, as characterized by the uneven distribution of knowledge production and innovation and by different development trajectories.

First, for those countries in which the characteristics highlighted in these models cannot be observed, such features turn into an *ex ante* condition to stimulate industrial and economic development. This becomes more critical for developing countries, once the *ex ante* matter is treated as if such countries were in a previous development stage in which there is no innovation, although in some cases there would be technical change. Second, the explanation in both approaches for the differences observed among countries is generally based upon arguments such as "market failures," "lack of institutions," or "lack of linkages." This is rather limited, once for these countries to catch up, and to foster innovation and development, it would be necessary to "correct" those failures and create an environment more favorable to dynamic interaction among agents and to innovation, following the path and the already observed features of more developed countries. In our opinion, this is limited because such arguments result from the fact that in both approaches, geopolitical aspects such as the international division of innovation and productive labor along industrial evolution and within the capitalist interfirm and interstate competition are treated in an incipient manner. Thus it is argued here that understanding the differences between developed and developing countries, in respect to knowledge production and innovation, needs a specific approach.

In order to discuss these issues, the following sections will focus upon the dynamics of knowledge production and innovation in the pharmaceutical industry. This is an innovation-intensive sector, characterized by a high degree of internationalization and a well-defined international division of productive and innovative labor, and in which the differences in knowledge production have important implications for public policies concerning socioeconomic, technological, and industrial development. Particular attention will be given to the Brazilian case, where there is a well-established pharmaceutical industry and some nucleus of knowledge production in the life sciences.

Our argument is that, in the case of the pharmaceutical sector, the observed geography of production and innovation, which poses many challenges for developing countries, reflects three intertwined dynamics. First, it reflects the dynamics of evolution of such sectors (market structure, conduct, and performance of firms) in each national context along the changes in knowledge production. Second, it reflects the world industry evolution in terms of interfirm and interstate capitalist competition. Third, it also reflects the evolution of regulation, industrial and science and technology (S&T) policies and intellectual property rights, in both national and international levels.

THE PHARMACEUTICAL INDUSTRY

In the pharmaceutical sector, generally large, integrated, and international or multinational pharmaceutical companies carry out R&D, production, and marketing of medicines. However, not every company is involved in all activities. These companies are present in almost all national markets, reflecting the industry's high degree of internationalization, mainly because their domestic markets may not always be large enough for such companies to recover development/introduction costs (Commanor, 1985; Halliday et al., 1997; Gambardella et al., 2000). The world pharmaceutical market, in terms of sales, is also concentrated in few developed countries with some developing countries lagging behind. This sector is also characterized by a clear division of labor in terms of production and innovation, with concentration of production, trade, and R&D in developed countries (Michaud and Murray, 1996; CIPR, 2002).

At the world level, in developed countries, especially those in which leading firms are located, the industry generally presents a vertical structure, with large research-based firms acting in R&D, production of active principles and finished medicines, and marketing.[2] In most developing countries, large multinational firms dominate domestic markets, accomplishing only production of finished medicines and in very few cases the production of active principles for their own consumption; national domestic firms are of a smaller size and devoted mainly to production of medicines.[3]

The resulting geographical concentration of production and R&D efforts in the pharmaceutical sector has severe implications for developing countries. Such countries play a marginal role in the sector's industrial and innovation dynamics. In terms of industry organization, the great majority of these countries are dependent upon imported active principles (with exceptions such as India) and imported finished medicines (with exceptions such as India and Brazil). The same is true in the case of knowledge production and innovation, once in developing countries R&D activities related to the sector are very incipient. Besides, although there are some developing countries with capabilities in health research, these are concentrated in public research universities and research institutes with few linkages to the industrial sector. Furthermore, although these features bear on a national imprint in terms of the sector's evolution within each country, it has to be considered that the pattern of insertion of these countries in the world economy has been different from developed countries.

The Evolution of Knowledge Production and Innovation in the Pharmaceutical Industry

The evolution of knowledge production and innovation in the pharmaceutical industry will be addressed here using as reference a recent study of Achilladelis and Antonakis (2001) about 1,736 innovations from 1800 to 1990. According to these authors, the evolution of the innovation dynamics in the sector could be analyzed through five main periods: 1820-1880, 1880-1930, 1930-1960, 1960-1980, and 1980 onward.

In the early 19th century, when the pharmaceutical industry was in its infancy, the search for medicines was based upon extraction of active substances from plants, though without proved efficacy once neither the scientific content nor quality-control procedures had been established in the sector. Most of this work was done by pharmacists and physicians in apothecaries, in some cases also working as academic researchers. At that time there was little difference among countries in terms of the knowledge involved in such activities and in terms of production once manufacturing of medicines was based on manipulation and artisanal methods. Also, markets for those products were generally of a national scope. Although most of the former apothecaries were overcome by the emerging pharmaceutical laboratories at the end of the 19th century as the source of medicines, there were many new pharmaceutical laboratories that emerged from former apothecaries belonging to academic pharmacists (like Merck AG), and from apothecaries that commercialized only imported medicines, like some U.K. and U.S. laboratories (Liebenau, 1985).

According to the study by Achilladelis and Antonakis (2001), the geography of innovation between 1820 and 1880 was the following: 14 academic innovations in France; 13 academic innovations and 1 industrial innovation in the United Kingdom; 10 academic innovations and 3 industrial innovations in Germany; 3 academic innovations and 1 industrial innovation in the United States; 2 academic innovations in Switzerland; 2 academic innovations in Brazil; and 1 academic innovation in the Netherlands. It was also possible to observe that since the beginning of the industry's evolution, Germany was the locus of knowledge production and innovation in the pharmaceutical sector. The explanation for such predominance seems to lie in the fact that organic chemistry was well developed in Germany and that policies were oriented to industrial and technological development, especially in respect to technology and knowledge transfer between technological research institutes and industry. Also it has to be considered that after Germany was unified at the end of the 19th century, the first patent law allowed only patents for process but not for products, either in chemicals or pharmaceuticals.

At the end of the 19th century the development of new disciplines such as pharmacology and physiology, the development of chemical synthesis, and the discovery by German chemical firms that some synthetic dyes also had therapeutic effects introduced a new pattern in knowledge production and innovation. This new dynamic involved the institutionalization of R&D by the first chemical and pharmaceutical firms, and ties were established between industry and academic research institutions, especially in Germany, and that became, later on, important features in the sector's dynamics.

German leadership in knowledge production in the pharmaceutical sector, especially knowledge economically useful, became more visible between 1880 and 1930, when, according Achilladelis and Antonakis (2001), from 145 innovations, 82 (56%) originated in Germany of which 62 (42.76%) were industrial innovations and 21 (13.79%) academic innovations. From the 62 industrial innovations, 21 (33.87%) became market successes, and from the 21 academic innovations, 4 (20%) became market successes in the period. The consequences of the German leadership were felt in several countries during World War I, especially regarding access to medicines and inputs for the final production. In turn, some countries began to develop strategies to foster the development of their own chemical industry, as was the case of the United States, which during World War I tried to use the expropriation of German patents as a technology transfer mechanism (Steen, 2001).

During World War II, the German predominance was still felt in many countries (developed and developing), which, among other policies, used the nonrecognition of intellectual property rights to foster the national pharmaceutical sector. While most developed countries began to recognize patent rights until the end of the 1970s and beginning of the 1980s (Switzerland and Italy, for example), most developing countries continued to neglect the granting of patents in the pharma sector. However, after the war efforts for producing large-scale penicillin, the 1950s saw the emergence of a strong North American industry. This can be observed in the innovative performance of U.S. laboratories in the period 1930-1960 studied by Achilladelis and Antonakis (2001). Between 1930 and 1960 the United States originated 270 (50.47%) of 535 innovations in the period. From the 270 innovations, 267 were industrial innovations of which 37.08% were market successes, and 3 academic innovations of which 1 was a market success. The U.S. pharmaceutical industry was followed by a large distance by Switzerland with 87 innovations (86 industrial innovations of which 22 were market success), Germany with 56 innovations (54 industrial innovations of which 23 were market successes), United Kingdom 54 (15 industrial innovations of which 13 were market successes), and France 31 (11 industrial innovations of which 11 were market successes).

This period was also marked by the development of antibiotics and new classes of medicines, especially for infectious diseases, and random screening became the main process of search for new molecules. In the case of the United States, besides the role played by war efforts, the size of the U.S. market and changes in regulation acted also as an important element for the development of the North American pharmaceutical industry. In addition, it has also to be noted that the health sector became the second major depository of the North American government's resource allocation in science and technology.

In the following period, between 1960 and 1980, Achilladelis and Antonakis (2001) report only innovations introduced by pharmaceutical firms. In this respect, among the 126 innovating firms observed in the sample, the

United States accounted for 31 innovating firms, Germany for 22, Switzerland for 3, the United Kingdom for 14, and France for 18. These 88 companies not only represented 70% of all companies in the sample but also 78% of all 672 innovations introduced in the period (43% by U.S. firms, 13% by German firms, 10% Swiss firms, 7% U.K. firms and 5% French firms). These firms were also responsible for 87% of the 137 radical innovations introduced in the period and for 91% of the 143 innovations that became market successes. In terms of radical innovations and market success, around 50% of the 137 radical innovations and 55% of the 143 innovations that were market successes were introduced by U.S. firms. In the case of the German, Swiss, and U.K. firms, their performances were quite modest in relation to the United States: Firms from each of these countries accounted for around 11% of the radical innovations and around 11% of the innovations became market successes. The performance of French firms was quite below the others, 4 of the 137 radical innovations were introduced and 2 of the 143 market successes were introduced by French firms.

Therefore, since the consolidation of the chemical synthesis as the main route for the search for molecules with therapeutic effects, one observes a group of countries in which production and innovation have concentrated: United States, Switzerland, Germany, United Kingdom, and France. This distribution changed only after 1980 when Japan replaced France as the fifth location of innovations in the pharmaceutical sector. The geography of innovation in the 1980-1993 period was the following: of the 380 innovations, the United States accounted for 39%, Germany for 12%, Switzerland for 12%, the United Kingdom for 11%, Japan for 8%, and France for 6%. In respect to the 82 radical innovations of the period, the United States accounted for 63%, Germany for 10%, Switzerland for 10%, the United Kingdom for 7%, and France for 4%. In respect to the 67 innovations that became market successes, the United States accounted for 54%, Germany for 7%, Switzerland for 13%, the United Kingdom for 13%, and France for 3%.

In respect to the period that begins in 1980 some remarks must be made. First, an important feature in the dynamics of innovation in the pharmaceutical sector according to Achilladelis and Antonakis (2001) is that after 1960 the number of radical innovations declined from 166 in 1930-1960 to 137 in 1960-1980 and to 82 in 1980-1993. Also the number of innovations that became market successes declined.

Second, the sector has been going through important transformations, which have had major impacts upon the dynamics of knowledge production and have also had important implications to developing countries. Some of these transformations include: the intensification of mergers and acquisitions, especially among leading firms, with impact upon the international division of labor in the sector; the emergence of new technologies and transformations in the knowledge base; as demand for pharmaceuticals increased around the world and governments also began to implement more stringent regulations not only in terms of safety and efficacy of medicines but also in order to control pharmaceutical expenditures; pressures for firms to increase the number of new medicines introduced in the market; a greater awareness of the society; the emergence of new diseases and reemergence of old diseases; a greater competition due to the strengthening of the generics markets; and a growing concern of different governments and civil society about neglected diseases in terms of R&D efforts.

Another important source of pressure involves the disputes concerning the Trade-related Aspects of Intellectual Property Rights (TRIPS) among those World Trade Organization member countries, resulting from the discussions during GATT's Uruguay Round. This is related to the fact that it has been acknowledged that intellectual property rights, besides other factors of analogous importance, affect countries in different ways, not only in terms of access to medicines, but also in terms of access to knowledge and technological development.

Third, although the competition environment may seem to have become less favorable to national policies, these, however, have had a preponderant role, especially in terms of international commercial disputes. As Gambardella et al. (2000) observed, the pharmaceutical sector is important for public policy not only because it is an innovation-intensive sector, with high growth potential, but also because the industry has proven highly internationalized and the transformations by which it has been going through (with the introduction of new technologies or changes in the industrial structure at a world level) have affected the dynamics of competition. These features pose important issues for national public policy, from the point of view either of countries individually or of regional economic agreements.

Most Important Aspects of Current Dynamics in the Sector

Besides the overall features of the pharmaceutical industry observed in the beginning of this section, and taking into account the evolution of knowledge production and innovation in the sector presented above, it is possible to highlight the most important aspects of the current dynamics of knowledge production and innovation in the sector.

First, there are few large research-based pharmaceutical firms located in developed countries that have been responsible for a great part of the R&D effort related to the development of new or improved medicines. These firms tend to locate innovative activities in their home countries and/or in other developed countries, because of access to a high-qualified scientific infrastructure, and proximity of the firm's headquarters. These companies have also established strong R&D facilities that allowed them for a long period to lead technical change and innovation. And, although the biotechnology revolution has shifted the locus of knowledge production and innovation from large firms to small biotech companies, large firms have moved toward establishing linkages with these firms (Galambos and Sturchio, 1998). According to a recent study, the majority of the biotechnology drugs approved by the U.S. Food and Drug Administration (FDA) have been commercialized by large firms (Gray and Parker, 1998).

In terms of sales, pharmaceutical markets are concentrated in the developed countries. In 2002, North American countries (United States and Canada) accounted for around 50% of world sales, the European Union accounted for 22%, and Japan for 12%. The rest of the world market was divided among developing countries as follows: 4% within Latin American countries; 8% among Asia, Africa, and Australia; and 3% for the remaining European countries (IMS, 2003).

Second, although pharmaceutical firms have been relying mainly on their internal R&D efforts, it is worth highlighting the role of universities and research organizations, and knowledge produced by the public research system. For example, Mansfield (1991) observed that, between 1975 and 1985, 27% of products and 29% of processes could not have been developed by the pharmaceutical industry without substantial delay in the absence of academic research. Also Cockburn and Henderson (1997 and 2000) observed that some firms have recently began to reward their scientists based on peer review process, stimulating them to participate in congresses and seminars, where they could be in permanent contact with the current state of the art of academic research once they are generally based in new fields or fields in which there is no consolidated scientific knowledge. These authors also observed that from 21 medicines with high therapeutic impact, introduced between 1965 and 1992, 14 were based upon researches funded by government resources.

Third, governments have, accordingly, played an important role in this dynamic, either through direct support to R&D within the research system; through the establishment of regulatory standards that have had important impacts upon the innovation process and decision taking to invest in R&D by firms; or through their purchasing power, mostly in developed countries, stimulating the growth of the pharmaceutical sector as well as the development of indigenous technological and innovative capabilities in some countries.

Fourth, another interesting side of this dynamic of knowledge production in the pharmaceutical sector was analyzed by Mariani (2000). This author studied the pattern of relations among inventors in partnerships—in biotechnology and pharmaceuticals—that resulted in patents and the mechanisms through which collaborative research was coordinated. This study corroborates the geographical concentration of innovative activities in triad countries as observed by Archibugi and Michie (1995). Moreover, the distribution of patents by the assignees' country of origin indicates that inventors both in biotechnology and pharmaceuticals tend to work with inventors from their own country and then with U.S. inventors. In European countries, inventors tend to work with U.S. inventors more in pharmaceuticals (11.5%) than in biotechnology (5.1%), while U.S. inventors work with Europeans more in biotechnology (16.6%) than in pharmaceuticals (13.3%). Inventors in Japan tend to work mainly within the country both in biotechnology (94.9%) and pharmaceuticals (95.2%).

Last, most pharmaceutical R&D efforts focus upon markets and diseases of more developed countries that have higher market potential (Commanor, 1996). Pharmaceutical manufacturers have always claimed that the lack of patent protection in some countries does not stimulate R&D activities among their subsidiaries. But for Pecoul et al. (1999: 365), "it is unlikely that western manufacturers will devote much of their effort to insolvent populations, with or without patents ... [and], tropical research may not have a more promising future, even if patents are widely enforced." Furthermore, the pressure for strengthening intellectual property rights since the beginning of the 1990s, culminating in the TRIPS agreement, has reduced the space for technology and industrial policies in developing countries, which are still in a marginal position in the international division of innovative labor. Such features have led to a regrettable outcome for developing countries that account for the larger share of the world disease burden, especially because of the lower market potential of the needed medicines, and the poor purchasing power of the ill population (Commanor, 1996). These countries also have to face other problems such as reducing disparities in wealth distribution, leveraging socioeconomic development and health conditions, among others.

Considering the facts highlighted in this section concerning the evolution of the pharmaceutical sector at world levels, it can be argued that the observed concentration of knowledge production and innovation in few developed countries results from the interplay of several economic and political dynamics, and is best expressed by the international division of innovative and productive labor. This division emerges as a result of how the sector evolved in each country, in terms of the interrelation among those involved in knowledge production and decision taking in relation to R&D, innovation, and production along the process of technical change and industrial evolution. Such dynamics are also influenced by the role of public policies and their linkage with the productive sector; by the structures of learning, training, and research; and by how each country is inserted in the overall interstate and interfirm capitalist competition. It also has to be observed that the effect of this geographical concentration has been different not only between developed and developing countries, but also among developing countries.

Although among developed countries the concentration of R&D, innovation, and production is also an important concern in terms of industrial and technology policy, the impact upon access to new or improved medicines is less critical than in developing countries. Developed countries face better socioeconomic conditions and R&D targets of leading firms that also participate in the epidemiological table set by authorities. In contrast, the concentration of knowledge production in few developed countries has had major effects for developing countries, where access to pharmaceuticals is a problem because of the low overall access to new or improved medicines but also because R&D targets of leading firms are not connected with their epidemiological table. In this sense, it has to be observed that while some developing countries' morbidity and mortality patterns are becoming similar to those in developed countries, some diseases still present public health problems and to which incipient R&D efforts has been directed.

THE BRAZILIAN CASE: AN EXPLORATORY OVERVIEW

There are in Brazil approximately 400 pharmaceutical firms, of which 20 are multinational companies accounting for 75% to 80% of the domestic market. While the domestic market of finished medicines has been supplied since the 1960s by local production, production of active principles is very incipient, mostly supplied by importation. There are also 15 public laboratories producing finished medicines specifically for the Ministry of Health (MoH) pharmaceutical assistance programs. These laboratories represent around 3% of the total supply of medicines (Hasenclever, 2002).

Multinational firms operating in the country have their overall strategies determined by their headquarters. In terms of R&D this means that such activities in Brazil are very incipient compared to what is observed in developed countries and are mainly devoted to clinical trials phases III or IV,[4] and quality control. The picture is not different in the case of national-owned firms that are generally limited to imitative efforts in formulation of imported active principles. There are a few exceptions to this rule, with a small number of firms conducting R&D activities in the search of new formulations or new molecules. One limiting factor for national-owned firms to conduct R&D has been the fact that they are generally of a smaller size (Hasenclever et al., 2000).

The Evolution of the Brazilian Pharmaceutical Industry

The emergence of a pharmaceutical industry in Brazil seems to have followed, at least in its initial phase, the same pattern observed in the United States—that is, it evolved from the manipulation and marketing of natural extracts and of imported medicines. Notwithstanding, it is also possible to observe a few cases in which the emergence of pharmaceutical laboratories in Brazil in the late 19th century seems to have followed a pattern closer to the specialized German pharmaceutical firms, through the initiatives of academic pharmacists. But it is important to note that in Germany, the pharmaceutical industry emerged from the diversification of chemical firms involved in the production of organic products, mainly dyestuffs, once these companies identified through R&D activities that some of these products also had therapeutic effects. Thus, while it can be said that the pharmaceutical industry emerged in Brazil between the ends of the 19th and 20th centuries, the first initiatives in the chemical sector seem to have happened only after the mid-1950s and mostly after the 1970s (Fontoura, 1938; Scheinkmann, 1965; Bertero, 1972).

In addition it has to be noted that the transformations through which the industry went at the end of the 19th century posed several challenges for the first Brazilian pharmaceutical firms. At the end of the century, the situation of the infant Brazilian pharmaceutical industry was not very different from that observed, for example, in the United States. But, while some North American firms seemed to have been capable of making the transition from artisanal or semi-industrialized production of medicines based upon plant extracts to chemical synthesis and industrial scale, Brazilian firms seem to have been incapable not only of making the necessary investments but also of failing to successfully integrate qualified human resources, a scientific infrastructure, and governmental support (Bertero, 1972; Frenkel et al., 1978).

As well as in other industrial branches, the pharmaceutical industry emerged in Brazil in the 19th century when the country's economy was based on primary products within a slavery system, thus there was a mismatching between the political, economic, and institutional exigencies of the industrial entrepreneurship and the Brazilian primary exporting periphery economy insertion pattern. In the United States, on the other hand, the industrial structure was more consolidated, the development of a scientific infrastructure was devoted to narrowing the technological gap in relation to Europe, and public policies intended to promote national industry.

Also in the last quarter of the 19th century some foreign laboratories were already marketing their products in Brazil, especially German firms, but also English and North American firms (Fontoura, 1938; Frenkel et al., 1978). The entry of foreign pharmaceutical firms was initially based upon importation but then some firms also began to produce locally, especially in the 1930s and after the 1950s, through the acquisition of local producers, basically those more dynamic laboratories that could not make the transition to chemical synthesis (Frenkel et al., 1978). Also, the pharmaceutical sector seems not to have been an agenda for public policy in terms of industrial and technological development until the 1970s. While after the 1950s it could be said that most of the domestic market of finished medicines was supplied by local production, it was already observed that the country was dependent upon importation of active substances and upon knowledge produced outside the country, especially knowledge economically useful to follow or lead technical change (Scheinkmann, 1965).

Between the 1950s and 1970s several authors (Bertero, 1972; Frenkel et al., 1978) pointed out different factors that, for their absence or inadequacy, would explain the dependent character of less-developed countries in the dynamics of the pharmaceutical sector at a world level: availability of qualified human resources; socioeconomic factors; development of chemical sciences; the existence of correlated industries; lack of government policies, like those put forward by Germany and United States that allowed them to establish positions of industrial leadership; the weak linkages between firms and universities; and the specificity of the sector's evolution with the emergence of large integrated firms operating at the world level, which led to an international division of productive and innovative labor. In most developing countries these differences crystallized mainly after the 1960-1970 decade. In the case of Brazil, at that time, the structure of the pharmaceutical industry had already acquired one of its current main features: more than 70% of the domestic market was dominated by multinational firms accomplishing only formulation, that is, the last phase of production of pharmaceuticals in the country (Bermudez, 1995).

In the 1970s the Brazilian government implemented some policies intended to foster development and to narrow the technological gap in the pharmaceutical sector, among which were the exclusion of pharmaceutical process and products of the Patent Law of 1971;[5] the creation of a Central Medical Store, the Central de Medicamentos (CEME), which was intended initially to act both as a public procurement agency and to support technological development in the sector;[6] the implementation of programs to stimulate the development of the chemical industry in the country within the Planos Nacionais de Desenvolvimento (PNDs); and of incentives by the Programas Brasileiros de Desenvolvimento Científico e Tecnológico (PBDCTs) in the 1970s and the Programas de Apoio ao Desenvolvimento Científico e Tecnológico (PADCTs) in the 1980s. However, although successful in some areas, the Brazilian pharmaceutical industry remained dependent upon importation of raw materials, specially active substances, and also dependent upon knowledge produced elsewhere, confirming the country's marginal role in the international division of productive and innovative labor of the sector.

After the 1990s the Brazilian pharmaceutical industry experienced significant changes in its structure: opening of the Brazilian economy; the end of price control in 1990; the new Patent Law 9279 in 1996 that reintroduced pharmaceutical process and products as objects of intellectual property rights; the successive mergers and acquisitions of firms at a world level and their strategies of location in developing countries; and the introduction of a specific law regulating the market for generic medicines (Hasenclever, 2002).

In terms of the effects of market liberalization, there was a sharp increase of importing trade principles after 1990 and the dismantling of the existing Brazilian fine chemical sector (producing active principles and synthesis intermediates). It has also been observed, recently, an increase in importation of finished medicines, and more recently of finished generic medicines (Hasenclever, 2002). In respect to intellectual property rights, until 1996 the Brazilian patent law excluded pharmaceutical processes and products, but in the course of the Uruguay Round of GATT, Brazil and other developing countries, in order to become members of the World Trade Organization, had to sign an agreement to comply with TRIPS until 2005. Differently from the other developing countries, Brazil introduced changes in its intellectual property regime in 1996, among which the granting of patents for pharmaceutical products and processes onward and the granting of *pipeline protection* for those deposits between January 1, 1995, and the implementation of Patent Law 9279 (Bermudez et al., 2000). In the case of the concentration trend among firms at the world level, some multinational firms decided to transform their domestic operations—production or packaging of finished medicines—into an exportation platform for Latin American countries. Other companies had decided to interrupt local production although maintaining marketing activities, and others decided to license products to some Brazilian firms. Last but not least, the introduction of Generics Law 9787 of 1999 has been expected to introduce significant changes in terms of greater competition among firms, especially in order to expand access to medicines in the country, because these medicines are sold at lower prices (Hasenclever, 2002). However, in these first years it has been seen as a substitution effect within the higher income classes and not necessarily a strong expansion of access to lower income classes.

In the case of R&D, while initial R&D efforts accomplished in developed countries by firms or government research institutions in chemical synthesis in areas like antibiotics were also connected to the epidemiological table of developing countries, like Brazil, there were still problems of access to medicines by the larger part of the population. And the improvement of sanitary conditions was very slowly. Also, while there are still problems of access to medicines, R&D efforts of the leading companies and main health research institutions of developed countries are no longer connected with diseases still prevailing in developing countries, which have low market potential, as observed above. Also, even despite some important achievements in the improvement of sanitary conditions and income level, many problems were not overcome, because in many parts of the country sanitary conditions are still poor and many people still live in misery.

It has been suggested, however, that Brazil would be in a favorable position for the location of knowledge production in the sector due to the potential represented by the biodiversity of its fauna and flora. Moreover, the government has signaled the will to promote technological development and innovation through new incentive mechanisms such as the establishment of a sectoral fund for health research.

The Brazilian Situation Revisited

Having the overall account of the evolution of the Brazilian pharmaceutical sector and the world pharmaceutical industry evolution, we are particularly interested in one question: In what degree could the country benefit from a "possible" window of opportunity like biodiversity for the location of knowledge production by already established large multinational firms or by local firms or new firms specializing in research and development of new or improved medicines that are public health problems in the country or medicines for other diseases? Based upon the theoretical approaches reviewed in the first section, these research questions would probably be explained in the following terms.

According to the national systems of innovation (NSI) approach, these questions would be addressed in terms of the existence or nonexistence of such systems in Brazil and in what degree the existing system seems to be more or less dynamic in terms of industrial and technological leadership or at least in terms of a *catch-up* level. In what concerns the study of the Brazilian pharmaceutical industry, Albuquerque and Cassiolato (2000) using the NSI approach concluded that it would be possible to speak of an "immature" national system of innovation in the health sector, although once there is a well-established pharmaceutical sector and there are important health research institutions, R&D efforts are still incipient in relation to countries like the United States. However, in a previous work, one of us argued that, considering the foundation of the NSI approach relies upon those configurations that proved to be more dynamic in one or more sectors from the point of view of innovation, it would not be possible to speak of such a system in Brazil and thus the NSI approach would not be applicable to the Brazilian context (Fialho, 2001).

In the triple helix model such issues would be investigated in terms of the roles of universities, academia, and government in knowledge production and their transformations over time. Following such an approach the Brazilian pharmaceutical sector should set up a system in which there are clear roles for each group without overlaps or permeability of borders, and also that the linkages between industry and academia would be very fragile or nonexistent. Such a picture would be an obstacle to creative destruction or innovation and technological development.

As seen above, although relaxing some of the prerequisites of both approaches could allow for the understanding of some characteristics of the UIG relationships and of the knowledge production that could be economically useful, they are still limited when they just explain part of the puzzle once it resembles a static photograph of the current situation.

Therefore, it is needed to take into account the international division of labor issue—which is an institution of the capitalist economic development dynamics and a source of competitive advantage for those countries in industrial leadership positions—how this is related to the dynamics at the national level, and how this reflects the different development trajectories and has impact upon the development of indigenous industrial and technological capabilities, especially in respect to innovation.

In this sense, in respect to the Brazilian case, it has to be taken into account that the evolution of the pharmaceutical sector in Brazil and its insertion into and relation to the world industry's dynamics—in production and knowledge generation—present several challenges for both industrial and technology policy and public health policy. In this sense, using the NSI or the triple helix as a guideline for policy design would not fully address the structural problems of the Brazilian pharmaceutical industry either in respect to the dynamics of knowledge production or in respect to the industry's dynamics. Therefore, in what concerns the study of the Brazilian pharmaceutical industry and the possibility of benefiting from a "window of opportunity," the current research considers that it is necessary to understand not only the dynamics of innovation per se but how countries are inserted in this dynamic. This is possible only through an evolutionary account of industry evolution and of knowledge production in each particular context and how this relates to the dynamics at the world level.

CONCLUSIONS

From the study presented above, four main conclusions can be drawn. First, the differences in the dynamics of knowledge production in the pharmaceutical sector reflect the divergent patterns of industrial and technological development, mainly in respect to research efforts and to generation of innovations, both among firms and among countries.

Second, the geographical concentration of production and R&D observed in the pharmaceutical sector reflects: (a) the dynamics of the emergence of such sector in each national context and how it evolved along the changes in knowledge production; (b) the dynamics of this evolution within the evolution of the world industry in terms of trade and R&D (both at macro and micro levels); (c) the evolution of regulations and industrial and S&T policies, in relation to the industry's dynamics at national and international levels.

Third, it is important to incorporate these three aspects in understanding the differences in development trajectories among countries.

Fourth, policies oriented to stimulate innovation and technological development in the health sector in developing countries shall consider that demand and supply sides, market structure, firms' decisions, and relations among knowledge producers are influenced by the dynamics of knowledge production at a world level, incorporating the roles of interfirm and interstate competition and the institutional environment in which these actors are embedded.

Thus, this overall picture has to be considered in order to evaluate the possibility of benefiting from the windows of opportunity that may turn out to be a way out for the spurious effects of concentration of knowledge production in the pharmaceutical sector. An important result of the ongoing research is that it may improve a better understanding of how developing countries can or are able to foster innovation in the health sector and to understand what mechanisms are available in order to balance public health goals with industrial and technological development.

NOTES

1. The development of a new medicine starts with the search for biologically active compounds, which then are submitted to early preclinical toxicology and pharmacology (mechanism of action) tests, to formulation (selection and stability), and chemical development (selection of synthesis routes). A handful of compounds (drug candidates) selected from these tests are submitted to four phases of clinical trials in healthy subjects and patients. The wining compound is then submitted to registration requirements, production, and launching (see Halliday et al., 1997; Chiesa and Manzini, 1997). In terms of production, this can be divided into production of active principles—those substances that are responsible for the therapeutic effect—and formulation of active principles and other synthesis intermediates into medicines that can be administered by physicians to patients for given illnesses.
2. According to Gambardella et al. (2000) one can distinguish at the national level, in the European pharmaceutical industry, research-based and innovative firms from the large majority of national-owned companies, frequently of a smaller size and mainly devoted to imitative production and/or commercialization. Even in developed countries, research-based firms tend to locate innovative activities close to their headquarters or in a few developed countries, such as the United Kingdom, the United States, France, or Germany, for example.
3. As observed, it could be said that this characteristic would not be exclusive of developing countries, but for them such industry organization and knowledge production geography have more serious implications, especially in terms of access to medicines.

There are some exceptions within developing countries such as India, where the domestic market is dominated by national-owned firms, which accomplish both production of active principles and medicines (Felker et al., 1993).

4. Clinical trials are intended to evaluate efficacy, effectiveness, and safety of medicines, and they are divided into preclinical (at laboratory) phase I for clinical pharmacology and toxicity (involving 20-100 individuals), phase II for investigation of the treatment effect in patients, phase III for large-scale evaluation of the treatment involving several individuals, and phase IV for postmarketing surveillance (Halliday et al., 1997).

5. It is important to mention that in 1945 the government had already abolished patent rights for pharmaceutical products.

6. This last function, however, did not survive for long.

REFERENCES

Achilladelis, B. and Antonakis, N. (2001). "The Dynamics of Technological Innovation: The Case of the Pharmaceutical Industry," *Research Policy*, 30 (4): 535-588.

Albuquerque, E.M. and Cassiolato, J.E. (2000). *As Especificidades do Sistema de Inovação do Setor Saúde: Uma Resenha da Literatura como Introdução a Uma Discussão sobre o Caso Brasileiro*. Belo Horizonte: FeSBE.

Archibugi, D. and Michie, J. (1995). "The Globalisation of Technology: A New Taxonomy," *Cambridge Journal of Economics*, 19 (1): 121-140.

Bertero, C.O. (1972). "Drogas e Dependência no Brasil, Estudo Empírico da Teoria da Dependência: O Caso da Indústria Farmacêutica," Ph.D. Thesis, Cornell University.

Bermudez, J.A.Z. (1995). *Indústria Farmacêutica, Estado e Sociedade*. São Paulo: Hucitec.

Bermudez, J.A.Z., Epsztejn, R., Oliveira, M.A., and Hasenclever, L. (2000). *The WTO TRIPS Agreement and Patent Protection in Brazil: Recent Changes and Implications for Local Production and Access to Medicines*. Rio de Janeiro: Fiocruz.

Chieza, V. and Manzini, R. (1997). "Managing Virtual R&D Organizations: Lessons from the Pharmaceutical Industry," *International Journal of Technology Management*, 13 (5/6): 471-485.

CIPR (2002). *Integrating Intellectual Property Rights and Development Policy*. London: Commission on Intellectual Property Rights.

Cockburn, I. and Henderson, R. (1997). *Public-Private Interaction and the Productivity of Pharmaceutical Research*, NBER Working Paper no. 6018. Cambridge, MA: National Bureau of Economic Research.

Cockburn, I. and Henderson, R. (2000). "Publicly Funded Science and the Productivity of the Pharmaceutical Industry." In A. Jaffe, J. Lerner, and S. Stern (eds.), *Innovation Policy and the Economy*. Cambridge: MIT Press, pp. 1-34.

Commanor, W. (1996). "The Pharmaceutical Industry and the Health Needs of Developing Countries." In World Health Organization (ed.), *Investing on Health Research and Development*. Geneva: WHO, pp. 205-212.

Etzkowitz, H. and Leydesdorff, L. (eds.) (1997). *Universities and the Global Knowledge Economy: A Triple Helix of University-Industry-Government Relations*. London: Cassel.

Etzkowitz, H. and Leydesdorff, L. (2000). "The Dynamics of Innovation: From National Systems and 'Mode 2' to a Triple Helix of University-Industry-Government Relations," *Research Policy*, 29: 109-123.

Felker, G., Chaudhuri, S., and György, K. (1993). *The Pharmaceutical Industry in India and Hungary: Policies, Institutions and Technological Development*. World Bank Technical Paper no. 392. Washington, DC: The World Bank.

Fialho, B. C. (2001). *Sistema Nacional de Inovação em Saúde*. Final Research Report. Rio de Janeiro: Fundação Oswaldo Cruz.

Fontoura, C. (1938). *Pharmacia e Pharmaceutico no Brasil*. São Paulo: Instituto Medicamenta.

Freeman, C. (1987). *Technology Policy and Economic Performance: Lessons from Japan*. London: Frances Pinter.

Frenkel, J., Reis, J.A., Araujo Jr., J.T., and Naidin, L.C. (1978). *Tecnologia e Competição na Indústria Farmacêutica Brasileira*. Rio de Janeiro: Financiadora de Estudos e Projetos.

Galambos, L. and Sturchio, J.L. (1998). "Pharmaceutical Firms and the Transition to Biotechnology: A Study in Strategic Innovation," *Business History Review*, 72: 250-278.

Gambardella, A., Orsenigo, L., and Pammolli, F. (2000). *Global Competitiveness in Pharmaceuticals: A European perspective*. Report prepared for the Directorate-General Enterprise of the European Commission. http://europa.eu.int/comm/enterprise/library/enterprise-papers/pdf/enterprise_paper_01_2001.pdf

Gray, M. and Parker, E. (1998). "Industrial Change and Regional Development: The Case of US Biotechnology and Pharmaceutical Industries," *Environment and Planning A*, 30: 1757-1774.

Halliday, R., Drasdo, A., Lumley, C., and Walker, S. (1997). "The Allocation of Resources for R&D in the World's Leading Pharmaceutical Companies," *R&D Management*, 27 (1): 63-77.

Hasenclever, L. (coord.) (2002). *Diagnóstico da Indústria Farmacêutica Brasileira*. Brasília: UNESCO/FUJB/IE-UFRJ.

Hasenclever, L., Wirth, I., and Pessoa, C. (2000). "Estrutura Industrial e Regulação na Indústria Farmacêutica Brasileira." In *Anais do XXI Simpósio de Gestão da Inovação Tecnológica*. São Paulo, USP, CD-ROM.

International Medical Statistics (IMS) (2003). "2002 World Pharma Sales Growth: Slower, but Still Healthy," *IMS Market Insight News*. London: IMS.

Liebenau, J. (1985). "Innovation in Pharmaceuticals: Industrial R&D in the Early Twentieth Century," *Research Policy*, 14 (4): 179-187.

Lundvall, B.-Å. (ed.) (1992). *National Systems of Innovation: Towards the Theory of Innovation and Interactive Learning*. London: Pinter.

Mansfield, E. (1991). "Academic Research and Industrial Innovation," *Research Policy*, 24 (2): 1-12.

Mariani, M. (2000). "The Location of R&D and the Networks of Inventions in the Chemical and Pharmaceutical Sectors." European Pharmaceutical Regulation and Innovation Systems (EPRIS) Working paper.

Michaud, C. and Murray, C. (1996). "Resources for Health Research in 1992: A Global Overview." In World Health Organization (ed.), *Investing on Health Research and Development*. Geneva: WHO, pp. 213-234.

Nelson, R.R. (ed.) (1993). *National Innovation Systems: A Comparative Analysis*. New York: Oxford.

Pécoul, B., Chirac, P., Trouiller, P., and Pinel, J. (1999). "Access to Essential Drugs in Poor Countries. A Lost Battle," *Journal of the American Medical Association*, 281 (4): 361-367.

Scheinkmann, J. (1965). "A Indústria Farmacêutica no Brasil," *Revista Brasileira de Farmácia*, 6, November/December: 323-353.

Steen, K. (2001). "Patents, Patriotism, and 'Skilled in the Art': USA v. The Chemical Foundation, Inc., 1923-1926," *Isis*, 92 (1), March: 91-122.

21

Success Factors of the CDMA R&D Project in Korea

Joong Ick Ryu and Heung Deug Hong

INTRODUCTION

Numerous national large-scale R&D projects have been established to promote national competitiveness and economic growth through the commercial introduction of new technologies (Brown and Karagozoglu, 1989). Most policymakers in the world have been devoting increasing attention to projects intended to promote research and industrial innovation (Nelson, 1984). Technology is considered to be a major competitive factor for countries at the macro level and for individual firms at the micro level (Georghiou, 1998).

Large-scale R&D projects have become the subject of attention, with discussion in relation to many of their aspects. Large-scale R&D projects have been viewed as a technology development method that has distinctive characteristics. The framework of large-scale national R&D projects—in which universities, national laboratories, and companies in competition participate to develop common technology—contains considerably different aspects from those of the traditional framework within which technology innovation depended solely on individual companies. In these large-scale national R&D projects, much attention has accordingly been paid to how to adjust technology development among competing firms.

This chapter is aimed at ascertaining the success factors of a national large-scale R&D project in Korea, and at developing policy directions for the management of national large-scale R&D projects. What are the key factors for success in national large-scale R&D projects that include various heterogeneous groups? What are the roles of the government in the process? These are questions that have not yet been clearly answered. The study investigates individual polices for mobile telecommunications technology and analyzes the digital CDMA (Code Division Multiple Access) Mobile Telecommunications Technology R&D Project, which was formulated to secure the competence of the Korean mobile telecommunications industry.

The technology policy for mobile telecommunications, including CDMA technology development, encompasses a wide range of purposes. In formulating the policy, multiple ministries in the Korean government and several organizations need to collaborate. In particular, the arena and scope of the policy have continued to expand in line with technological innovations in information technology as well as in wired and wireless telecommunications technology combined. In the light of this, this chapter covers the process of CDMA mobile telecommunications technology acquisition, development, and commercialization as well as the process of policy formation. At the same time, an analysis is conducted from a macroscopic viewpoint of how the characteristics of the mobile telecommunications policy have been utilized in the process of policy formation. In particular, the process of technological cooperation and technology learning, which was present during the technology acquisition stage, is observed. Another factor to be reviewed is the evolution process and dynamics of the mobile telecommunications industry, where Korea has caught up with advanced countries.

NATIONAL INNOVATION SYSTEM AND LARGE-SCALE R&D PROJECT

Definition and Structural Elements

A national innovation system is a nationwide network of innovators working together as a cohesive entity (DeBresson and Amesse, 1991; Freeman, 1991). The focal point here is that technology learning, or the process of creation, dissemination, and application of technological knowledge, is patterned and systemized (Lundvall, 1992).

As shown in Figure 21.1, the process of technology learning in enterprises and that of technology creation between enterprises or the interaction between enterprises, universities, and government research institutes constitutes a uniquely systemized pattern in a nation.

Four typical structural elements exist in a national innovation system (NIS). First, it is national R&D capacity that has grown on the basis of universities and public institutes funded mostly by government, public foundations, and sometimes nonprofit organizations. Official R&D elements come next: the R&D capacity of enterprise research institutes included in structural elements of innovation systems accumulated by enterprise. This also includes core engineering design and other types of innovational know-how on the formational basis of enterprises. Third, there are educational organizations that provide not only scientists and engineers, but also technicians who possess appropriate skills. Last, it is a science and technology policymaking organization that supervises the facilitation of R&D in the public sector and enables some coordination with business-sector R&D (OECD, 1997). In this respect, a national innovation system is the collective and comprehensive definition that includes related actors participating in R&D as depicted in Figure 21.1.

Figure 21.1
Elements of a National Innovation System

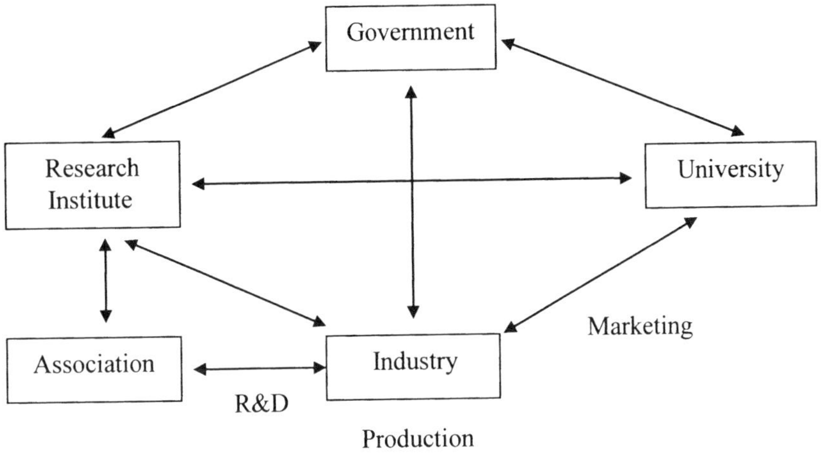

Source: Carlsson and Jacobsson (1994).

The large-scale R&D project is an instrument of an NIS for accomplishing technology innovation with a specific purpose through R&D or technology induction performed by each main member with collective efforts, technology, and resources. We need to define the characteristics of a large-scale R&D project from the viewpoint of resource investment, the management process, and production efficiency. In terms of resource investment, large-scale R&D projects requiring large amounts of resources (manpower, facility, time, technology, information) (Kerzner, 1989; Sykes, 1990) and the issues of resource allocation, technology ability, management ability, and government support are reviewed at the global level (Horwitch, 1990). In general, there are not many experienced experts in the government and industry because there are only a few successful cases of such projects (Sykes, 1990).

In terms of the management process, many people and organizations with specific goals are involved in large-scale R&D projects (Horwitch, 1984). However, any such project should be verified in advance through a thoughtful review process, because it takes a long time to complete a project and uncertainty exists in resource supply and market needs (Sykes, 1990; Shibata, 1984; Sayles and Chandler, 1971). Many large-scale R&D projects fail to enter the market due to the large scale and complexity of the knowledge and technology, as well as political and environmental difficulties. Many large-scale projects that have succeeded in entering the market have failed to get satisfactory results. Due to the difficulties in the management of large-scale organizations, political methods are required to adjust the benefits of the participating organization.

The objectives of national large-scale R&D projects are various and cannot be generalized due to their different characteristics. The objectives of large-scale R&D projects carried out can be summarized country by country. The large-scale industrial technology R&D system that started in Japan in 1966 had four objectives: advancement in the country's industrial structure, the strengthening of international competition, the rational development of natural resources, and the prevention of industrial pollution (Tanaka, 1989). In addition, it stated that R&D for national security, the protection of the environment, the medical system, and welfare should be promoted as a role at the national level.

Success Factors of Large-scale R&D Projects

As we have seen in large-scale research projects, there are many management factors to be considered because of the participation of many organizations. First we consider the success and failure factors of large-scale research projects, because the starting point of this chapter was to research the success and failure factors of technology innovation. Sayles and Chandler (1971) demonstrated the controllability of large and complex R&D projects through the arrangement of organization management characteristics within the success of NASA's important technology program. They said that the key factors for the success of projects are the ability of the project manager, effectiveness of schedule management, the control system, the responsibility system, communication, monitoring, feedback, and continuous participation. Afterward, many researchers studied large-scale project management (Baker and Wilemon, 1977; Quinn, 1979; Horwitch, 1984, 1990; Sykes, 1990; Morris, 1990). They categorized success factors of large-scale technology research projects into characteristics of technology (project), movement/ structural factors, and environmental/cultural factors. However, the more recent studies of large-scale research projects have emphasized the strategic utilization of technology and project characteristics (Morris, 1990; Sykes, 1990). In the management of small-scale technology innovation, the importance of strategic management was not fully understood. However, large-scale research projects are concerned with the importance of the strategy, compatibility with the environment, and the arrangement of the management system. So, in order to improve the result of technology innovation, it is important to take a systematic view when setting up and executing strategies. However, this systematic view has not been examined in detail in the aforementioned studies. Thus we will introduce strategic factors, including the characteristics of technology and projects based on the previously presented factors of technology strategy in this research. Movement/structural factors are classified as managerial factors and environmental/cultural factors are classified as environmental factors, as shown in Table 21.1.

Table 21.1
Success Factors of Large-scale R&D Projects

Classification	Strategic Factors	Managerial Factors	Environmental Factors
Sayles & Chandler (1971)		- Ability of project manager, Schedule control, Control and responsibility system, Communication, Monitoring and feedback	- Continuous participation in project
Baker & Wilemon (1977)		- Leadership style of project manager - Complication adjustment, - Determination style, - Variation of organization design and project authority	- Superior organization of project team, Customer relationship with outside organization
Sykes (1990)	- Preparation of situational plan - Project execution by several experienced specialists - Insurance of appropriate members	- Enough preliminary examination, - Observability of whole project, - Activation ability and evaluation of availability	- Maximum observation of objectivity
Horwitch (1984, 1990)	- Valuable project (effect to expense)	- Effective management activity	- Right relationship among organizations, Political environment, Active management of resource supply
Morris (1990)	- Effective management of strategy and technology plan	- Active development and activation - Accurate plan of project period and steps	- Effective bureaucracy and political motion

Conversely, Shibata (1984) identified a set of obstacles to large-scale R&D projects as uncertain resources, supply and market demands, diversity of sense of value and regulation, lack of preparation of cooperation between government and nongovernment organizations, rupture of communication, rigidity of organization, insufficient data and information, vague policy, and difficulty in agreeing.

The Research Framework

The overall research framework to be used in this chapter is shown in Figure 21.2. Drawing on Pinto and Slevin's (1989) four stages model of the research process, this chapter uses the four stages process model of the formation of concept, planning, the execution stage, and the completion stage. These are revised in relation to the CDMA case and the Korean policy structure and situation. The architecture of the framework is divided into three stages based on the project development process: the policymaking stage of the project, the R&D execution stage, and the commercialization stage as depicted in Figure 21.2. The detailed research factors will be discussed in the next section.

Figure 21.2
Research Framework

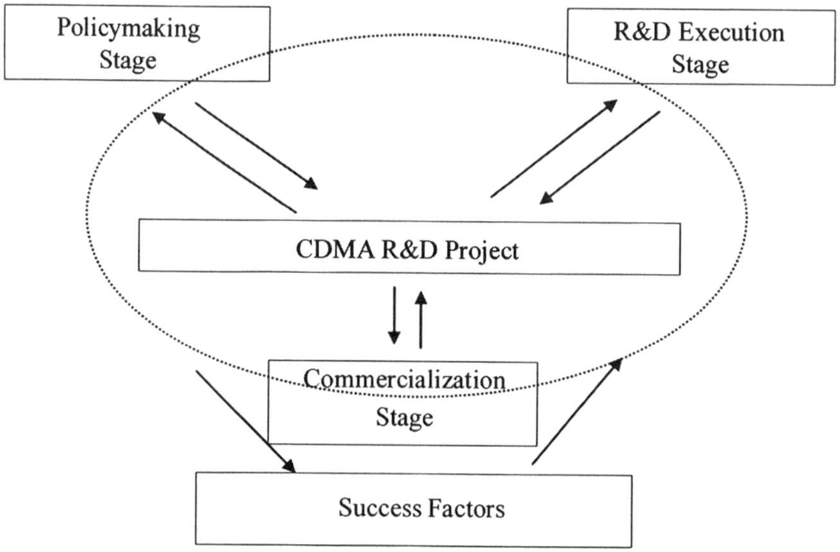

On the basis of both the in-depth analysis and the results analysis, a strategic model for a large-scale R&D project is presented. This is then applied to R&D projects in the high-technology area so as to assist its successful progress. This result can also be used for the selection of research projects within limited resources, for the establishment of a support policy to promote novel technology development, or for the driving of national large-scale R&D projects by other technology-developing countries.

Taking these factors into consideration, we asked the experts in the government several questions: Why was the CDMA technology adopted? What were the difficulties faced in the policymaking and problem-solving processes? What was the role of the government and the success factors identified by government? We also asked experts from various research institutes the following questions: What was the process of the development of the prototypes? What were the difficulties faced developing prototype systems? What was the problem solving process? and What were the success factors identified by research institutes?

CDMA PROJECT AND THE BACKGROUND

Why CDMA?

The project was one of the most significant large-scale and successful projects in the history of Korean R&D policy. It was also the result of the collective efforts made by the country as a whole, where various R&D organizations such as universities, government-supported research institutes, industries, and several ministries of the Korean government participated. For the areas where domestic R&D capacity was lacking, international cooperation was actively pursued. Moreover, its technological and economic outcomes were clear and evident.

The CDMA R&D project is a good example because it contains all the elements that need to be considered in developing R&D projects. These elements are: (1) the CDMA project is a national large-scale R&D project that has been successfully completed; (2) it comprises the whole cycle from research to commercialization; (3) R&D principals from industry, university, and public research institutes joined together and worked effectively from the very early stages of R&D; (4) the key basic technology was secured by technology transfer from a foreign firm; and (5) based on the real need for a national telecommunications system, the government induced the participation of system suppliers in the project. Accordingly, the view was taken that if the CDMA R&D project, a representative Korean large-scale R&D project, was selected and its success factors analyzed in such areas as the policy decision process, the R&D process, the R&D management method, the technology acquisition process, the technology transfer process, and commercialization and servicing, this would provide effective policy alternatives on future large-scale R&D projects. On the basis of these viewpoints, the CDMA project was selected as the case for study.

The Background of the Project

Starting from the end of 1989, the basic policy in relation to the telecommunication industry has dramatically changed. In 1988 the Ministry of

Communications (whose name was changed to the Ministry of Information and Communication in 1994) presented a policy that adopted free competition and privatization of the telecommunications industry. This meant that the government monopoly was surrendered and the telecommunications industry was rebuilt as a private management system. In 1989 the Ministry of Communications (MOC) officially announced the conversion of the government monopoly to a free competition system, and established the first plan for the structural reform of the telecommunications industry in July 1990 (Suck-Hwan Yoon, 1996). By adopting free competition in the fields of international communication services, mobile telecommunications systems, and valued communication, the structural reform allowed most sectors of the telecommunications industry to compete freely.

Through the new basic policies, the structural reform in the mobile telecommunications industry has been achieved. The Ministry of Communications announced a new policy to approve a new provider for the nonmonopoly system in the mobile telecommunications service by revising the Law on Information and Communications in 1991. Also it designated the second mobile service provider and put it into competition with Korea Mobile Telecom (KMT) in order to provide a better service to customers at a reasonable price, with its aim to strengthen the competitiveness of the domestic providers in the face of international communication service providers. However, in the structural reform of the telecommunications industry, it was decided not to allow the existing provider to set up the new service, as in the case of the international phone service of DACOM Corporation, but to select a new provider.

ANALYSIS OF THE SUCCESS FACTORS OF THE CDMA PROJECT

This research identifies the success factors that strongly influenced the success of the CDMA R&D project sequentially in the stage of policymaking, the R&D execution, and the commercialization as described in a previous section on Research Framework.

In the policymaking stage, the target of R&D activities and strategies for technology selection and technology acquisition and a plan for financial resource and human resources were established. In this stage, the Ministry of Information and Communication (MIC) played a major role in laying the cornerstone by establishing a plan for developing CDMA technology and creating the R&D environment.

In the R&D execution stage, the Electronics and Telecommunications Research Institute (ETRI), in charge of the project, played a key role. In this stage, the technology was introduced, studied, and researched, and a prototype system was developed.

In the commercialization stage, technology transfer, certification, verification, and field tests for commercial services were conducted. Commercial products were developed and produced. In this stage, designated manufacturers and wireless service providers took up a crucial role.

Policy Decision-Making Process of the Project

The choice of CDMA technology was made in Korea to satisfy the increasing demand for mobile phones, to construct the basis for advancement, to support domestic high technology, and to meet the demands of the changing global world communications market. For instance, it is essential to fit this technology to a global trend of better calling quality as compared with the analog method and the ability to accommodate new subscribers. Here, the selection of CDMA is understood as the challenge for the unknown world (Jung-Uck Seo, 1999). Furthermore, the choice of CDMA technology is explained as the outcome of the study of technology coping with a situation of uncertainty and crisis (Wi-Chin Song, 1999). Although the ETRI had the initiative in developing a digital mobile telecommunications system, it did not have any basic technology related to wireless communications, so it was in a difficult situation. After consulting with the MOC, the ETRI changed its strategy from self-development to co-working with a foreign company. Thus, it acquired the core technology for CDMA from Qualcomm and effectuated the technology with domestic manufacturers in order to co-develop new systems. The introduction of the technology and its development took place concurrently. This parallel strategy is one of the major success factors in the development of CDMA.

For TDX, the Korea Institute of Science and Technology (KIST) research group developed a private exchanger and applied it practically but the result was negative.[1] Technology acquisitions in technologically developing countries that have no core technology must keep pace with the introduction of technology and self-development. Preparing the ability to absorb new technology is important as well. The effort to absorb new technology must be exerted in various ways (Gug-Hyeon Cho, 1996). After conclusion of the co-development contract with Qualcomm in 1991, the ETRI sent its researchers to Qualcomm and made an effort to acquire the new technology as soon as possible. After the researchers gained the core technology from Qualcomm, they returned to Korea to develop CDMA and helped to spread the technology.

In addition to these factors, technology ability and positive attitude were also determinants of the result of technology acquisition. Here, time was conceived of as an important factor. For example, shortening the two-year period of development for service by using domestic production was analyzed as a kind of devised crisis, which reflected a strategic intention to acquire technology through renovating the technology development process (Wi-Chin Song, 1999). The U.S. selection of a CDMA standard, and Korea's decision to select the CDMA design as well, were seen as related actions. Specifically, CDMA had a positive effect on Canada, South America, and Australia, countries politically influenced by the United States. In Korea, the invisible and large power of U.S. influence gave underlying meaning to the success of CDMA (interview with Han-Ju Kim, 2001). Of course, these matters can be understood as the effects that have arisen from the introduction of foreign technology because the CDMA R&D project had little connection with existing business

(Chul-Won Lee, 1994). In addition to these points, the control and cooperation for technology acquisitions and self-development among driving members, research institutes, enterprises, and the like are conceived as an important factor in CDMA development.

The purpose of this section is to understand the strong motivation of the MOC and the ETRI, the major research organization. There were many objections and problems, and much resistance to the policy—for example in relation to the direction and process of technology selection, R&D, and commercialization. The MOC, however, settled the direction and standard of policy and maintained it in a consistent manner. We can see the strong will of the government that was ready to cancel a business license of Shinsegi Telecommunication for apparently trying to collude with the foreign company and stockholders (Wi-Chin Song, 1999; Geum-Ae Jung, 1999[2]). The strong will of the government to improve domestic industry using core technology worked.

When the Ministry of Information and Communication decided to develop the mobile telecommunications technology by using CDMA, the core organization was the ETRI. This organization had experience of TDX development and large manpower in the information and communications field. The ETRI made policy decisions for the Ministry of Information and Communication and performed practical and core duties in project management, technology, development, and testing. The president of the ETRI at that time checked the rate of development every week. Responsible researchers were highly involved in researching and reporting the progress and solving the problems at the beginning of every week. The complete support and encouragement from CEOs increased researchers' morale so that they were further motivated. This encouragement motivated researchers to concentrate on their duty to develop the CDMA successfully. From these facts, we know that the responsibility and willingness of the ETRI as a policy conductor was a principal factor in the successful development of CDMA.

The goal must always be clear in order to perform a national R&D project successfully. Horwitch (1978) said that both "unfocused growth" and "uncontrolled growth" gave rise to problems and only "controlled growth," like the Apollo program, which considered the life cycle would succeed (Gug-Hyeon Cho, 1996).

As explained previously, the CDMA project was established and promoted based on the development program. In the early research and technology stage, when the method of mobile telecommunications had not been decided, there were uncertainties about research on the mobile telecommunications concept and the promotion of wireless-method technology research, because the background of self-researched technologies concerning wireless communication had not yet been established. However, after the CDMA method was decided upon in conjunction with Qualcomm, the goal was decided based on the users' requirements. At this point, the goal of the CDMA project was clear and this fact affected the result.

R&D Execution Process of the Project

In order to complete R&D for the new technology of information communications like CDMA, there was a need not only for the development of core technology, but also implementation of tasks to support effective R&D management including research management, schedule management, and the management of the results at every stage of the research. Also, systematic management of laboratory equipment for R&D was required. So, for the joint development project to achieve the best result, strategic management was essential, and a proper solution and response to various problems occurring in the process of the project were needed.

In the management of the CDMA project, the methodology obtained from the TDX R&D experience was systematically utilized. The CDMA project was based on this experience, and system development was promoted systematically and effectively. The CDMA development system is similar to the TDX development methodology in that they both emphasize the configuration management system, the organization of document creation, the management guide, and test activity.

The project research management was divided into several components, such as research performance management, which aided joint research environment, technology management, which analyzed and maintained the R&D progress and research results, and post management, which preserved final research results and led to commercialization. First, the objective of research performance management was to predict a series of processes that were necessary for the joint research and development performance, to prepare various systems for research performance, and to solve problems occurring in the course of R&D. Research performance management consisted of determining annual R&D projects, the establishment and execution of various guiding principles, and the management of research funds. Second, the objective of technology management was to understand the present performance of the joint research development project, to check actual performance results, to maintain configuration management of intermediate and final research results, and to produce in print the relevant information, knowledge, and know-how embedded in all researchers' brains. Technology management consisted of technology document management, the management of research project results,[3] and the holding of technology interchange meetings. Third, the final assessment of research results, which was performed at the level of postmanagement for domestic security of technology and mutual technology exchange, was divided into intermediate, final, and total assessment and was performed by objectively reproducing the results of the development of joint research. In the CDMA project, the expected results of the research goal were made as clear as possible in preparation for the assessment, and the intermediate target was established from the research goal. In the assessment, the first verification team was brought together and they presented technology documents to the assessment committee concerning specific problems after checking them with related documentary evidence. Then the assessment committee estimated the research results objectively by comparing the initial development goal and the research results.

CDMA research and development required life-cycle management, including the digestion of introduced technology, the development of prototypes, test evaluation, and mass production, as well as quality warranty and operation. The ETRI prepared various guiding principles when developing the digital mobile telecommunications system and has promoted effective and systematic R&D in relation to large-scale projects.

In the development process, the ETRI and three companies obtained the technology from Qualcomm in 1992, and conducted product development separately, so the direction of development varied from company to company. For strategic management of the CDMA project, initially incentives were created for the participation of companies and secure competition. This followed the government's policy that mobile telecommunication technology development was to be a pan-national government-run project, utilizing all domestic technologies and resources. In the selection of participating companies and the scope of their participation, a regular incentive system was used, so that a system of cooperation was devised, which was an important factor in the good results of the project because it encouraged a continuous cooperative evaluation system and the rapid reflection of its results. Also, the incentive system could stimulate competition. Through this process, intended technology development could be accomplished rapidly.

Table 21.2
Mobile Telecommunication Technology Development Policy Participants

<table>
<tr><th>Division</th><th>Organization</th><th colspan="2">Interest-related participants</th></tr>
<tr><td rowspan="3">Policy Decision</td><td>Subject</td><td colspan="2">Ministry of Information and Communication</td></tr>
<tr><td>Cooperation</td><td colspan="2">Central government departments such as the Ministry of Commercial Industry, the Ministry of Science & Technology</td></tr>
<tr><td>Related</td><td colspan="2">The Blue House, Congress, the Board of Audit and Inspection</td></tr>
<tr><td rowspan="3">Project Execution</td><td>Quasi-public</td><td>Project Management/ Volunteer</td><td>ETRI/Project Management Organization</td></tr>
<tr><td rowspan="2">Private Company</td><td>Communication Company</td><td>Communication Company (KMT, Shinsegi communication)</td></tr>
<tr><td>Equipment Manufacturing</td><td>Samsung, LG, Hyundai</td></tr>
</table>

As shown in Table 21.2, groups participating in the policy process had various viewpoints relating to institutes and organizations. As a result, CDMA technology development policy was a pan-national policy; decision making was spread horizontally, and each participating group participated based on its organizational interest. Thus, mutual application based on the cooperation of participating groups was needed, and this was a clear example of mutual operation to achieve the common aim (Kyung-Ha Seo, 1996). In particular, government, project management organizations, research institutes, and companies were all awareness of common problems, and endeavored to achieve the common aim (Geum-Ae Jung, 1999).

Here the government decided the principles and policies to reduce uncertainty and danger, and to play a supporting role. The ETRI managed cooperative research by private companies and prepared the environment in which knowledge and technology could be integrated. Designated manufacturers took part in the technology development with supplied funds, human resources, and market information.

As we can see from Table 21.3, the policy coordinators in the course of the development of the mobile telecommunications technology were technical bureaucrats and experts who had special ability and the power to promote policy, and they played the main role as decision makers and coordinators. It was known that successive ministers of information and communication recognized the mobile telecommunications technology development policy as having the greatest priority and placed emphasis on it.

Table 21.3
Coordinators of the CDMA Project

Year	Policy Decision	Implication	Policy Execution	Implication
1991			Head of ETRI	Expert knowledge
1993	Minister of Communications	Leadership of information	The head of project management organization/ Head of ETRI	Experience, promotion force/expert Knowledge and Experience
1994		Excellent insight and Expert knowledge	The head of business management body/ Head of ETRI	Expert knowledge and driving force
1995	Minister of Information and Communication	Political leadership		
1996		Administrative management	The head of the business management	

In addition, a joint information utilization system was constructed. This can be justified by the fact that the infrastructure was constructed so that it could perform common operations in mobile telecommunications technology-related organizations. This also had far-reaching effects in practice. Consequently, personal exchanges and information sharing were possible through the joint information utilization system.

Commercialization Process of the Project

The performance process and research results of the joint research development project were important, but postmanagement activities including management, maintenance, and commercialization of developed technology must also be emphasized.

Now let us examine the process of commercialization of CDMA. The ETRI and Qualcomn agreed on the joint development of CDMA in April 1991 and took the first step in joint development in August 1991. The second step in joint development started in July 1992. In December 1992, the ETRI chose Samsung, LG, Hyundai, and Maxon as the designated manufacturers and proceeded with the development of the technology. In June 1993, a new mobile telephone provider, Shinsegi Telecom, Inc., selected the CDMA system as a medium of communication. In August of the same year, Qualcomn entered into a technology execution contract with local companies, and in November a CDMA-type technology standard was achieved. In February 1995, Samsung Electronics Co., Ltd. finished its first commercialization test (108 items); and in May 1995, LG Information & Communications Co., Ltd. finished its commercialization test for 830 items. Also in November, the MIC through public discussion determined that CDMA should be the standard technology of Personal Communication Services (PCS), because a single standard is more desirable than multistandards when researching ability is restricted. Finally, in January 1996, in the Incheon and Bucheon regions, commercial CDMA service was started. The important aspects in this process were that, in the commercialization process, the end-provider, KMT, joined in the system developing process with the research institutes and the producing companies. Through these, the CDMA project changed into a self-controlled competitive developing system. In addition, KMT acquired the specific contents of the CDMA system and its operating staff were confident of their work. So the operating know-how that the staff of KMT brought to the CDMA commercialization process helped the three late-starter PCS companies. It played a decisive role in allowing them to smoothly operate their service system without great problems (Jung-Uck Seo, 1999).

Four manufacturing companies were in competition to supply the products to KMT, the end-provider (Geum-Ae Jung, 1999). In addition, the situation where Korea Mobile Telecom and the secondary telecommunications provider, Shinsegi Telecom, Inc., each chose one company as a supplier worked to create competitive pressure, thereby preventing a "free ride" in the public technology development. At this point, each system manufacturing company tried its best to be chosen as a supplying company to telecom service enterprises, and the technology learning effect was maximized. As there were technology developments based on purchasing power, it was estimated that the commercialization of CDMA could be viable. It was possible to confirm through a series of policies in relation to such aspects as know-how about consuming, the end-providers' specs (needs), certification and verification about system commercialization, and the introduction of a competition system.

SYNTHESIS AND CONCLUSION

The CDMA R&D project mobilized all research actors in the field of mobile telecommunications in a national innovation system in Korea. It was a network of various public and private organizations and systems in which new technology was introduced, modified, and changed through the individual activity and interactive activities of CDMA-related R&D innovation subjects.

In the CDMA project, both methods were employed. That is, core technology was introduced from Qualcomm and self-study was performed at the initial stage, but, in the second half of the development, the ETRI and each company conducted self-development. Third, structural selection can be performed in a relationship of cooperation or separately according to the cooperation levels between involved organizations, government, research institutes, universities, and private enterprises. The combination of competition and cooperation was very important for the successful introduction and R&D of the technology, which had been selected through careful survey. In the process of the CDMA R&D, competition and a cooperation system were established and affected the results of the project both directly and indirectly. In other words, the CDMA development process began with a separated relationship, but with the progress of R&D activities, it was transformed into an interrelated relationship. In particular, the establishment of the Mobile Telecommunication Technology Development team to manage various R&D projects and organizations created an atmosphere of free competition between participating organizations.

In the processes of technology selection and the development and commercialization of the CDMA system, the ETRI played a prominent role in the technology innovation system. The ETRI, a government-supported research institute, is a unique system in Korea.

Based on such a discussion, the conclusion can be drawn that the process of the acquisition of CDMA technology is unique. As shown in Figure 21.3, the CDMA technology acquisition model suggests that it well reflects the situation in which policymaking on technology transfer should be done, the structure and circumstances of the R&D system, characteristics of the executive officers and executive systems, and the manner of technology implementation. In addition, the factors that produced the result of the R&D are interactive with each other in the processes of application and modification of policy and the adaptive adjustment of the policy setting. The process of the adaptive adjustment of policy setting eventually produced the intended results, technology, products, and service, and then demonstrated the possibilities of the Korean CDMA technology acquisition model to other countries.

Figure 21.3
The Success Factors of the CDMA R&D Project in Korea

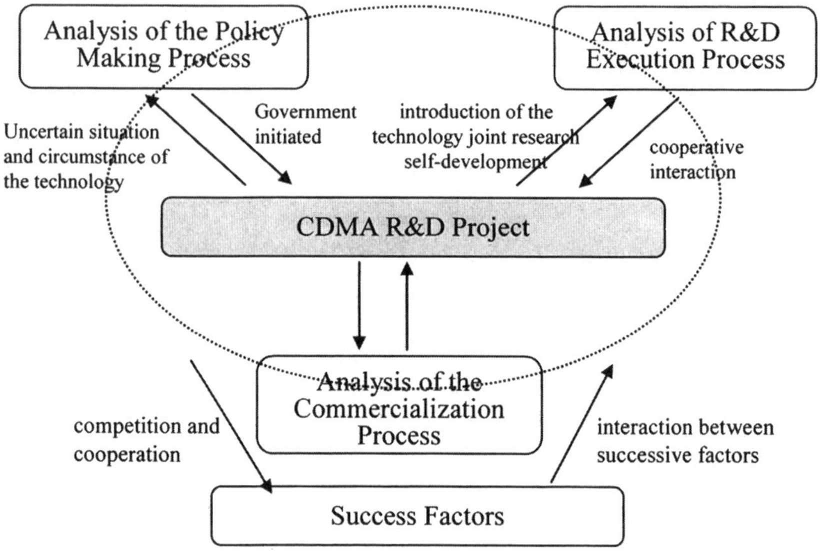

As described in the above case study and analysis of the Korean CDMA R&D project, the categories of successful factors in technology development were expanded from previously defined categories obtained in case studies of developed countries' R&D projects. The success factors of existing models do not fit the R&D environment well, even though they are determined depending on research objectives and conditions. In addition to the success factors identified in previous research, other success factors were identified through this research: (1) the strategy for technology selection and technology acquisition; (2) the strong motivation of the policymakers; (3) the clarity of goal setting; (4) strategic project management; (5) coordination and cooperation among operative actors; (6) the experience and know-how gained from the TDX project; (7) the strategy for commercialization; and (8) the participation of consumers and the principle of competition. In the course of the project development, the factors interact with each other leading to successful implementation. This can be understood as a unique aspect discovered in successful R&D projects of a developing country, namely Korea.

NOTES

1. The TDX development also succeeded by pursuing a parallel development strategy of technology introduction and self-development for TDX through trial and error.
2. Geum-Ae Jung published a novel *Our Mobile Phone is a Slam-dunk*, based on the fact that it contains an account of the CDMA development success process.

3. The management of research project results is divided into monthly result management, semiannual result management, and annual result management, and includes the inspection of other R&D promotion results and of results of activities financed by research funds and others.

REFERENCES

Baker, B. N. and Wilemon, D. L. (1977). "Managing Complex Programs: A Review of Major Research Findings," *R&D Management*, 8 (1): 23-28.

Brown, W. B. and Karagozoglu, N. (1989). "A Systems Model of Technological Innovation," *IEEE Transactions on Engineering Management*, 36 (1), February: 11-16.

Carlsson, B. and Jacobsson, S. (1994). "Technological Systems and Economic Policy: The Diffusion of Factory Automation in Sweden," *Research Policy*, 23 (3): 235-249.

Chul-Won Lee (1994). *Technology Acquisition Strategy of Designated Manufacturer for Commercialization of Result of Collaboration R&D*. STEPI, Seoul.

DeBresson, C. and Amesse, F. (1991). "Networks of Innovators: A Review and Introduction to the Issue," *Research Policy*, 20 (5): 363-379.

Freeman, C. (1991). "Networks of Innovators: A Synthesis of Research Issues," *Research Policy*, 20 (5): 499-514.

Georghiou, L. (1998). "Global Cooperation in Research," *Research Policy*, 27 (6): 611-626.

Geum-Ae Jung (1999). *Shot the Dunk-shot of Korean Mobile Phone*, Soochaewha Publisher.

Gug-Hyeon Cho (1996). "Analysis of Critical Factors in National R&D Projects Performance—With a Focus on the TDX R&D Project," unpublished Ph.D. Thesis, Department of Administration, Korea University.

Horwitch, M. (1978). *Uncontrolled Growth and Unfocused Growth.* Cambridge, MA: Harvard Business School Press.

Horwitch, M. (1984). "Managing Large-Scale Programs: The Managerial Dilemma," *Technology in Society*, 6 (2): 161-171.

Horwitch, M. (1990). "From Unitary to Distributed Objectives: The Changing Nature of Major Projects," *Technology in Society*, 12 (2): 173-195.

Jung-Uck Seo (1999). *History of CDMA Mobile Communications—History of Korean Information and Communication, Those who are involved in opening the future world*, SK Telecom.

Kerzner, H. (1989). *Project Management: A Systems Approach to Planning, Scheduling and Controlling*, 3rd ed. New York: Van Nostrand Reinhold.

Kyung-Ha Seo (1996). "A Study on the Strategic Alliances of the Digital Cellular Communication System: On the CDMA System Development," unpublished Masters Dissertation, Yonsei University.

Lundvall, B.-Å. (1992). *National Systems of Innovation: Towards a Theory of Innovation and Interactive Learning*. London: Pinter Publishers.

Morris, P. W. G. (1990). "The Strategic Management of Projects," *Technology and Society*, 12 (2): 197-215.

Nelson, R. (1984). *High Technology Policies: A Five-Nation Comparison*. Washington, DC: American Enterprise Institute.

OECD (1997). *National Innovation Systems*. Paris: OECD.

Pinto, J. and Slevin, D. (1989). "Critical Success Factors in R&D Projects," *Research-Technology Management*, 32 (1): 31-35.

Quinn, J. (1979). "Technological Innovation, Entrepreneurship and Strategy," *Sloan Management Review*, 20 (3): 19-30.

Sayles, L. R. and Chandler, M. K. (1971). *Managing Large Systems: Organizations for the Future*. New York: Harper & Row.

Shibata, Y. (1984). "Toward a Policy Guidance System for Complex Innovation." In H. Eto and K. Matsui (eds.), *R&D Management Systems in Japanese Industry*. Amsterdam: North-Holland Publishers, pp. 35-41.

Suck-Hwan Yoon (1996). "A Study on the Policy Network in the Information and Telecommunications Field," unpublished Ph.D. Thesis, Choongnam University.

Sykes, A. (1990). "Macro Projects: Status, Prospects and the Need for International Cooperation," *Technology in Society*, 12 (2): 157-172.

Tanaka, M. (1989). "Japanese-Style Evaluation Systems for R&D Projects: The MITI Experience," *Research Policy*, 18 (6): 361-378.

Wi-Chin Song (1999). "Technology Innovation Pattern of Mobile Communications Industry," *Policy Research*, 99-40. STEPI, Seoul.

Index

About the Contributors

Philip E. Auerswald is Professor at the School of Public Policy, George Mason University. Contact: auerswald@gmu.edu

Ioanna Boulouta is a MPhil student in Technology Policy at Cambridge University, United Kingdom. Contact: ib235@cam.ac.uk

Cristina Casanueva is a researcher at the Research Institute for the Development of Education (INIDE), Universidad Iberoamericana, Mexico. Contact: cristina.casanueva@uia.mx

Jae-Yong Choung is Professor in the Faculty of IT Business, Information and Communications University, Korea. Contact: jychoung@icu.ac.kr

Pedro Conceição is Assistant Professor at the Instituto Superior Técnico, Technical University of Lisbon, and a researcher at the Center for Innovation, Technology and Policy Research at IST. He is also a Deputy Director and Senior Policy Analyst, Office of Development Studies, United Nations Development Programme (UNDP), New York. Contact: pedroc@dem.ist.utl.pt

Gabriela Dutrénit is Full Professor of the Master and Doctorate in Economics and Management of Technology at the Metropolitan Autonomous University, Campus Xochimilco (UAM-X), Mexico.
Contact: dutrenit@correo.xoc.uam.mx

José Rui Felizardo is CEO of Inteli—Intelligence for Innovation.
Contact: jrf@inteli.pt

Beatriz C. Fialho is a researcher and graduate student at the Technological Innovation and Industrial Organization Area at COPPE, the Federal University of Rio de Janeiro, Brazil. Contact: beafialho@yahoo.com.br

Pliny Fisk is Co-Director of The Center for Maximum Potential Building Systems, Austin, Texas, and Fellow for Sustainable Urbanism joint appointment in Architecture, Landscape Architecture and Planning, Texas A&M University. Contact: pfisk@cmpbs.org

Rashmi A. Gehani is Professor at the College of Literature, Science and the Arts, The University of Michigan. Contact: rgehani@umich.edu

R. Ray Gehani is Associate Professor of Management and International Business at the College of Business Administration, The University of Akron. Contact: rgehani@uakron.edu

David V. Gibson is Associate Director and Nadya Kozmetsky Scott Centennial Fellow at the IC^2 Institute, The University of Texas at Austin. Contact: davidg@icc.utexas.edu

Carlos A. Góngora-Caamal is co-founder and regional director of Innovateur Capital at Monterrey, Mexico. He has worked as an independent consultant for diverse innovation and private equity consulting firms, government agencies, tech-based ventures, and business development entities in Mexico. Contact: carlosagc@exatec.itesm.mx

Lia Hasenclever is Professor at the Institute of Economics, Federal University of Rio de Janeiro (UFRJ), Brazil. Contact: lia@ie.ufrj.br

Lauro Noboru Hassegawa is Senior FAE Consultant for Advanced Micro Devices, Inc. in Brazil, and has worked in international projects such as the Brazilian Voting Machine and Motherboard Design in the Silicon Valley, California. Contact: lhassegawa@yahoo.com.br

Manuel V. Heitor is Secretary of State for Science, Technology, and Higher Education at the Portuguese government since March 2005. He is Full Professor at the Instituto Superior Técnico in Lisbon, and was the founding director of the Center for Innovation, Technology and Policy Research, IN+. He is a Senior Research Fellow of the IC^2 Institute, The University of Texas at Austin. Contact: mheitor@ist.utl.pt

Salvador Padilla Hernández is Full Professor at the Faculty of Economics Vasco de Quiroga, Universidad Michoacana de San Nicolás de Hidalgo, Mexico. Contact: spadilla@zeus.umich.mx

Heung Deug Hong is Assistant Professor at Miryang National University, Korea. Contact: hdhong@mnu.ac.kr

Hye-Ran Hwang is with the Daejon Development Research Institute, Korea. Contact: hrhwang@djdi.re.kr

Alejandro Ibarra-Yunez is Professor of Economics and Public Policy at the Graduate School of Business and Leadership (EGADE), Instituto Tecnológico y de Estudios Superiores de Monterrey (ITESM), Mexico.
Contact: aibarra@itesm.mx

Matti Lähdeniemi is Professor at the Satakunta Polytechnic, Finland.
Contact: matti.lahdeniemi@samk.fi

Kari Laine is Professor at the Satakunta Polytechnic, Finland.
Contact: kari.laine@samk.fi

Enrique Díaz de Léon López is the Director of Innovation and Technology at the Institute of Technological Entrepreneurs. He is also an MBA professor of Strategic Management and Entrepreneurship at the Instituto Tecnológico y de Estudios Superiores de Monterrey (ITESM), Campus Guadalajara, in Zapopan, Jalisco, Mexico. Contact: ediazdeleon@itesm.mx

María de la Luz Martín is a Ph.D. student in Social Sciences at El Colegio Mexiquense, A.C., Mexico. Contact: luzmc@yahoo.com

José M. C. Mello is Visiting Professor in the Production Engineering Graduate Program and associate researcher at the Nuclei of Studies in Innovation, Knowledge and Work, both at the Federal University Fluminense, Brazil.
Contact: josemello@aol.com

William H. Melody is Managing Director of Learning Initiatives on Reforms for Network Economies (www.lirne.net), and the World Dialogue on Regulation for Network Economies (www.regulateonline.org). He is a Visiting Professor at the Center for Information and Communication Technologies, Technical University of Denmark; the Media @LSE Programme, London School of Economics; and the LINK Centre, University of Witwatersrand, South Africa. He is Emeritus Professor, Economics of Infrastructures, Delft University of Technology, the Netherlands. Contact: melody@cti.dtu.dk

John Motloch is Professor in the Department of Landscape Architecture and Director of the Land Design Institute at Ball State University.
Contact: jmotloch@bsu.edu

Pedro Pacheco is Associate Professor in the Department of Architecture, Instituto Tecnológico y de Estudios Superiores de Monterrey (ITESM), Campus Monterrey, Mexico. Contact: ppacheco@itesm.mx

Carlos Quandt is Professor at the Graduate Program of Business Administration, Pontifícia Universidade Católica do Paraná (PUC-PR), Brazil.
Contact: quandt@rla01.pucpr.br

Rodolpho Ramina is the coordinator of the Ecodesign Program and the Kyoto Protocol Initiative at the Industrial Federation of Paraná State, Brazil.
Contact: rhramina@uol.com.br

Apiwat Ratanawaraha is a Ph.D. candidate of the International Development and Regional Planning Group, Department of Urban Studies and Planning, Massachusetts Institute of Technology. Contact: rapiwat@mit.edu

Pablo Rhi-Perez is Associate Professor at the School of Business Administration, The University of Texas at Brownsville.
Contact: rhiperez@utb.edu

Victoria E. Rodriguez is Vice Provost and Dean of Graduate Studies at The University of Texas at Austin and holds the University's Ashbel Smith Professorship at the LBJ School of Public Affairs.
Contact: v.rodriguez@mail.utexas.edu

Joong Ick Ryu is with the Korean Ministry of Science and Technology.
Contact: jiryu@most.go.kr

Roberto Sbragia is Full Professor in the Business Administration Department of FEA/USP and Scientific Coordinator of the Center for Technology Policy and Management, University of São Paulo, Brazil.
Contact: rsbragia@usp.br

Tobias Schauf is managing director of the Institute for World Economics and International Management, University of Bremen, Germany.
Contact: iwim@uni-bremen.de

Luiz Márcio Spinosa is Professor in the Graduate Program of Production Engineering, Pontifícia Universidade Católica do Paraná (PUC-PR), Brazil.
Contact: spinosa@rla01.pucpr.br

João Pedro Taborda is with the Brazilian aerospace and defense corporation Embraer, as Director of External Relations in Europe. Before joining Embraer he worked as an aerospace and defense analyst at Inteli, a Portuguese think-tank in Lisbon concerned with industrial policy.
Contact: joao.taborda@embraer.fr

Alejandro O. Vera-Cruz is Full Professor in the Master and Doctorate in Economics and Management of Technology programs at the Metropolitan Autonomous University, Campus Xochimilco (UAM-X), Mexico.
Contact: veracruz@correo.xoc.uam.mx

Kaja Wendt is a Researcher at the Norwegian Institute for Studies in Research and Higher Education, Oslo. Contact: kaja.wendt@nifu.no